Intelligent Networks and New Technologies

IFIP – The International Federation for Information Processing

IFIP was founded in 1960 under the auspices of UNESCO, following the First World Computer Congress held in Paris the previous year. An umbrella organization for societies working in information processing, IFIP's aim is two-fold: to support information processing within its member countries and to encourage technology transfer to developing nations. As its mission statement clearly states,

> IFIP's mission is to be the leading, truly international, apolitical organization which encourages and assists in the development, exploitation and application of information technology for the benefit of all people.

IFIP is a non-profitmaking organization, run almost solely by 2500 volunteers. It operates through a number of technical committees, which organize events and publications. IFIP's events range from an international congress to local seminars, but the most important are:

- the IFIP World Computer Congress, held every second year;
- open conferences;
- working conferences.

The flagship event is the IFIP World Computer Congress, at which both invited and contributed papers are presented. Contributed papers are rigorously refereed and the rejection rate is high.

As with the Congress, participation in the open conferences is open to all and papers may be invited or submitted. Again, submitted papers are stringently refereed.

The working conferences are structured differently. They are usually run by a working group and attendance is small and by invitation only. Their purpose is to create an atmosphere conducive to innovation and development. Refereeing is less rigorous and papers are subjected to extensive group discussion.

Publications arising from IFIP events vary. The papers presented at the IFIP World Computer Congress and at open conferences are published as conference proceedings, while the results of the working conferences are often published as collections of selected and edited papers.

Any national society whose primary activity is in information may apply to become a full member of IFIP, although full membership is restricted to one society per country. Full members are entitled to vote at the annual General Assembly, National societies preferring a less committed involvement may apply for associate or corresponding membership. Associate members enjoy the same benefits as full members, but without voting rights. Corresponding members are not represented in IFIP bodies. Affiliated membership is open to non-national societies, and individual and honorary membership schemes are also offered.

Intelligent Networks and Intelligence in Networks

IFIP TC6 WG6.7 International Conference on Intelligent Networks and Intelligence in Networks, 2–5 September 1997, Paris, France

Edited by

Dominique Gaïti
University Pierre and Marie Curie - Paris
and University of Technology - Troyes
France

SPRINGER-SCIENCE+BUSINESS MEDIA, B.V.

First edition 1997

Originally published by Chapman & Hall in 1997
MyCopy version of the original edition 1997

DOI 10.1007/978-0-387-35066-0

A catalogue record for this book is available from the British Library

Printed on permanent acid-free text paper, manufactured in accordance with ANSI/NISO Z39.48-1992 and ANSI/NISO Z39.48-1984 (Permanence of Paper).
www.springer.com/mycopy

CONTENTS

Preface vii
Background vii
Guide to the book ix
Programme committee x
Referees xi
About the editors xii

Part One Market Issues

1 Telecommunication trends
Olli Martikainen 3

2 Implications of technology and policy for intelligent network research priorities
William H. Melody 7

3 On intelligent network certification: performance issues
Manfred Schneps-Schneppe 16

4 Application of intelligent networks in banking
Troels Schmidt Jensen 27

Part Two Service Creation

5 The P103 service creation environment model
Carla Capellmann, Kjell Hermansen, Parminder Mudhar and Jørgen Nørgaard 37

6 The feature interaction problem in the IN: in search of a global solution
Nadir Belarbi and Dominique Gaiti 51

7 Formal criteria for feature interactions in telecommunications systems
Jan Bredereke 68

8 Open service node for intelligent networks
Pasi Kemppainen and Olli Martikainen 84

9 Advanced concepts in an IN-switching platform: standards and IN
Jørgen Dyst and L. M. Ericsson 95

10 Nokia SCE – an architecture for a lightweight SCE
Mikko Kolehmainen 107

Part Three Database issues

11 Design issues and experimental database architecture for telecommunications
Juha Taina and Kimmo Raatikainen 121

Part Four Performance

12 Reusable simulation models for performance analysis of intelligent networks
Claes Wohlin, Christian Nyberg and Anders Larsson 143

13 Methods to synchronize the IN SCP's overload protection mechanism
Philip Ginzboorg 155

14 The optional utilization of multi-service SCP
Tang Haitao and Olli Simula 175

Part Five Standards and IN

15 The telecommunication standardisation process: can it be 'reformed' to support 'de-regulation'?
Richard Hawkins 191

16 Formal description of services
Yuzhang Liu, Fangchun Yang and Junliang Chen 201

17 Intelligent network service creation: an ITU-T CS-1 view
Sehyeong Cho, Choongjae Im and JeongHun Choi 213

Part Six Multimedia and Mobility Services

18 IN and mobility
Terje Jensen 225

19 Call processing model for multimedia services
Olli Martikainen, Valeri Naoumov and Konstantine Samouylov 241

20 A generic service management architecture for multimedia multipoint communications
Constant Gbaguidi, Simon Znaty and Jean-Pierre Hubaux 252

21 The multimedia reference model: a framework facilitating the creation of multi-user, multimedia applications
Stephan Abramowski, Karin Klabunde, Ursula Konrads, Karl Neunast and Hermann Tjabben 265

22 Call processing architecture and algorithms for future network
Geonung Kim and Sunshin An 281

23 An overview of the telecommunications information networking architecture
TINA Consortium, USA 296

Index of contributors 308

PREFACE

The *IFIP International Working Conference on Intelligent Networks '95* was held at the Technical University of Denmark (DTU) in Copenhagen. The conference was organised jointly by the Center for Tele-Information at DTU and Tele Danmark Research[1] during August 30-31 1995 and was hosted by the Center for Tele-Information. The conference was sponsored by IFIP TC6.

The papers contained in this book are edited versions of the papers from the conference and some of the tutorial presentations that were arranged prior to the conference. The mixture of tutorial papers and conference papers gives a broad and yet thorough coverage of activities and central areas of development in the IN world at this point in time. Future developments, especially in the area of extended mobility support and multimedia capabilities, are also presented.

BACKGROUND

The theme of the conference was the two major trends in Intelligent Networks (IN) development:

- ITU and ETSI based IN standards, such as CS-1 and the upcoming CS-2 and
- long term development as undertaken by, for example, the TINA consortium and the European RACE and ACTS programmes.

The present IN technology based on *capability sets* is on its way into the networks, and operators are pushing for open platforms to be deployed in the networks in the near future in order to obtain greater freedom in selecting platforms for service provisioning.

At the same time, new entirely software-based platform technologies are being developed in the TINA consortium and the European RACE and ACTS programmes to answer this demand. The goal is to enable *Open Network Provisioning* on the one hand and to get flexibility on the other, both goals aiming at a better and more versatile communication infrastructure.

Concurrently, migration paths from IN to TINA-based architectures are being studied by a number of EURESCOM and ACTS projects. For the evolution of today's CS1-based IN networks, it is important to know the capabilities of the future technologies, potentials and integration strategies in order to plan for long-term network upgrade paths.

The trend is a distinct move towards more open platforms for the functional entities, such as SSP, SCP, SDF, etc. supported by, for example, the core INAP protocol while a number of research projects are defining an IN architecture to be used in broadband networks as well as in present networks. The first demonstrations of these concepts took place at *Telecom '95*. The integration of intelligence in broadband and mobile networks and the way in which the service logic is going to be distributed in the future may stir the industry into action and result in an entirely new structure of the communication industry.

---

[1] Now part of Tele Danmark Research and Development

GUIDE TO THE BOOK

In his opening paper, Olli Martikainen, Chairman of IFIP TC6 Taskforce on Intelligent Networks, presents the current trends in the telecommunication business and the market place transition from offering generic services towards offering services based on digital media, which is a new growth area. Following that line, the paper *Implications of Technology and Policy for Intelligent Network Research Priorities* analyses the impact of the information infrastructure initiatives on the organisation of the telecommunication sector. For telecommunication operators in a liberalised environment this new situation also has impacts on how R&D is to be carried out, focusing on customer demand and business opportunities rather than technology as it has traditionally been done in the industry. Looking at the market from a different perspective, the next paper in the *Market Issues* chapter discusses certification of IN equipment for open networks in general and the Russian market in particular. A number of issues that need to be considered are presented and the author asks for standardised IN certification methodologies suggesting a Memorandum of Understanding in Europe on this. The final paper, *Application of Intelligent Networks in Banking*, discusses the opportunities of using IN in the banking sector and is an example of innovative uses of IN.

Service creation is the starting point for rapid introduction of services in the networks and it covers a number of different activities. These activities and their relationship are described in the paper *The P103 Service Creation Environment Model*, which establishes a context for comparing service creation environments. A particular and important aspect, feature interaction, is explored by surveying work in the area in the paper *The Feature Interaction Problem in the IN: In Search of a Global Solution.* A formal definition of interaction is given in *Formal Criteria for Feature Interaction in Telecommuncation Systems* as well as methods for detection. The *Open Service Node for Intelligent Networks* paper addresses the issue of executing services by describing a platform that aims specifically at supporting third party service creation. Finally, two specific environments are described in order to give an understanding of current commercial possibilities as well as providing a view on possible evolution strategies.

Designing databases for use in IN networks presents challenges in the form of strict real-time requirements and handling of massive numbers of requests to be served. These requirements are being analysed in the paper *Design Issues and Experimental Database Architecture for Telecommunications*, which also proposes a design fulfilling the requirements.

In a commercial environment, performance of the IN network is of the utmost importance. Being able to control load and prevent overload is a necessity because the telecommunication network is expected to be available at all times. The chapter *Performance* investigates this topic, starting with work that aims at simulating network models which will enable prediction as network properties. This is the topic in the paper *Reusable Simulation Models for Performance Analysis of Intelligent Networks.* A number of specific overload prevention techniques are described in the paper *Methods to synchronize the IN SCP's overload protection mechanism*, which examines different techniques and their properties. Methods based on adaptive techniques, as described in *The Optimal Utilization of Multi-Service SCP*, aim both at maximixing the through-put of a node and preventing overload.

Standards pose a particularly difficult question in the changing world of telecommunication. Asking how standards should be used, the paper *The Telecommunication Standardisation*

Process: Can it be 'Reformed' to Support 'De-Regulation'? surprisingly concludes that standards are needed now more than ever before even if this is not yet realised. From a technical viewpoint, standards based on service creation are presented in the papers *Formal Description of Services* and *Intelligent Network Service Creation: an ITU-T CS-1 view*, describing the merits and drawbacks of standards and suggesting improvements.

The final chapter *Multimedia and Mobility Services*, starts by giving an overview of the different types of mobility, Terminal, Personal and Service mobility and their characteristics. The possible strategies for supporting mobility services are then examined in the context of IN. One platform could be the lightweight *Call Processing Model for Multimedia Services*, which describes a specific implementation of an execution platform for IN and multimedia services. Moving on to the natural extension of services in terms of capabilities, the multimedia services, the next group of papers present proposals for flexible network handling, which is needed to enable these new, complex and promising services. A proposal for the definition of a conceptual model for multimedia services inspired by the standardised IN model is presented in *The Multimedia Reference Model: A Framework Facilitating the Creation of Multi-User, Multimedia Applications*. The papers *A Generic Service Management Architecture for Multimedia Multipoint Communications* and *Call Processing Architecture and Algorithms for Future Network* are inspired by the work in the TINA consortium at Bellcore, adding levels of details which are required but not yet found in the TINA documentation, as well as addressing issues which are not being considered by TINA. Concluding the chapter is an overview presentation of the TINA Consortium project and its central areas of work and current status. This paper will help understanding the work and the very important links to present state of IN, which is the prerequisite to begin developing future-proof plans for upgrading and evolving today's network towards mobility, multimedia and information service capable networks.

PROGRAMME COMMITTEE

The following people participated in the IFIP IN '95 programme committee and secured a successful conference:

Olli Martikainen, Chair of IFIP TC6 Taskforce on Intelligent Networks, Telecom Finland, Finland
Kimmo Raatikainen, Department of Computer Science, University of Helsinki, Finland
James White, AG Communication Systems, USA
Carla Capellmann, Deutsche Telekom, Technologiezentrum, Germany
Dominique Gaîti, Columbia University, USA
Raymond Schlachter, EURESCOM, Germany
Caroline Knight, Hewlett-Packard Labs, UK
Andy Bihain, GTE, USA
Guy Pujolle, University of Versaille, France
Jørgen Nørgaard, Tele Danmark Research, Denmark
Villy Bæk Iversen, Technical University of Denmark, Denmark

REFEREES

The editors wish to thank the referees for their effort in evaluating the papers submitted for publication

- Kimmo Raatikainen, Department of Computer Science, University of Helsinki, Finland
- Olli Martikainen, Telecom Finland, Finland
- Manfred Schneps-Schneppe, R&D Center KOMSET, Russia
- Carla Capellmann, Deutsche Telekom, Technologiezentrum, Germany
- Karl Neunast, Philips Research Laboratories, Germany
- Philip Ginzboorg, Nokia Research Center, Finland
- Terje Jensen, Telenor Research and Development, Norway
- Jørgen Nørgaard, Tele Danmark Research, Denmark
- Jens Kristensen, Tele Danmark Research, Denmark
- Flemming Andersen, Tele Danmark Research, Denmark
- Valeri Naoumov, Russian University of Peoples' Friendship, Russia
- Konstantine Samouylov, Russian University of Peoples' Friendship, Russia
- Jan Bredereke, University of Kaiserslautern, Germany
- Dominique Gaîti, Center for Telecommunications Research, Columbia University, USA
- Villy Bæk Iversen, Technical University of Denmark, Denmark
- Tang Haitao, Laboratory of Computer and Information Science, Helsinki University of Technology, Finland
- Raymond Schlachter, EURESCOM, Germany

Jørgen Nørgaard and Villy Bæk Iversen

Copenhagen, March 26, 1996

Addresses

Jørgen Nørgaard

Tele Danmark Research
Lyngsø Allé 2
DK-2970 Hørsholm,
Tel: +45 4576 6444, Fax: +45 4576 6336
E-mail: jnp@tdr.dk
URL: http://www.tdr.dk/~jnp

Villy Bæk Iversen

Department of Telecommunication
Technical University of Denmark
Building 343,
DK-2800 Lyngby
Tel: +45 4525 3648, Fax: +45 4593 0355
E-mail: vbi@tele.dtu.dk

ABOUT THE EDITORS

Villy B. Iversen received the M.Sc. degree in electrical engineering in 1968 and the Ph.D. degree in teletraffic engineering in 1976, both from The Technical University of Denmark, where he is associate professor at the Department of Telecommunications. He is Professor Honoris Causa at Beijing University of Posts and Telecommunications and vice-chairman of the International Advisory Committee of the International Teletraffic Congresses. He is also Danish member of IFIP TC6. His research interests include stochastic modelling, communication systems and teletraffic engineering.

Jørgen Nørgaard graduated with a M.Sc. in computer science from the Computer Science department at Aarhus University, Denmark. After joining Tele Danmark Research he has been working at the software engineering department and participated in the RACE Cassiopeia project and in the EURESCOM P103 project in which he led the task on activities for service creation and methodology work. At the computer science department at Aarhus University he co-designed and developed an open, distributed hypermedia system on object oriented database system and the X window system. At Nokia Telecommunications he worked with object oriented techniques and their application in telecommunications software, using the notation SDL-92 and OMT.

Part One
Market Issues

1
TELECOMMUNICATIONS TRENDS
Opening speech 30.8-95

Olli Martikainen
Telecom Finland Ltd.
P.O.BOX 106
FIN-00511 Helsinki, Finland
Phone:+358 2040 3503, Fax:+358 2040 3251
E-mail: olli.martikainen@tele.telebox.fi

1 INCREASING ROLE OF INFORMATION

In addition to the basic factors: capital and labor, information is becoming a factor of increasing importance. Following many others we claim that the role of information - more closely its share of the value products create - has been increasing rapidly during the past decades. In our terminology we say that information has become a product element. Information can be considered to be present in products in three major roles, information of markets and technology (know how), information as embedded elements in products (software) and information as a product itself (media products). When information is created and processed in digital form, it can also be transferred via digital networks. During the next ten or twenty years remarkable changes in industries and in the society will take place, these changes being a result of the increasing value of digital information and its new infrastructure: the modern broadband networks (Figure 1).

2 GROWTH IN MOBILE AND VALUE ADDED SERVICES

The distribution of value in the value chain will also change. In Figure 2 there are two basic types of value chains, a generic one and a differentiated one. In the left hand side of a value chain we can see the accumulated value of raw materials and technology, in the middle the product itself, and in the right is the value which customer pay for the product. Basic telephony products have a generic nature: They are based on expensive high quality technology, and there is little customer oriented value from differentiation; there are only local, long distance and international calls. Their value chain resembles the value chain of air traffic: high technology production with little differentiation, only business, tourist and charter flights. In value

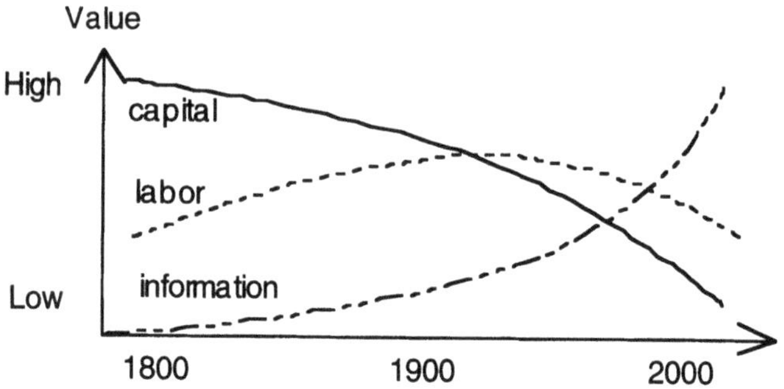

Figure 1 Relative value of Capital, Labor and Information.

added products there are more customer oriented value and often the basic technology has become more cost effective thus enabling more value adding capability. What can we say about the value added telecommunications products with low cost broadband infrastructure. To continue our traffic analogy, in the road traffic it is free to use the roads, there are hundreds of thousands of service providers transporting goods and people, and then there is the expensive terminal equipment which people are willing to buy: the private car. The concept Information Superhighway clearly represents the similar development in telecommunications, the mobile terminal being the private car.

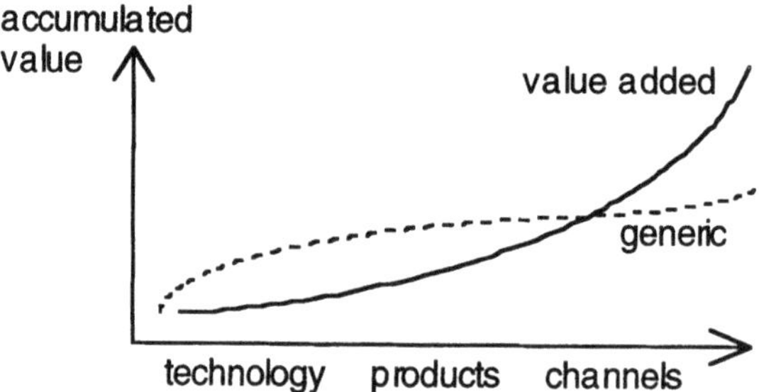

Figure 2 Accumulated value in Generic and Value Added value chains.

The changes created by the modern broadband networks will be as profound as those caused by the development of physical traffic networks during this century. The transition from trains and railways to roads and motorcars changed the structures of society and industries. The competing network operators will merge or build alliances, a comprehensive broadband infrastructure will be built, thousands value added service providers will be established with lots of added value from mobile and intelligent services. The multimedia products will be available via the broadband networks, the content often being more valuable than the container.

3 NEW NETWORKS WILL BE BROADBAND AND INTELLIGENT

In telecommunications technology substantial new breakthroughs are going on. The introduction of cellular radio networks and mobility is probably the most influential one in the next few years. The broadband transmission and switching technology is also maturing and will provide a cost effective platform for service provision. The shift from basic telephony to mobile and value added services, and the increasing value of digital multimedia services, is depicted in Figure 3.

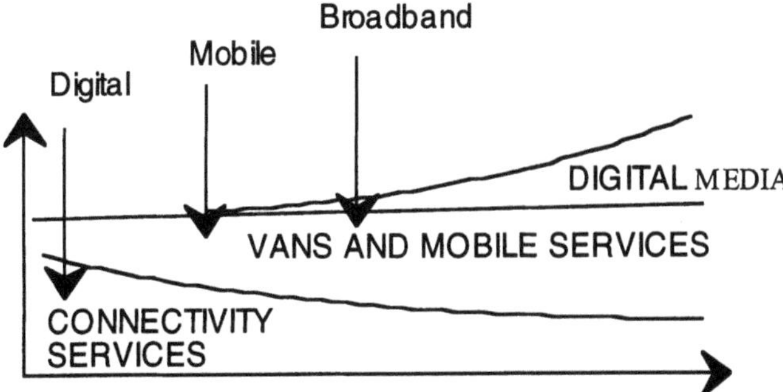

Figure 3 Telecom services value shift to Mobile, VANS and Digital Media.

When the broadband customer access will be available, interactive business and consumer services based on video and multimedia will become possible. The increasing role of service logics and multimedia software represents the computer controlled structure of modern telecommunications, where protocols, application technology and resource management are key factors. In modern telephony the most promising computer control and management concept is the Intelligent Network *(IN)*. It facilitates the development and management of new services by service providers. Intelligent networks can be considered as the bridging technology between the customer based service management, the service application execution and the connection control in the digital voice services. Considering the mobile and broadband technologies it is a natural question, how this concept of intelligence will develop. It seems to be clear that the service applications will be distributed over terminals and different service control and provider nodes. Examples of distributed intelligence can be found in Internet services *(WWW)* and in the specifications of Universal Mobile Telecommunications Services *(UMTS)*.

4 INTELLIGENT, CUSTOMER ORIENTED SERVICES

Modern computer controlled service creation is capable in customer oriented mass tailoring. It may even be so that the user will be able to configure the communications environment in the way she or he likes it most. This could be called the personal sphere or the information and communications landscape. It will be the user that makes the choices between services. The coming standards will be so called dominant designs which are evaluated and chosen by the market. Several experts currently believe that the dominant design for multimedia services will grow from Internet services and that the value added voice services integrated to them will evolve from Intelligent Networks.

We are entering into an era of competitive telecommunicatoíons services market, where the competencies available in the companies will be of the most important value creating factors. By developing competencies and new technology the research and development activities (R&D) have an increasing role in modern telecommunications. R&D has also an important mission in providing education and skill building for young professionals. New technology experts are of great value to business units. Enthusiastic teams of experts in relevant technology areas form the best basis of new technology applications and innovations.

5 BIOGRAPHY

Olli Martikainen, Ph.D., from Helsinki University and M.Sc. from Helsinki University of Technology. He has been doing research in several positions in Helsinki University of Technology (1976 - 1982), Oxford University (1980 - 1981) and Technical Research Centre of Finland (1982 - 1985). Later he has been R&D manager at Nokia Electronics (1985 - 1986), department manager at Nokia Research Centre (1986 - 1988), professor and head of Datacommunications Institute at Lappeenranta University of Technology (1988 - 1989) and professor at Technical Research Centre (1989 - 1991). Since 1991 he has been research director at Telecom Finland. Currently he is vice president, R&D, at Telecom Finland Ltd., and professor at Helsinki University of Technology. His main areas of interest are telecommunication software methods and tools, network architectures, performance analysis and new industrial and economic structures in telecommunications.

2

Implications of Technology and Policy for Intelligent Network Research Priorities

William H. Melody
Center for Tele-Information
Technical University of Denmark
Building 371,2nd fl.
2800 Lyngby, Denmark
Tel: +45 4525 5174
Fax: +45 4506 3171
E-mail: melody@cti.dtu.dk

Abstract

Windows 95 may provide a major stimulus to Intelligent Network (IN) development, to real technological convergence between computing and telecommunication, and to a major reordering of research priorities in the field. The paper examines IN in the context of the evolution of convergence and the larger framework of information superhighway, infrastructure and society developments. It shows how the pace and direction of IN development will be driven by demand associated with organisational and institutional change, not the simple supply of greater capacity or the discovery of 'killer applications'. This will be associated with a shift in priorities to distributed networking, information-based service development and a diversity of very specific new applications.

Keywords

Intelligent network development, convergence, information infrastructure, IN demand, IN applications.

1 INTRODUCTION

The launch of "Windows 95" may turn out to be a defining moment in the evolution toward a future information society. After years of discussion about the *convergence* of computing and telecommunication, Windows 95 may prove to be a bridge, or the first of many bridges that begin the process of real technical convergence. Windows 95 is a major attempt to integrate telecommunication network capability into PC operating systems. As

Windows is already on 80% of PCs, its implications for intelligent network development could be profound. Among other things, it could contribute to a major reordering of research priorities in the field. This paper examines recent developments relating to convergence in light of the economic and policy trends that are moving many countries toward an information society.

2 PREPARING THE GROUND FOR INTELLIGENT NETWORKS

We can now look back about a quarter century to some key policy decisions that redirected the course of development of the computing and telecommunication industries. In computing, IBM was forced by court order to unbundle its hardware and software. This fostered a rapidly growing independent software industry that passed the hardware industry in market size during the 1980s. Now hardware is increasingly becoming a commodity item, while software opportunities appear to have no limit.

In telecommunication, in 1968 a change in US policy forced AT&T to permit the network attachment of terminals produced by independent suppliers. Interconnection policy and standards development became subject to the direct influence of independent equipment suppliers and large customers. It led to the general unbundling of customer terminal equipment from network services and permitted the attachment of computers to the telecommunication network. This stimulated the process of converting telecommunication equipment from analogue to digital standards, and then the gradual conversion of telecommunication networks to the digital standards of computing that is now underway.

It is widely accepted that the cornerstone of future telecommunication network design, provision, service development and applications will be new software. In keeping with the grossly exaggerated terminology adopted to describe information society developments - such as 'information superhighways', 'global information infrastructures', and 'killer applications' - telecommunication software developers now claim to be creating 'intelligent networks'. As if this hyperbole isn't enough, the term 'advanced intelligent networks' already has arrived in the literature. The new communication and information networks are becoming more sophisticated and complex, requiring software development by very intelligent people. But I doubt if this adoption of the language of marketing hype will help either the research and development initiatives or public understanding.

3 CHANGING PRIORITIES FOR RESEARCH & DEVELOPMENT

The history of telecommunication has been characterised by continuous improvements in technology. For many years the driving thrust of research and development was directed to conquering distance. Then the emphasis shifted to expanding capacity, and later to providing new and more diversified types of communication, such as data and video. In recent years centre stage of the telecommunication technology drama has been taken by the concept of integrating the diverse range of telecommunication services into a single network, the Integrated Services Digital Network (ISDN), with open network provision

(ONP) expanding the ubiquity and flexibility of the network. The current popular term 'intelligent networks' apparently encompasses virtually all software development for network design, provisioning, management, service provision and applications. It covers both communication network software and computing information processing software.

But these two areas of software development have never been comfortable bedfellows. Information processing software and telecommunication software have been quite distinct historically and they remain so today. They reflect the very different heritages of the two industries and technology systems from which they have been developed. Computing has reflected a focus on information processing in a culture dominated by fierce competition, rapid improvements in technology and obsolescence of equipment, and close attention to customer demand and marketing as the primary driving force for industry development.

Telecommunication has reflected a focus on the incremental expansion of stable physical networks in a culture dominated by monopoly, technological improvement in long-lived facilities, and much closer attention to government policy than demand as the primary force driving industry development. Although there have been significant strides taken in the direction of convergence and integration over the years, the very different cultures of these two industries have kept real convergence more on the horizon than within our grasp. The attention being directed to the potential future information society by leading government policy makers in many countries has provided a stimulus toward convergence and integration, and therefore highlighted the range of problems yet to be solved on future intelligent networks.

4 COMPONENTS OF THE INFORMATION INFRASTRUCTURE[1]

The information infrastructure of a future information society has a number of fundamental components. The *telecommunication facility system*, which is currently being upgraded and/or extended in all countries is the cornerstone of the infrastructure. A maximally enhanced 'broadband' telecommunication system with advanced interactive capabilities has been labelled the "information superhighway". However, the *information content* sector and a newly developing *value-added communication services* sector are equally important in the development and application of new services. The communication/information *equipment* sector (hardware and software) is the primary driving force behind most of the new technological developments that are opening new opportunities for both industry and individual end users. These are illustrated in figure 1.

It is apparent that the information infrastructure requires the co-ordinated development of all its components. The information superhighway is an important part of it, but only a part – "the plumbing" as the *Economist* magazine once called it. Appropriate terminals and terminal software are essential for users to obtain access to value added services. This is where Windows 95 hopes to find a market. Ultimately, the major benefits must be found in the

---

[1] See Melody, W.,*Toward a Framework for Designing Information Society Policies,* CTI Working Paper No. 5, June 1995, for a more detailed exposition and analysis of the issues summarized in this section.

Figure 1 – **Creating the Networks for an Information Society**

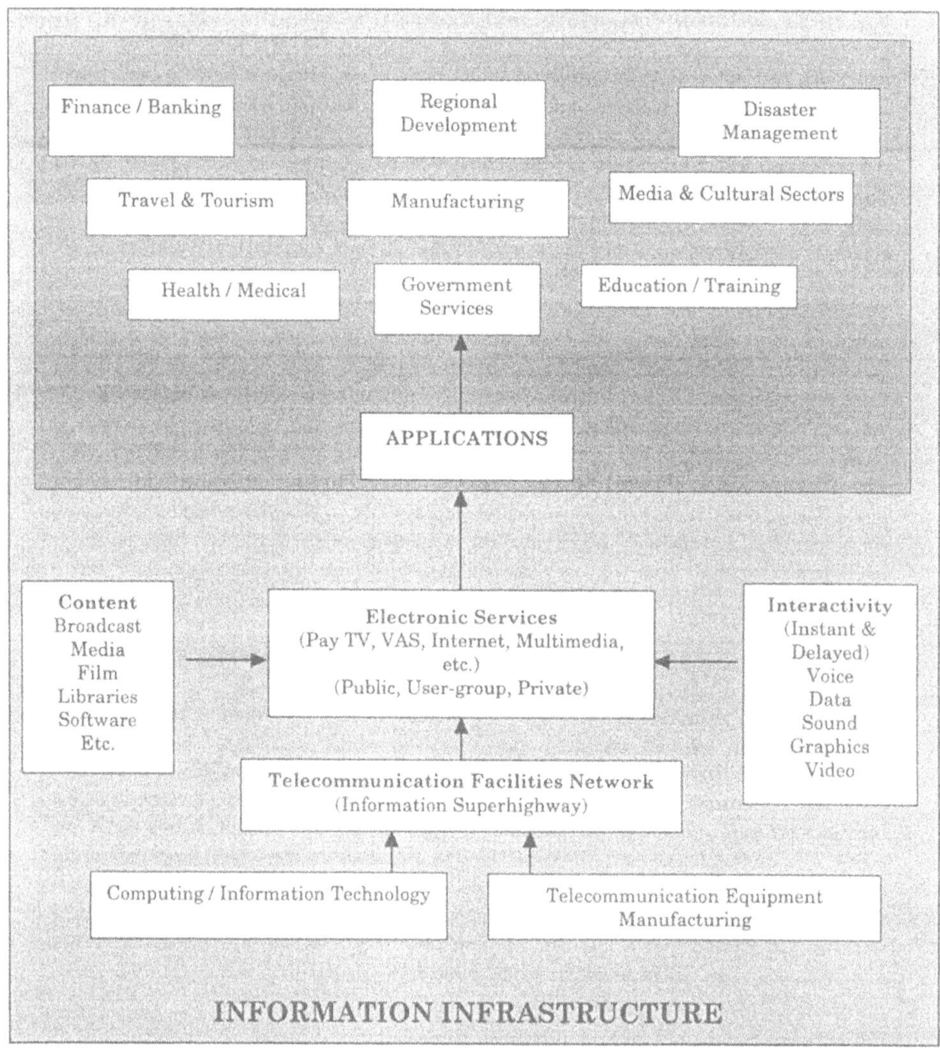

content and services. This raises a number of fundamental issues with respect to the design, management and governance of the information infrastructure, ranging from compatible technical standards development across all its components to issues of intellectual property rights, information security, privacy, universal access, pricing, monopoly and competition, etc.

A crucial element for success is the parallel development of the skill-base. New skills are needed for employment in the production sector – equipment supply, information creation and new services development. But most importantly the cultivation of user skills in the work-force attempting to apply the new services, and in potential household users will be essential to participation and market growth. The potential diffusion of new services throughout the population will be heavily influenced by the diffusion of PCs and the skill base necessary to use them most effectively in the new environment. Achieving universal service coverage for the information infrastructure will be a difficult task, and will be achievable only over a long period of time even in countries with highly developed telecommunication systems.

The information society discussion to date has been preoccupied with the enhancement of the "Telecommunication Facilities Network" (identified in the bottom portion of Figure 1) to the standards of a broadband information *superhighway*. The information technology and telecommunication equipment manufacturing sectors are providing the hardware (e.g. transmission, switching, terminal equipment), and increasingly the software to the telecommunication network operators (PTTs, PTOs, Telecoms, Telcos) as well as corporate, government and individual users.

An increasing number of electronic communication and information services supplied on the telecommunication facilities network are being generated by service suppliers outside the traditional telecommunication sector, and even outside the information technology sector. This has been made possible by telecommunication reform policies which have permitted access to the network by new firms to sell value-added services (VAS) directly to customers. Figure 1 illustrates this growing separation between the telecommunication facilities network, which provides the physical capacity to communicate, and the "Electronic Services" which reflect the design of special communication services that use the facilities network. As shown in Figure 1, this includes such services as Pay TV, VAS, the Internet, Multimedia Services, and others. Electronic news services and database services are now wide-spread. The services can be public, closed user groups (e.g. the banking industry), or private (e.g. a government agency or corporate network).

The growth of electronic services as a distinct component of the information infrastructure has provided an avenue for the design of new services that are more responsive to the specific needs and demands of particular users. It reflects a shift away from the almost total supply-side orientation of service development that has characterised the traditional approach of telecom operators toward a more demand-side orientation that pays more attention to specific customer needs. These service providers typically acquire more detailed knowledge of specific customer needs, which then represents the basis of the "value-added" they provide. Their entry to the market also has stimulated the telecom operators to offer their own value-added services and improve their responsiveness to customer needs.

5 APPLICATIONS OF INTELLIGENT NETWORK SERVICES

It is doubtful that this very expensive and comprehensive upgrading of the entire electronic information infrastructure could be justified in any country simply in terms of the supply of a range of new services to business, government and household users. Although entertainment services in the form of expanded options for viewing television and playing interactive games at home are expected to provide significant markets by the 21st century, the major benefits of the upgraded information infrastructure must arise as a result of applications of new services throughout the economy and society. These applications are expected to transform the traditional ways of operating for large and small businesses, government agencies, education and health organisations, and other institutions. It is anticipated that applications of the new electronic services will permit a major restructuring of all organisations so they can be more efficient and more responsive in their markets. The anticipated benefits will arise from the integration of the new information/communication services into the operations of each major sector of the economy, and of society. This is illustrated at the top of Figure 1.

The beginnings of such changes have been seen in the global restructuring of banking and finance, in major changes in travel and tourism, in the early applications of electronic funds transfer, just-in-time management systems, and electronic document interchange in a variety of different industries and types of organisations. It has facilitated the transformation of the music and media industries to global dimensions.

Successful applications of new services that go beyond home entertainment also will require changes in the structure of the organisations adopting them. Many large corporations are realizing that information and communication networking services justify (or require) changing their traditional hierarchical organisational structures to more horizontal structures based on distributed communication networks - providing improved management quality at reduced cost. Teleworking requires significant changes in the structure and even the culture of organisations. Similarly, self-directed education and training based upon new network services, the creation of new integrated health and medical networks, and most other potential major applications require quite fundamental institutional changes within each applications sector. The organisations applying the new services will have to be restructured so as to be able to take full advantage of the new services, and they will have to ensure the skill levels of employees, customers and suppliers are appropriate for the new structure as well as the new technologies and services.

Clearly a successful transformation of any major sector of society will have to take place over a considerable period of time. The field of education and training is a good example. The increasing requirements for "learning" in the 21st century will mean that such activities as distance education, computer assisted learning (and eventually interactive multimedia) must be moved from the fringes of national learning systems, where they are now, to the centre. Societies will have to move the education and training system on to the electronic information infrastructure. But to date there has been little co-ordination, let alone convergence of policy development and planning in telecommunication and in the education/training sectors in any country.

For the future, governments in the most developed countries should not invest primarily in physical facilities for stand-alone educational institutions. Rather, investment should focus on the most efficient and effective use of a variety of new communication and information services, and new forms of content. This should permit more and more diversified opportunities for access to learning by people of all ages and circumstances. It should permit more effective use of existing educational resources. The role of teachers in this new environment would shift from being suppliers of information (filling empty vessels) to more highly valued facilitators of access to information and knowledge (guiding and collaborating in the learning activity).

The application of the full range of new electronic information and communication services is expected to permit the redesign and transformation of the education/training/learning system for the 21st century. This will require the development of a series of intelligent networks designed to meet the very specific needs of different kinds of education in the new environment. Clearly this is a tall order and will require major reallocations of resources, both within the education/training sector and between this and other sectors, before it can be achieved. The transformation will not be an easy one and is likely to be characterized by many missteps and dislocations along the way. Figure 1 illustrates just some of the major areas of application of information infrastructure services.

6 DIRECTIONS OF CHANGE FOR INTELLIGENT NETWORK RESEARCH PRIORITIES

To date the research priorities for intelligent network development have been driven primarily by the public telecommunication operators and their equipment suppliers. The primary thrust has been to enhance the technical capabilities of the telecommunication system. These telecommunication sector priorities have been supported by government policy in research and development funding and other areas, particularly in Europe. A major limitation of this supply side emphasis in research has been its relative inattention to demand, and in particular its failure to uncover new demands that are necessary to justify the intelligent network enhancements now under development.

Windows 95 launches the Microsoft Network on-line service. It is expected to attract up to 3 million subscribers the first year, as many as America On-line, currently the largest service. Microsoft's goal is to make networking, and access to networking simpler and cheaper through its network, and eventually to merge it with the Internet. Whether or not this is achieved, Windows 95 has stimulated all Microsoft's current and potential information sector competitors into expanded efforts directed to ensuring Microsoft does not control key network software standards, as it has done with PC software. Developing software for improved networking is now the top priority for the computing industry.

This industry response could signal a major shift in influence over the future direction of network development and intelligent network research priorities. This shift in influence is from the telecommunication to the information processing industries, from supply side to demand side issues, from centralised to distributed networking, and from telecommunication network enhancement to new information-based service development. It should also stimulate an increase in research associated with future service applications. In addition, it could signal a major opening of small business and residential user service opportunities, reducing the pervasive influence of the largest corporate users on telecommunication network development. As a result of Windows 95, the proportion of household PCs connected to networks can be expected to increase rapidly from its current level about 10% in the leading countries. The market for software to run PC networks is forecast at US$ 30 billion for 1996, nearly ten times what it was in 1987.

Undoubtedly Windows 95 is as much a product of the trend of events as a cause. Perhaps the fundamental issue for the future with respect to software development is competition over the distribution of the network "intelligence" among telecommunication operators, specialist network managers, service suppliers, applications specialists in particular industry sectors and individual end users. This could represent one step in the direction of turning the telecommunication infrastructure into a commodity and forcing the public telecom operators to reassess their comparative advantages in this new market environment. For research priorities, it marks another significant step in the shift from supply to demand related issues. The telecommunication network eventually will become a great information processor with characteristics much closer to computing than to telecommunication as we know it today.

7 REFERENCES

Arnbak, J. (Editor). *Innovative Services or Innovative Technology?* IFIP/ICCC 1989, Elsevier Science Publishers B. V. (North-Holland).

Austin, Marc T. *Europe's ONP bargain: What's in it for the user?* Telecommunications Policy 1994, Vol. 18, Issue 2, p. 97-113.

Baker, Dan. *Intelligent networks: A report from the trenches.* Telephony, 1994, Vol. 227, Issue 15, p. 40-46.

Cane, Allan. *Era of intelligent networks.* Financial Times, London, March 1, 1995, p. VIII.

Commission of the European Communities. *Recommendation of the Co-ordinated Introduction of the ISDN in the European Community.* Brussels 1986, COM(86) 659 final.

Commission of the European Communities. *Present Status and Future Approach for Open Access to Telecommunications Networks and Services.* Brussels 1994, COM(94) 513 final.

Commission of the European Communities. *Green Paper on the Liberalisation of Telecommunications Infrastructure and Cable Television Networks. Part II. A Common Approach to the Provision of Infrastructure for Telecommunications in the European Union.* Brussels 1995, COM(94) 682 final.

Commission of the European Communities. *Proposal for a European Parliament and Council Directive on the Application of Open Network Provision (ONP) to Voice Telephony.* Brussels 1995, COM(94) 689 final.

Downs, Stephen J. *Asynchronous Transfer Mode and public broadband network: The policy opportunities.* Telecommunications Policy, 1994, Vol. 18, Is. 2, p.114-136.

Higham, Nicholas. *Open Network Provision in the EC: a step-by-step approach to competition.* Telecommunications Policy 1993, Vol. 17, Is. 4, p. 242-249.

Melody, William H. *Telecommunication: Policy Directions for the Technology and Services.* Oxford Surveys in Information Technology, 1986, Vol. 3.

Melody, William H. *Policy Issues in the Evolution of ISDN*, in Arnbak J. (Editor), Innovative Services or Innovative Technology? IFIP/ICCC 1989, Elsevier Science Publishers B. V. (North-Holland), p. 53-62.

Melody, William H. *Toward a Framework for Designing Information Society Policies.* CTI Working Paper No. 5, Center For Tele-Information, Technical University of Denmark, June 1995.

Reid, A. H. *The Integrated Services Digital Network: A Presentation of Related Policy Issues.* Trends of Change in Telecommunications Policy, OECD, Paris, 1987, ICCP 13, p. 99-115.

7 BIOGRAPHY

William H. Melody is Guest Professor and Chairman of the International Advisory Board, Center for Tele-Information, Technical University of Denmark. He was Founding Director of the Centre for International Research on Communication and Information Technologies (CIRCIT), Melbourne, Australia, 1989-94; and the Programme on Information and Communication Technologies (PICT), Economic and Social Research Council, London, UK, 1985-88. He has held appointments at St. Anthony's College, Oxford University; Simon Fraser University, Vancouver; University of Pennsylvania, Philadelphia; Iowa State University, Ames, and is a former Chief Economist, US Federal Communications Commission, Washington.

He has authored five articles on communication and information technology issues in the *Canadian Encyclopedia* as well as the article on "Telecommunication" in the *International Encyclopedia of Communications*. He has been a consultant and advisor to universities, national and international government organizations, and corporations on all continents.

3
On Intelligent Network Certification: Performance Issues

Manfred Schneps-Schneppe
R&D Center KOMSET
7 Zeleny prospect, Moscow, 111141 Russia
Fax +7 095 3060104

Abstract

With the rapid deployment of multi-vendor equipment on Russian telecommunication market (in accordance with Russian state and private investment plans) comes the major challenge of preserving network integrity. The more it concerns intelligent networks exploiting the SS7 network - the most sensitive part of telecommunications system. The paper discusses the foreign experience (US, Germany, France, etc.) in service provider certification (qualification), as well as some proposals in development of the standard ITU-T version of the IN concept, the revision of existing ITU-T recommendations and the challenge for memorandum of understanding for implementation of IN CS1.

1 ON THE RUSSIAN TELECOMMUNICATION MARKET

Russian state investment plans have a habit of beginning nationwide projects only to become far more modest when constraints on money and time catch up with planners. Yet, ever since its inception, the 50x50 project has grown and grown. At the outset it was an ambitious plan to connect 50 Russian cities with 50,000 km of fiber optic cable. In the middle of 1994, project was to include at least 100 cities and 100,000 km fiber optic cable. How did it begin? The Russian Federation property committee sold off 30 percent of the former Ministry of Communications (state-owned cables and trunk exchanges and upper level switching nodes) and formed ROSTELECOM in 1993, which is now leading a consortium financed by foreign telecom operators which will undertake a massive construction project called the 50x50 project [1]. Russian officials confirm these ambitions [2]. The most active companies on the Russian telecommunication market are the following: Nokia, Ericsson, Alcatel, AT&T, Italtel, Siemens, etc.

The Ministry of Telecommunications is the regulator exploiting its R&D institutions for certification (qualification) process, e.g., ZNIIS (Moscow) is testing toll exchanges, LONIIS

(St. Petersburg) is with urban and rural switches, NIIR is responsible for radio facilities, KOMSET has experience in billing system certification and network design, etc. Certification procedure contains detailed specifications for hardware and software, factory tests, line tests. Field tests are carried out in the actual Russian public network or in the LONIIS Certification Center (for rural and urban applications) by simulation and protocol-testers as well as through real trunks to existing step-by-step and crossbar switches [3].

In the paper we should like to discuss some difficulties arisen at implementation of intelligent networks on Russian telecommunication market. Intelligent Networks (IN) throughout the world offer the opportunity for telecommunication administrations as well as third-party service companies to provide application development and control of services. These INs are generally composed of systems from many different suppliers, and the services offered may need to function in cooperation with other carriers. In an environment with such distributed and shared responsibilities, it is important to ensure that each part meets expected levels of service integrity, performance, and quality. A certification process for telecommunication administrations, equipment providers, and other service providers may help identify serious vulnerability and network integrity concerns early, so that corrective or compensatory actions can be taken.

We are looking to IN implementation process in Russia as a next step of the now going Russian ISDN pilot project based on the Moscow toll and international network (MMT) having several connections inside Russia and a few international routes. We would like to restrict our discussion on IN certification by the INs built in accordance with ITU-T standards. Considering the standard SS7 as an IN backbone, we have several challenges put by INs:

- IN services lead to a significant increase in the volume and complexity of the SS7 messages exchanged between network nodes. New strategies for overload control must be defined, e.g., different internal overload control strategies for MTP and SCCP levels of the SS7 protocol led to unfair treatment of different message classes in an uncontrolled fashion.
- When telecommunication administrations open their networks to third-party service companies, additional issues of application protection, security, network integrity, network performance and overload must be addressed.

2 IN CERTIFICATION: THE US EXPERIENCE

Nowadays the service evolution is mainly supported by the introduction of the SS7 network. Several well publicized SS7 outages occurred in the US during the last years (in January 1990, the outage of AT&T network had nationwide impact and involved the loss of 65 million calls, the other (in 1991, Washington, DC) involved entire cities and impacted 10 million customers)are of great importance for Russian network designers. After the hearings at the US Congress (in 1989 and 1991) a Network Reliability Council has been formed by FCC [4]. Nevertheless, there is growing dissatisfaction with public network and a crisis is developing.

Why did the recent failures occur? McDonald [4] goes through many driving forces: fiber technology, digital switches, common channel signaling, competition, divestiture of AT&T, customers requirements for faster introduction of new services - all these forces that shaped network evolution had caused a concentration of network assets and this had increased the customer impact of a given network element failure. Technology to maintain adequate integrity has not been deployed. Forces of technology, regulation, and customer demand have conspired to push the network away from acceptable levels of integrity.

Qualification (certification) procedure as of Bellcore [5] is a three-step process. The first step (so called white box step) is devoted to design review and contains a paper analysis of the supplier's design specifications and focuses on potential vulnerabilities (e.g., overload control algorithms) identified by previous studies and/or network failures. In a design review, internal behavior is examined in detail (internal architecture, functionality, interfaces in reference to generic requirements).

The second, black box step, is devoted to conformance analysis: the independent assurance of stand-alone product conformance to industry standards and telecommunication administrations requirements. A black box's responses to stimuli generated by other simulated network elements are studied. In the case of IN SCPs and switching systems, particular attention is given to testing feature interactions between IN and non-IN features, as well as among various IN features.

The third step is product to product compatibility analysis. As part of the qualification program, Bellcore's laboratory hub concept makes it possible to test IN network elements and systems in an environment that closely resembles an actual network. From a network integrity perspective, real world failure scenarios are introduced in the network (thirty extensive test scenarios until now) and high traffic loads are simulated.

To assure the network integrity, FCC suggested that third parties may provide new services through carriers' IN via mediated access [6]. However, third-party access to INs raises many problems, relating, for example, to incorrect messages, protection from unauthorized users, non-robust products, impacting on others' INs, etc. Third-party IN service providers having their own SCPs can make use of the proposed certification process for suppliers to have their equipment certified. In particular, the certification process should emphasize software integrity, software quality, service introduction and service interaction.

3 WHILE WAITING FOR OPEN EUROPEAN NETWORKING

The European telecommunication market will almost completely be liberalised at the beginning of 1998. The German Telecom announced a "Concept for provision of open access to telephone network functions" [7]. This concept contains a detailed description of how Telecom plans to provide new, non discriminatory accesses to the Telecom telephone network for its competitors (Open Network Provision, ONP). Foreseeing the customer/market needs, the offensive strategy of Telecom in the ONP area is based on Telecom's corporate policy interests: making optimal use of resources, making its services more flexible with regard demand, and stimulating new effective demand while satisfying regulatory requirements.

For provision of different access types, ONP interface protocols have to meet three basic requirements: 1) to be standardised internationally, 2) to provide powerful compatibility mechanisms and 3) to provide functions to protect all interconnected networks.

To discuss the network integrity comprising different aspects, a classification into three areas of risk is applied by Telecom: 1) a disruption of the network (could lead to a complete loss of the signaling network or a complete network outage), 2) disturbances which could cause a degradation of the quality of service, 3) misuse of network functions without the risk of outages or even a degradation of QoS.

At ISS'95, Kung [8] of France Telecom has discussed the third-parties' requirements under question-mark: is an open networking technically possible? His discussion is based on simple model of two players: on the one hand, a network operator who owns a network that provides some "reserved" telecommunication services, and on the other hand, third parties that are able to provide services that are under competition.

The requirements are the following:

1) Independence: Third parties should be able to develop their services independently from other third-party services without rising service-interaction problems.
2) Restricted competition: It is necessary that third parties develop their services jointly or with the network operator.
3) Full life cycle: Rules are to be followed by third parties exist at all stage of the "service life cycle" (from service preparation till service deployment).
4) Guaranteed QOS: It is important to check that all relevant aspect of the calls (busy call, no answer, congested network, call abandoned, etc.) are handled by the service.
5) Hidden basic architecture: Introduction of a new system by the network operator should be unknown to third parties.
6) Standard physical interfaces: Systems should be available from several vendors to be able to shop for the best systems.
7) Global services: It should be possible to provide more global services on several networks. There is the need to take into account all relevant levels (circuit, data bases, OA&M).

4 SIEMENS PROPOSALS FOR THIRD PARTIES

For provision of open network concept, some considerations of SS7 in context of network integrity are needed. The MTP constitutes the signaling network. The network management functions are powerful tools and, if misused, can cause significant problems within a signaling network. To avoid such dangers, the concept for provision of SS7 access products contains as an important requirement that no access is connected directly to the national gateway signaling network. Each network provider has its own signaling network interconnected by the gateway network. This is the essence of proposal by Sevcik and Lueder [9]. They have pointed out some weak points of classical IN scheme:

1) Changes in the service logic by service provider/subscriber are difficult in many practical cases since they might affect and incorrectly interact with other services in the common centralized SCP (e.g. due to common rooting databases, common SS7 SCP-SSP protocol data, etc.)

2) Service providers/subscribers may not be willing to store and manage their proprietary corporate data in a SCP/SMP owned by network operator and shared by other competing service providers.

They proposed a new IN architecture supporting private SCP (PSCP) in addition to the SCP owned by a network operator (NSCP) and using TCP/IP protocol for interworking between PSCP and NSCP. A private SMP exists also. There is clear isolation between IN service logic in the NSCP and the IN services in PSCP (no common data, no common access to the SS7 signaling).The biggest potential for a PSCP owner lies in the opportunity to introduce the sophisticated service features like:

1) Personal screening.
2) The access to customer data base records on a workstation for each call.
3) Individual billing.
4) Voice and fax mailbox controlled by the PSCP.
5) Multimedia server containing product advertisement (video, voice).
6) Private video on demand server.

Leconte with co-authors [10] have offered a way of adding value through the deployment and implementation of innovative Intelligent Peripherals (IP)/Service Node (SN) services. The SN combines: the IP, a service control function, a switching function, a service creation/customization function, all under one physical (or logical) node.

According to the standard IN architecture, intelligent peripherals provide the service resource function for SCP, for example, send information to users participating in a call (prompts, announcements), receive information from users (e.g. authorization codes), modify user information (text to speech synthesis, protocol conversion), provide special connection resources (e.g. audio conference bridge, information distribution bridge).

Whereas an IP's main function is to support an IN SCP, the SN may have its own service logic and may support advanced services such as voice activated dialing, voice/fax messaging and other types of services. The IP/SN may also be directly connected to an SCP via an SS7 Transaction Capability Application Part (TCAP) interface.

5 STANDARD IN CS1 ARCHITECTURE: UNSOLVED PROBLEMS

As a compulsory requirement for certification we propose that the IN facilities follow the standard IN architecture in accordance with ITU-T Recommendations Q.12xx, more precisely in accordance with IN Capability Set 1 defined by Q.121x. Therefore, we would like to discuss the problem in the strong framework of the IN functional model defined by Q.1204 “IN distributed functional plane architecture (DFP)” for CS1 services.

Recall that the DFP architecture:

- provides the support for a large variety of services,
- is vendor/implementation independent,
- accommodate service creation.

For service certification procedure, we are to study relationships between functional entities related to IN service execution. There are as many as 12 types of IN control/management relationships that may be established between functional entities in the DFP (see Figure 2-1 in Q.1204).

The contents of these functions are defined by Q.1204. Each interaction between a communicating pair of functional entities is termed "an information flow".

Let us point out the call control function (CCF) providing call/connection processing and control. An important term here is the trigger mechanism acting in so called detection points of call model to access IN functionality. Triggering tables are located in SSP (Service Switching Point). Whenever a new service is to be introduced, or whenever a user wants to subscribe to a line-based service, it is necessary to update these tables. As SSPs belong to the network operator (otherwise, the third party would be the network operator), the problem is to decide how third parties should access and update triggering tables.

To say a few words about Intelligent Network Application Protocol (INAP). The SS7 INAP for IN CS1 as it was defined by ITU does not support the interconnection of IN network nodes in different networks, the INAP was explicitly developed for the operation within one network. There will be different kinds of INAP interfaces SSF-SCF, SCF-SDF and SCF-SRF. One possibility would be to introduce a new network node type in the IN architecture, the "Inter Network Unit". This unit would separate the links between SCF and SSF. The INAP dialogues would split into two parts. As a consequence, the Inter Network Unit would then have knowledge about the service logic of each SCP connected to it [7].

The next question worth mentioning regards the ISUP-INAP interconnection for correct charging procedures. Q.764 currently recommends that an originating exchange awaits the receipt of an answer message before it through connects its speech path in the forward direction. The procedure is intended to assist in the prevention of fraudulent use of the network, as it is common practice to link the start of charging for call at an originating exchange with the receipt of the answer message. In order to meet the requirements of IN for in-band dialog with the calling party to collect information necessary to complete the call set up to the called party, it is necessary to through connect the speech path at the originating exchange before the called party. This will require a revised protocol that will be included in a future release of ISUP recommendations. It will be necessary for the IN to send an "early" answer message to the originating exchange before the call is established to the called party. Such an early answer message as a procedure in ISUP-INAP interactions recommendation would be specified for the case of calls originating in exchanges that use existing ISUP protocol.

Such revision of ISUP-INAP interaction procedure can be followed by negative features: the operation of some ISDN supplementary services can be seriously jeopardized on calls for which an early answer message is used to ensure through connection at an originating exchange. This is because the change in ISUP procedures associated with the early answer message will prevent the transfer of ISDN supplementary service information from the destination exchange to the originating exchange. An early answer message to ensure through connection at an originating exchange may also cause the exchange to commence charging the call prematurely, or to charge for an unsuccessful call.

The implementation of any new service requires the updating of all triggering tables. How to do this procedure? As mentioned above, the network physical configuration is hidden to the third party, so intermediate management "mediation device" would be needed. Next question. Triggering tables are structured according to call states. Whenever the switch enters a new call state, the switch looks at the relevant part of the triggering tables to check if any triggering condition is met. This particular way of implementing services may raise some service interaction issues. Sometimes, it may happen that several conditions are met and several services may be triggered at the same time. Service interactions problems may arise because parameters in INAP may not exist in ISUP or DSS1 and vice-versa.

It is necessary to provide some management capabilities to third parties. We have already identified the need of filling SSP trigger tables. There are many other data and many other systems than the SSP that need to be updated by third parties: creation of new subscribers, modification of subscriber's service profile, etc. As the exact physical configuration should be hidden to the third party, some single entry point for the management by the third party is needed to manage, e.g., routing tables of switches, triggering tables of SSP, global title data updated in STP.

6 REVISION OF E. 411 AND E. 412 RECOMMENDATIONS

These recommendations are of special interest for IN performance analysis. What about E.411 "International network management operational guidance"? [11], the very meaning of network management is to be changed. Instead of traditional management control status indicators based on 30 second traffic measurement, in case of SS7 and SCP management we are to control the traffic intervals as short as 1 sec and less. SS7 system status provides information that will indicate failure or signalling congestion within the system. It includes such items as: receipt of a transfer prohibited signal; signal link unavailability; signal route unavailability; destination inaccessible. These items form the basis for IN performance management.

Effective intelligent network management requires good communications and cooperation between the various network management elements within an Administration and with similar elements in other Administrations as well as with IN service providers and users. This includes the exchange of real-time information as to the status and performance of circuit groups, exchanges and traffic flow in distant locations.

Such information can be exchanged in variety of ways, depending on the requirements of the Administrations. Voice communications can be established between or among network management centers using dedicated service circuit or the public telephone network. Certain signals may be related to the exchange status (as for the switching congestion indicators) and to the status of the destination as for Hard-To-Reach (HTR) information in particular.

It should be noted that in network management applications, the volume of data to be transferred can be quite large and its frequency of transmission can be as high as every three minutes for traditional voice teletraffic control and much higher (as every 1 sec) for IN services. When this data is transferred over signalling links which also handle user signalling traffic, stringent safeguards much be adopted to minimize the risk of signalling system overloads during busy periods when both user signalling traffic and network management data transmission are at their highest levels. These safeguards include, besides others, the following:

- limiting the amount of network management information to be transferred on signalling links which also carry user signalling messages;
- developing appropriate flow control priorities for network management information;
- equipping the network management operations system in such a way that it can respond to signalling system flow control messages.

As it is pointed out in the draft of revised E.411 [11], it is important that the network management controls should not become completely unavailable due to the failure or malfunction of the network management operation system or of its communications links with exchanges. Therefore, network management operations systems should be planned with a high degree of reliability, survivability and security.

There are many unsolved problems of intelligent network management. How to introduce IN call gapping for SS7 protection, when a service control function (SCF) communicates with a service switching function (SSF) via an overloaded SS7 line? Traditional congestion control works for an SCF communicated via circuit group. This is the case of SCF implemented in an adjunct, intelligent peripheral or service node that contacts to an SSP via circuit groups with e.g. ISDN interface.

What about new version of E. 412 "Network management controls", two problems are worth to be mentioned:

1) How to control HTR services. To explain the situation. An SCF detects that the destination of a dialled number, which is generally at customer's premises, receives a relatively high volume of ineffective call attempts. The SCF then issues a signalling message to an SSF to request for a call rate control. In return, the SSF activates a call rate control to reduce the rate of services requests that are sent to the SCF. Of special concern is how frequently the HTR information is updated in the receiving exchanges (taken into the mind several IN functions: SST, SCF and others, including service management agent function);
2) What method for IN traffic control should be implemented. With the call rate control method, an upper limit on the rate that calls are allowed to access the network is established (for example 4 calls per second). The leaky bucket counter it seems in here of special interest. Of course, the performance of call percentage control (comparing with call rate control) is to be considered also.

The revision of E.411 and E. 412 from the viewpoint of SS7 overload for IN use is to be carried out under requirements of Q.543 "Digital exchange performance design objectives". An exchange must continue to process a specified load even when offered call exceed its available call processing capacity. Two basic requirements for exchange performance during overload are: to maintain adequate exchange throughput in sustained overload; to react sufficiently quickly to load peaks and the sudden onset of overload.

It is well known that when traffic on real-time processing system rises beyond its capacity, the overall performance of the system degrades. In an ideal case, the throughput rises linearly

as a function of the input load until capacity of the system is reached and then levels off at the system capacity. In the case of actual system the throughput will rise linearly for light and moderate loads, but near the saturation the throughput will not increase as fast as the offered load and beyond saturation the throughput will usually decline. In most cases, if there are not adequate overload controls, this decline in throughput beyond saturation can be precipitous.
The goals of overload control can be stated as follows [12]:

1) Maintain throughput near system capacity, even under periods of extreme overload.
2) Ensure system sanity and perform all critical functions necessary for the system and network to work properly regardless of overload level.
3) The system should protect itself from becoming deadlocked.
4) The system should be able to protect itself if network congestion or flow controls do not work.
5) The system should be able to recognize and shed excessive network management work without violation goal 2.

Recommendation Q.542 "Digital exchange design objectives" contains the description of statistical indicators for HTR service detection. For IN management the HTR process should be modified in many aspects:

- HTR administration (codes of IN services, thresholds for detection of overload and load reduction conditions, procedure of HTR control list review) and
- methodology for HTR service determination, e.g. by SIB QUEUE mechanism.

7 NEW E.7INX SERIES

ITU-T Study group 2 is in a very starting point with drafts for this series. For our reason of IN certification the most useful should be E.7IN4 "Traffic and congestion control requirements for SS7 and IN-structured networks"[12]. The authors of E.7IN4 draft have stated four main purposes:

- to choose the performance characteristics implementable in different types of network;
- to avoid fault and congestion propagation, network instability;
- to provide the interconnecting networks for various vendor products;
- to provide high level requirements for traffic and congestion control.

IN control includes SS7 control as an element MTP traffic management procedures (according to Q.704) contain the transfer-prohibited, transfer-restricted and transfer-allowed procedures. Congestion control of a signalling route in under transfer-controlled procedure and traffic flow control actions. These MTP control procedures as an important part of SS7 are developed now in detail. SCCP management procedure for signalling point and subsystem status changes including broadcasting about subsystem prohibited conditions are less developed.

What about IN, only preliminary congestion control requirements are named. Part of them would be included in IN certification procedures. The question is - what part? A list of recommended high level requirements includes the following aspects:

- Share load. Load should be balanced over the possible serving element so that delay may be minimized. Thus congestion controls should automatically balance loading across more than two network elements, across elements of unequal capacity, across elements with distributed logic or data.
- Harmonize across levels and layers. The IN congestion and overload controls should be a harmonized set of functions providing overload control within each element to maintain sanity and call processing should higher level controls not work, as well as congestion control appropriate to the scope of data available and to the span of the control or network management system.
- Control effectively. The control process (the way a congestion is activated and modified) should be capable of taking the appropriate control action to handle traffic surges and peaks that cause congestion across network providers, across network boundaries (including Global Title Translation), across networks implementing different SS7/IN procedures.
- Control properly. A proper IN flow control is essentially a feedback control system. It should operate in stable, controlled manner, avoiding wide swings in the control parameters and actions. The controls should respond quickly to changing user traffic levels. Flow control is based on parameters such as: dialled number, originating switch, class of service of caller, IN-service requested.
- Overload triggers. Overload is usually caused by insufficient real-time in a processor to handle the workload, and it is usually recognized by certain buffer occupancies exceeding a threshold. Overload triggers should be set so that adequate time is available for the controls to take effect before the system performance degrades significantly. Unsolved problem here is how to set thresholds.
- Network robustness is concerned with the ability of a network to withstand both traffic overloads and failures of network elements. In the case of IN-structured networks, in addition to controls in the IN function layer, some control activation may be appropriate in the signalling and bearer portions of the network, such as call admission controls in the circuit-switched network.
- Requirements for screening messages. One of the potential dangers in interconnecting signalling network is that one network might send another message the receiving network considers invalid. One way to protect against this potential failure mechanism is to have a screening capability where the traffic enters the system (e.g., at the receive side of signalling link) and check messages to ensure all parts of the message meet the requirements for that network.

These are basic requirements be considered as topics of IN certification methodology.

8 SUMMARY

With the rapid deployment of intelligent networks comes the problem of presenting network integrity. The IN certification process is a systematic activity associated with a multi-supplier, multi-carrier environment, with new network systems, new operating systems, new interfaces, additional signalling traffic, and the emergence of third party service companies. Russian telecommunication system is in the very start point of this process. Therefore we are looking for the knowledge and existing experience.

The paper proves the emergency of standardised IN certification methodology. The main features of methodology might be the following:

- three-step approach (as of Bellcore);
- standardised subset of IN architectures (e.g. containing private SCP, enhanced SN);
- fixed ISUP-INAP interaction procedure for fixed subset of CS1 services (e.g. freephone, credit card, VPN);
- revised E. 412 (leaky bucket for HTR service control);
- developed E.7 IN4, contained minimised high level requirements, fixed overload triggers, developed concept of robustness (as for Q.543 defined overload control scheme based e.g. on SIB QUEUE mechanism);
- minimum subset of QoS parameters for IN throughput analysis.

Following the Memorandum of Understanding (MOU) for the implementation of an European ISDN service by 1992, the similar kind of document MOU IN'96 could be created.

9 REFERENCES

[1] Lyudi, D. (1994) Supplement Telecommunications, June , 3-8.
[2] Krupnov A. J., Elektrosvjaz, 1994, #8, 2-3.
[3] B. S.Goldstein, IEEE Journal on SAC, vol. 12, #7, Sept 1994, 1186-1191.
[4] J. C. McDonald, IEEE Journal on SAC, vol. 12, #1, 1994, 5-12.
[5] R.J.Baseil, and B.Hoang, ISS'95, Berlin, vol. 1, paper P.g5.
[6] FCC, Notice of proposed rulemaking, CC Docket #91-346, Aug 1993.
[7] R. Kickartz, and H. Gottschalk, ISS'95, Berlin, vol. 2, paper B5.4.
[8] R. Kung, ISS'95, Berlin, vol. 2, paper B5.3
[9] M. Sevcik, and R. Lueder, ISS'95, Berlin, vol. 2, paper P.y6.
[10] A. Leconte et al, ISS'95, Berlin, vol. 1, paper P.g3.
[11] Draft E.411. ITU-T COM 2-R 28E, 1995.
[12] Draft E.7IN4 ITU-T COM 2-R 32E, 1995.

10 BIOGRAPHY

Manfred Schneps-Schneppe has published numerous papers on economics, stochastic processes, and teletraffic engineering, and written several books on the same subjects. He has had the chair in Telecommunication at Latvia technical University in Riga and been visiting professor to The Technical University of Denmark. Now he is working with future telecommunication services and training at R&D Center KOMSET in Moscow (editors note).

4
Application of Intelligent Networks in Banking

Troels Schmidt Jensen
Department of Telecommunication
Technical University of Denmark
DK-2800 Lyngby
Fax (+45) 4593 0355

Abstract
The actual application of telecommunications in banking is described from a service point of view. The future possibilities of using intelligent networks services is disscused. In particular we have dealt with the possibilities of the concept of Virtual Privat Network, VPN.

Keywords
Virtual private network, Intelligent networks in banking

1 INTRODUCTION

This paper is a discussion about the potential advantage for banks by the application of intelligent networks, IN, in connection with communication. The discussion will primarily deal with the future use of an IN service, called Virtual Private Network, VPN.

2 QUANTITATIVE USE OF TELECOMMUNICATION IN BANKING

Today an intensive application of telecommunication is found especially within the financial sector. In the USA the transmission costs amount to approximately 40% of the total telecommunication costs in a typical financial institution. This may be converted into Danish conditions, which means that the total telecommunication costs amount to almost 10% of the operational costs of in banking. Therefore, the banks are very interested in teleservices, which can mean retrenchments of the costs of telecommunication.
The explosive development in the use of telecommunication is especially related to the growing use of datacommunication in connection with the introduction of technology for information and telecommunication within the financial sector. This development is reflected in the relative consumption of datacommunication of the sector compared to the consumption within other sectors. In 1989 the share of the financial sector in datacommunication costs for the telephone companies in Europe amounted to approximately 1/3, far exceeding the relative size of the sector as well as the share of the sector in the total communication costs. The

sectors share of the the total communication costs amounts to well over 10%. These figures are an argument for the operators to take a special interest in the financial sector by the development and marketing of teleservices.

3 THE GENERAL STRUCTURE OF IN

A general construction of IN, containing advanced services - such as VPN - can be described shortly as follows (ref. figure): When a user makes a call to an IN service, the local exchange of the user transfers the call to an exchange with a Service Switching Function, SSF. This exchange is called a Service Switching Point, SSP. When the IN call has arrived in SSP (containing a trigger criterion with the function for deciding whether it is a call for a Public Switching Telephone Network, PSTN, or an IN service), a message is sent from SSP to Service Control Function, SCF. (SCF can possibly be in the same exchange and in such case the exchange is called a Service Switching Control Point, SSCP). If SCF and SPP are placed in different exchanges, the exchange, where SCF is placed, is called the Service Control Point, SCP. In SCF an analysis of the A-number and the B-number is performed by means of a database. Thus SSF forms the communication part between the user and IN, and SCF is in control of the dialogue.

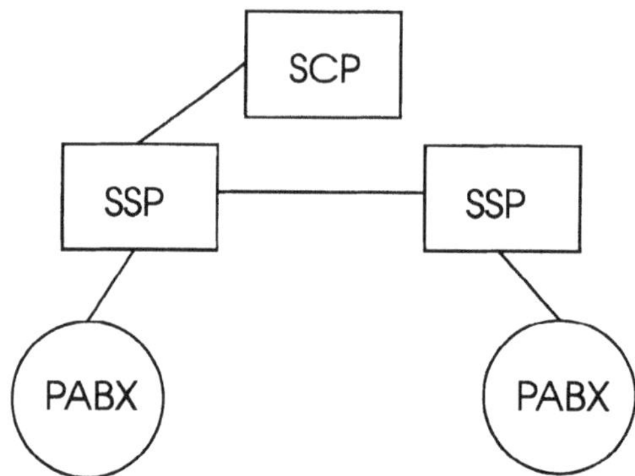

4 CAPABILITY SET NO. 1

The designation *Intelligent Network* was for the first time used by Bellcore in 1984. Since then the aim has been internationally to define an architecture of IN. This has resulted in some recommendations collected in one set called IN Capability Set no. 1, CS1. Two standardisation organisations have agreed upon CS1. These are ITU-T, The International Telecommunication Union, and ETSI, European Telecommunications Standards Institute. ITU's recommendations for IN are given by the Q.12xy-series. The number of CS has the index x, and the description of its recommendation part has the index y. CS1 is described by the Q.121y-series, and full standardisation has not been performed until 1993.

With a basic view CS1 gives an open interface between the Service Control Point, SCP, and the Service Switching Point, SSP. This contributes in making IN services independent of the individual provider of telecommunication equipment. CS1 supports a number of IN services. Here 25 services are included out of which a couple are only supported partly by CS1. A great part of the CS1 services may advantageously be applied within the banking sector. In the client-related bank activities (front-office) especially the services: Credit Card Calling, CCC, and Freephone, FPH, will be relevant.

5 IN AND CLIENT-RELATED BANK ACTIVITIES

CCC gives the subscriber the possibility of calling from and to any place whatsoever. When the subscriber wishes to make a CCC call, the first action will be to press a PIN-code (Personal Identification Number). When calling, the bank account of the subscriber will automatically be charged the costs of the call.

Freephone is a service allowing subscribers to call e.g. a bank branch free of charge, where the bank will be debited with the costs of the call. It will be possible to combine this service with routing of calls by bank clients so that a bank has only one main number. When a client makes a call, this call will be routed to the nearest branch, irrespective of the place in the world from which the client makes the call. Or possibly by pressing a user number the call will be routed to the client's own branch. Furthermore, the service can be extended by an automatic telephone answering device to be applied in connection with calls not falling within the opening hours of the individual branches.

Finally, the functions of homebanking and officebanking are both applicable combined with one or both of the services mentioned above. Homebanking is a way of offering bank services through the Public Switching Telephone Network, PSTN (or PSDN), direct to the home of the client. This facility gives the client the possiblity of performing bank transfers and different forms of information services. Just as CCC, homebanking requires the IN service to be connected to the edp-exchanges of the individual banks. From a financial point of view the costs in connection with telecommunication, when using homebanking, can form the barrier for deliveries over long distances. Most private clients will hardly acccept to pay more than the local charge. However, this problem can be solved by the establishment of a node, passing on conversations, so that the client pays the local charge, and the bank pays any long-distance charge.

Officebanking is the answer of the corporate clients to homebanking, but with a wider functionality. The communication will often be channelised through dedicated data networks or leased lines. A great part of the small, middle-sized, and partly the big companies communicate through the public telephone network. The facilities offered to the corporate clients include access to financial information services, information about accounts, and other electronic payment services. Trade in securities can also be performed by an officebanking system. Actually most routine transactions can be arranged for the corporate clients.

In the years to come all four IN services mentioned above will undoubtedly be far more in demand by bank clients in the part of the world which has an extended infrastructure for telecommunication. This means that the banks in the client-related operation will have an increased need of being able to offer such services, which again will mean an increased turnover for the operators.

6 VIRTUAL PRIVATE NETWORK

The sector-internal bank activity (back-office) means the activity performed internally in a bank company, for instance between the head office, subsidiary companies, branches, representative offices and agencies. By this facility a great deal of transmission of conversations and data in connection with e.g. trade in foreign exchange, trade in securities, clearing, booking, transactions, and credit information is made. Therefore, the CS1 service, VPN (Virtual Private Network), might possibly be applied with advantage in the internal banking operation. Besides, VPN's PABX (Private Automatic Branch eXchange) functionality may advantageously be applied for the construction of virtual dealing rooms. The PABX functionality may be relevant for both the client-related activity and the sector-internal activity.

Today the implementation of VPN services is solved by a switching-based VPN design. Here a relatively cheap starting VPN-service is offered which is quickly made more expensive by an increased number of switches. But by the development of IN, demanding large once-for-all investments, VPN operators can save costs as a centralisation of the information from the VPN configuration is possible. The use of software technology will furthermore mean an easier control of this information. Finally, there will be less likelihood of inconsistency between different nodes of the networks. Likewise it will be possible to extend the number of nodes without further investments.

In future the IN development will allow the VPN users - usually public network operators - to create client services which are independent of the sellers of VPN equipment. From the point of the public operators, VPN is a method of creating several private networks, using the common network infrastructure. This means that - with the same profit - the operators will be able to offer VPN services cheaper than fixed leased lines. Banks can remove some of or all their fixed leased lines to a VPN provider. Thus, the VPN service offers the banks to build virtual private networks by applying the resources of the public network. The lines of the subscribers = the bank departments, being connected to different switching nodes, hereby form a virtual Private Automatic Branch eXchange, PABX. This virtual PABX includes a private numbering plan, call transfer, call hold, routing information, all stored in a database at the VPN provider. Furthermore, the VPN-PABX may also include voice-response functions.

From any point whatsoever in the virtual private network it will be possible to obtain access to the network in accordance with authenticity. This authenticity gives the individual user access to a specific class of services providing specific rights and privileges. By change of places where private network facilities are desired or by changing capacity needs, it is easy for an operator to adjust by a software change the management configuration of the VPN services. (New design of modules named Service Independent building Block, SIB).

The public network can offer private customers a complementary supplement to transmission, called Value Added Service, VAS. One of the most important VAS is centrex. Centrex provides full PABX functionality. Private companies, such as small bank branches, can quite quickly establish or abolish PABX-similar bank facilities even at small places. This is possible without big investments for a PABX system at the individual place. In the strict sense of the word, Centrex is a forerunner of the VPN services. It contains the PABX functions of the VPN service, but not the intelligent functions of it. Thus, Centrex does not form part of the VPN architecture, but it can still be an important service, allowing the users - even at small places - to obtain access to VPN services.

VPN can be used both as complete network or as part of a network for complying with specific wishes. When calling a VPN service, extra signalling is required as it must be possible for the exchange to distinguish between PSTN and VPN calls.
As a backbone for VPN it can be advantageous to apply the signalling system "Common Channel Signalling System No. 7, CCS#7". CCS#7 provides the VPN users PABX similar facilities in the public network. The signalling between two PABXs can either be end-to-end or link-by-link signalling. End-to-end signalling is limited by the fact that it is not easy to extend the VPN service for offering Centrex service or for overflow of the offered traffic into PSTN.

VPN is a supplement both to the existing private networks and to the fixed lines in the public network leased by a company with own its management. Both types of networks are, as mentioned above, applied by big banks in the sector-internal operation. Leased fixed lines are not always an ideal solution. For safety reasons their transmission capacity must be calculated as based on the peak traffic. (The traffic in busy hours). However, this means a low average occupation per line. Alternatively, the overflow of the peak traffic can be led to PSTN, but this will limit the transparency between the PABC users. By internal bank operation, all things considered, a certain minimum of offered traffic will be required, before leasing of a fixed line can be financially profitable. Besides leased lines are characterised by a fixed allocated bandwidth, only changeable with relatively big extra costs in case of a changed traffic pattern. From the companies' point of view, having a present or future need for a high and safe transmission capacity of data, telephony, and possibly video and graphic arts, the VPN advantages can be summarised as follows:

- Reduced management costs as more parts will share the management of the infrastructure.
- Reduced capital costs for purchase of communication equipment.
- Savings of telephone charges and costs in connection with leasing of fixed lines.
- Increased operating safety as VPN is used through the public network, PSTN.

7 INTERNAL USE OF TELECOMMUNICATION IN BANKING

Here follows a description of the strategies of the big banks for use of primarily internal networks. Since the end of the eighties when there have only been three alternative strategies for the banks, "Global", has been the keyword in the financial world.

1. Banks can build their own private systems and use them as competition tools. This strategy is only possible for the big banks.
2. Big banks can build systems (overcharge-network services), which can be sold to banks being too small for building their own systems.

3. Banks can buy themselves into systems, established by others, or they can share systems based on public networks with other banks.

In principle, the existence of private financial service networks does not exclude the use of public switched networks. But in reality the banks operating on the international level have preferred to use privately leased fixed lines. Banks with bigger networks of these leased lines have built up centres, where management and supervision of the networks are being performed. Among others, this application of strategy number one is used by Citibank International Plc. Two main reasons for banks having developed such private networks during the seventies, the eighties, and partly the nineties, are the following:

- The increased data-service requirements from the banks were not generally available in the public networks. This includes problems with combining LAN's over public networks and often difficulties in setting the necessary services across national borders.
- The charge structure in the public network has made it advantageous for banks to lease fixed lines for high-capacity use.

These conditions gradually seem to be changed during the nineties. On the one hand, the new technology will reduce the costs for private networks and the European operators reduce the prices for leased fixed lines. But on the other hand, the public switched facilities will become more attractive for the financial institutions for two main reasons:

1. The net operators have started offering VPN, where lines are allocated dynamically in accordance with the needs of the clients. Besides the operator performs management and supervision of the virtually private network. This can mean a reliable and cost-effective solution for many big banks. VPN can advantageously be applied in connection with the third strategy of the banks.
2. Reason number 2 says that the banks have an increasing need for being connected directly to the clients.

8 ADVANTAGES IN CONNECTION WITH USE OF VPN IN BANKING

VPN will enable the banks to obtain access to advanced services, which today are not available from the established monopolists. In this connection John Sale, the leader of the group, European Virtual Private Network Users Association, has expressed: "Many of us are too eager to have audio conferences, telephone cards, common numbers and short codes".

A trade especially needing advanced telecommunication services, e.g. VPN services, are the banks. During recent years the banks have experienced big revolutions due to big trade slides, among other things in consequence of the strong development within the information and telecommunication technology. This gives the banks the possibility of entering into non-traditional bank areas such as credit secured by mortgage on real property, insurance and giro inpayment. But at the same time non-bank companies have entered into several bank services. Among other things, the insurance and finance companies have entered into banking, the retail trade offers its clients consumer loans, and telecompanies enter into various payment and credit systems. In the future it will be much easier to offer financial products by telephone, computer, or automatic teller machines. This means considerably less need for bank branches.

Furthermore, using the telecommunication facilities the back-office functions might just as well be removed to relatively cheaper addresses outside the expensive finance centres.

It is quite difficult to give a survey of the VPN services' dynamic allocation af bandwidth for a virtual private network as against banks' leasing of fixed allocated bandwidths. This is due to the fact that the prices differ greatly when a bank department is to establish or extend a fixed leased line. The price i.a. depends on whether extra two- or four-wired lines have to lead to the department, on how far the department is situated from the nearest exchange, the conditions for the final destination of the line and on how great a capacity per time is desired for transfer. As to the prices of the VPN services, these of course for the same reasons, differ greatly, and depend on how advanced the desired intelligent PABX functionality must be. However, in any case it is an absolute must for the banks that at any time a VPN service guaranties the agreed transfer capacity and security against hacking. For small and middle-sized banks the savings of costs for management and supervision of private networks will speak in favour of the use of VPN services. But for the big banks, such as Citibank International plc., having already spent big sums af money on construction of networks and communication centres, the advantages will be limited. Furthermore, there are also some strategic reasons for big banks preferring their own communication networks.

VPN's PABX functionality can be used for construction of logical Dealing Rooms. In geographically differently placed bank departments a relatively small number of physical dealing positions (terminal places for dealers) can be established at any place. By means of the PABX function these can be connected to *one* logical Dealing Room. Small bank departments can quickly establish and abolish dealing positions as required. Besides, it is evident that the client dealers will be placed in one department, and the interbank dealers in another.

A modern dealing position is constructed as a multi-media terminal. From this conversation to and from clients, brokers, and other dealers shall be transferred. The other dealers can then be physically placed in the same department or in another geographically placed department. Besides conversation it shall be possible to send and receive text, graphic arts, and telex as well as to receive information from financial news services (text, graphic arts, audio, and video). To this must be added the access to data systems of the banks, e.g. daily update databases with risk management information. This means information about limits for amounts to be traded in with the individual client, and the rate of interest to be demanded by the customer. Furthermore, booking and tape recording (confirmation) of transactions are performed. All this means a strong and manysided utilisation of the different teleservices. A combination of more dealing departments into *one* big department on top of a common PABX functionality will therefore require a considerably higher transfer capacity. In connection with the morning meeting of the dealers it will i.a. be advantageous if video conferences could be arranged between these geographically separated dealing departments.

9 CITIBANK INTERNATIONAL PLC.

Citibank International plc. has branches for corporate clients in the capitals of Copenhagen, Stockholm, Oslo, and Helsinki. Each of these branches has approximately 30 employees. Through leased fixed lines they are interconnected mutually and as well as connected to the headquarters in London. It could be imagined that the four Nordic branches by means of VPN will be connected - either all functions or only the dealing functions - as *one* big virtual department. VPN's PABX functionality will then give this Nordic Citibank's department a number of common telephone numbers and a kind of common secretariat. Contrary to ordinary telephony VPN will furthermore ensure that the necessary bandwidth, as agreed in advance, will always be available.

As partly mentioned before Citibank International plc., using the mentioned communication Strategy 1, has already built up a private network, consisting of its own communication centres with tandem, management and supervision functionalities, its own private cables and leasing of fixed lines at public operators. To this must be added the establishment of ISDN2 and ISDN30 connections, today primarily applied as backup for the fixed leased lines. In the Northern countries it is e.g. furthermore possible to connect the modern MD110 PABX'es of the individual branches to one big virtual Northern PABX. Therefore, it is quite certain that, in years to come the bank will prefer to upgrade its own network instead of applying the VPN services.

10 REFERENCES

Ericsson, L.M. (1994) *IN Survey*. 1994. p. 1.5 - 1.13, 1.31

Martikainen, O., Lipiöinen. J. & Molin, K. (1994) *IFIP TC6 Workshop on Intelligent Networks*. Lappeenranta University of Technology - Department of Information Technology. August 8-9 1994. pp. 53-55.

Schnepps, M. (1995) *Intelligent Networks*. Lecture Notes, Department of Telecommunication Technical University of Denmark. pp. 81-90.

United Nations. (1994) *The Tradability of Banking Service - Impact and implications*. Geneva. 148p.

Henten, A. (1995) *Impacts of information and communication technologies on trade in services*. Technical University of Denmark - Center for Tele Information. Januar 1995. 227 pp.

Falch, M., Henten, A. & Skouby, K. E. (1993) *DATE, Telematiserede bankydelser*. Department of Social Sciences, Technical University of Denmark, pp. 57-60 (In Danish).

Jensen, T. S. (1995) *Telekommunikation i Banker*. Department of Telecommunication, Technical University of Denmark. Master Thesis (in Danish).

BIOGRAPHY

Troels Schmidt Jensen got his M.Sc. in Telecommuncations in 1995. He has been working as a consultant in telecommucation and datacommuncation at Citibank CPH.

Part Two
Service Creation

5
The P103 Service Creation Environment Model

Carla Capellmann
Deutsche Telekom AG, Technologiezentrum Darmstadt
P.O.Box 100003, D-64276 Darmstadt
Tel.: +49 6151 83-3070, Fax: +49 6151 83-4221,
E-mail: capellmann@fz.telekom.de

Kjell Hermansen
Telenor Research and Development
Instituttveien 23, Postboks 83, N 2007 Kjeller, Norway
Tel: +47 63 80 91 00, Fax: +47 63 81 00 76
E-mail: Kjell.Hermansen@kjeller.fou.telenor.no

Parminder Mudhar
BT Laboratories, Martlesham Heath, Ipswich IP5 7RE, UK
Tel +44 473 642849, fax +44 473 637400
E-mail: parminder.mudhar@bt-sys.bt.co.uk

Jørgen Nørgaard
Tele Danmark Research
Lyngsø Allé 2, DK-2970 Hørsholm, Denmark
Tel: +45 4576 6444, Fax: +45 4576 6336,
E-mail: jnp@tdr.dk, URL: http://www.tdr.dk/~jnp

Abstract

One of the major objectives of the Intelligent Network concept is the rapid introduction of new services and features. The development of new services, from initial idea to implementation, called service creation, is therefore a vital process in the achievement of this objective. Within EURESCOM project P103 "Evolution of the Intelligent Network", the activity to support service creation, i.e. the service creation environment, was closely investigated. A generic model for the process of service creation was developed that covers a broad range of different service creation scenarios. This paper presents the P103 service creation environment model, thereby introducing important concepts of service creation and related aspects.

1 INTRODUCTION

The Intelligent Network (IN) is an architectural concept for the rapid introduction of new services and features into the network. For this, the process of developing new services, called service creation, is vital in order to reduce development costs and time-to-market of new services. It was one of the major objectives of EURESCOM[1] project P103 "Evolution of the Intelligent Network" [Vaar94] to investigate closely the complex task of service creation. A generic model, the so-called P103 service creation environment (SCE) model, was developed that describes the different service creation activities. Figure 1 illustrates the problem domain.

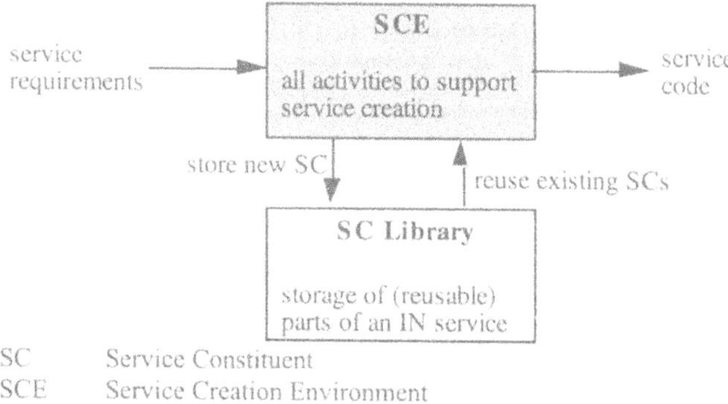

Figure 1 Problem domain.

Service creation is the transformation of often vague and imprecise requirements for a new service into code that implements the required service. To enable rapid and efficient service creation, the process is based on reuse. Existing service components, called service constituents in P103, are reused during the creation of a new service, whenever possible. *Service creation environment* is here defined as the encapsulation of all activities required to support service creation. The P103 SCE model specifies such an environment, i.e. it models the process of service creation.

In the following section the prerequisites for modelling the service creation environment, the P103 service life cycle and the P103 modelling technique, are introduced. Next the model and some of its major components will be presented. Finally, a summary will be given and the achieved results will be discussed.

---

[1] EURESCOM is a consortium formed by a wide number of European network operators in order to initiate, coordinate and supervise research and strategic studies in the field of telecommunications.

The aim of this paper, besides of course introducing the P103 SCE model, is to give a general overview on service creation and to highlight some important aspects in this area. The paper will not present any concrete or existing SCE tool though the model may serve as reference or basis for developing, assessing and/or comparing SCE tools.

2 PREREQUISITES

2.1 The P103 service life cycle

Starting point for the development of the P103 SCE model was the P103 service life cycle [P103TR3]. A service life cycle describes all the phases a service may be subject to through its entire life. Each phase defines the actions that can be prescribed to the service in that phase. Phases may be divided into sub phases called activities. The P103 service life cycle defines three phases: service creation, service deployment (including service withdrawal) and service utilisation. Figure 2 shows part of this life cycle, the service creation phase, which consists of six activities: requirements capture, analysis, specification, design, implementation and testing.

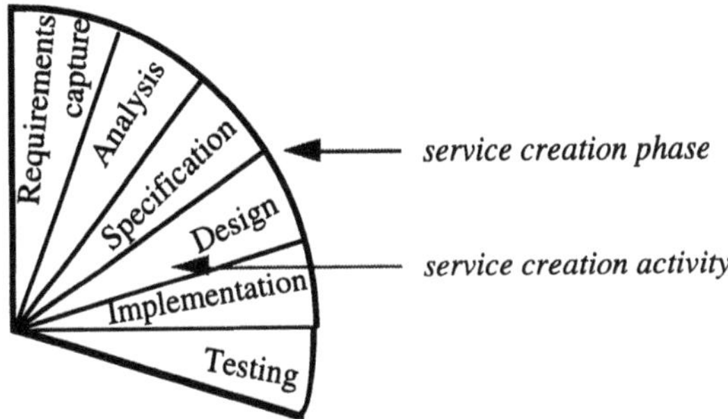

Figure 2 Service creation phase of P103 service life cycle.

In order to provide a structured description for each phase and activity, P103 developed a scheme for the description of the service life cycle activities. Each activity is defined in terms of its purpose, the required inputs and generated outputs, the tasks or actions to be carried out, the participants related to or involved and eventual requirements to the environment. Figure 3 gives an example of such an activity description.

2.2 The P103 modelling technique

The goal of the P103 service creation work was to describe the service creation process. The approach taken is based on the usage of an object-oriented method developed within the project, the P103 service composition technique [P103TR2]. This technique is based on OOram (object-oriented role analysis and modelling, [Reen92]), MSCs (message sequence charts, [Z.120]) and SDL-92 (not used in the SCE model, [Z.100]). Although the method was developed to model services, it is perfectly suited to describe processes. Role modelling which plays a central part in the method, is especially suitable to describe tasks or responsibilities to be taken during the different service creation activities, and MSCs provide a good means to visualise dynamic aspects within the service creation environment.

Acticity	Service Specification
Purpose	To derive a formal description of the service behaviour
Input	Service requirements
Output	Service specification
Actions	To specify the service using a formal description technique To validate the service specification ...
Actors	Service creator
Requirements to environment	Access to service requirements Access to existing specifications ...

Figure 3 Example for an activity description.

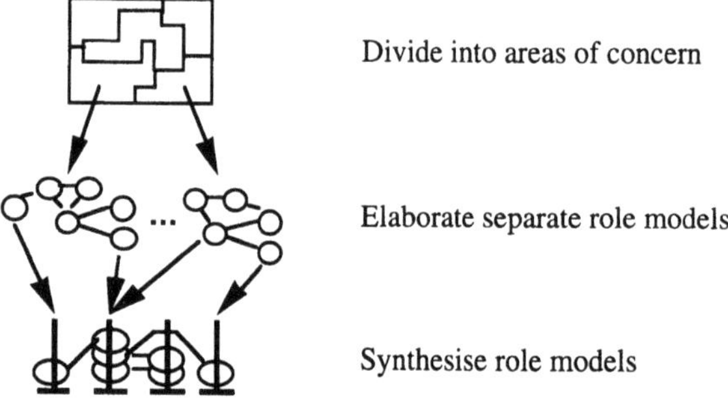

Figure 4 Main steps of the P103 modelling technique.

The main steps of the method are illustrated in Figure 4. First, the problem domain, here the service creation phase, is divided into different areas of concern, each of which can be treated independently of the others. Note that this step can be applied iteratively. The service creation phase is divided into its different activities, each of which again may be broken down into smaller tasks or areas of concern. For these areas of concern, so-called role models will be developed in the second step. A role model describes a problem domain in terms of roles and their collaboration structure. Thereby, a role defines the responsibilities of an object in the structure of collaborating roles. According to the method, the role behaviour is specified using SDL-92. This has not been done for the SCE model though; here the roles are described in prose. The messages that the roles can exchange are defined by contracts. Typical message exchanges may be described in form of MSCs. In a last step, the different role models are synthesised together by merging problem inherently related roles. Synthesis may also be applied iteratively.

3 THE MODEL AND ITS COMPONENTS

3.1 The SCE model

The P103 SCE model describes the service creation activities identified in the P103 service life cycle in terms of roles and relationships between them. Depending on the complexity of the activity at hand, several models describing its different aspects may exist. By merging roles the individual role models are synthesised in order to describe the complete activity and then the service creation process. Altogether the P103 SCE model contains 17 role models and 42 unique roles. An overview of this model is given in Figure 6, the complete model will not be included here. Besides the role diagram that graphically presents roles and their relationships, the model description contains textual descriptions of the area of concern, stimuli to and responses from the model[2], its roles and contracts (see Figure 5). In addition, many MSCs exist that describe typical interaction scenarios. An example for one such scenario is given in Figure 7. For a detailed description of the model, see [P103TR3].

The service creation process will be stimulated by a service creation *Client*, a role that requests a service to be created (see Figures 5 and 6). The request will be received and handled by the *Service Creation Coordinator* (see Figure 6). This role ensures that the *Client* has a single point of contact for meeting its service creation requests, and it takes care of the coordination of the creation process.

Although the main purpose of the service creation process is the creation of a service, it may also be possible for a *Client* to ask only for "intermediate" results, e.g. service requirements, specification, etc. The model thus covers the complete service creation process as well as the individual activities and any (reasonable) combination thereof.

In the course of its creation, the service will go through different activities within the service creation phase. Normally, the creation process will start with collecting the requirements for the new service (activity *requirements capture*). The collected requirements will be analysed with respect to consistency, completeness, service interaction, and realisation (activity *requirements analysis*). On this basis a consistent formal specification of the service will be derived (activity *service specification*). Note that the specification activity includes the vali-

---

[2] Stimulus/response pairs are part of a complete description of a role model. A stimulus is a message spontaneously created by one role of the model. It triggers certain events within the model and leads to a corresponding response.

Role Model — Service Creation Environment

Area of concern

The creation of a service

Stimulus	Response
create a service	created service
...	...

Roles

Client	Role that requests a new service
...	...

Contracts

Contract	From	To	Message
cs	Client	Service Creation Co-ordinator	createService (ServiceID) createSpec (ServiceID) ...
...	...	...	...

Figure 5 Part of the SCE model description.

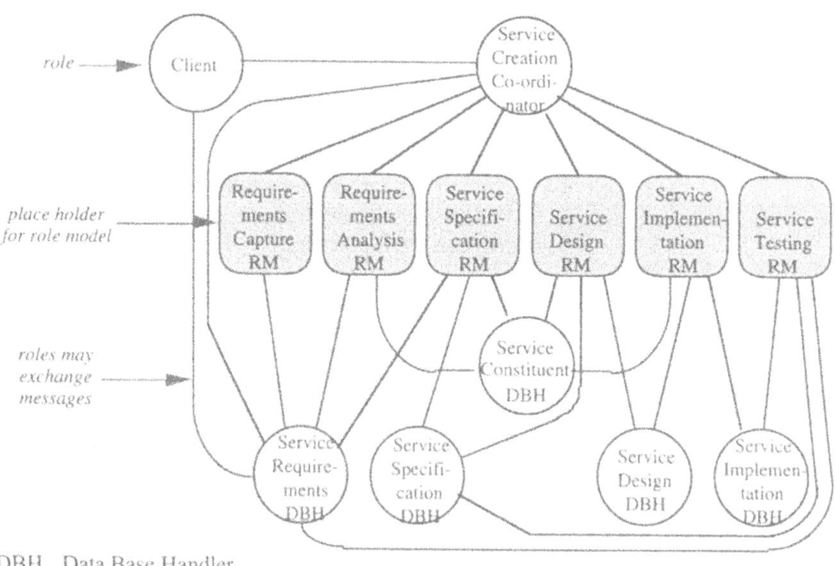

Figure 6 Overview of the P103 SCE model.

dation of the specification. From the specification the service structure to be implemented will be derived (activity *service design*), and code for distribution in the network will be generated (activity *service implementation*). Again it has to be mentioned that also these activities cover a validation of the results obtained. In the last activity of service creation, it will be checked that the service implementation fulfills its requirements and specification (activity *service testing*). Although this "normal" path through the service creation process is likely to occur often, it is not a must. Jump-backs to earlier activities are possible as well as forward-jumps to later activities in special cases where an activity might be omitted. The coordination of the sequence of the activities lies in the responsibility of the *Service Creation Coordinator*, who takes care of initiating activities, receiving the results and asking the *Client* for information or decisions, if necessary (see Figure 7 for a simple interaction scenario).

Each service creation activity is described in detail by separate role models, not included here. The overview diagram in Figure 6 only indicates these models (see rounded boxes in the figure).

During service creation a lot of information about the service (and its components) is created and used. This information is maintained by the different data base handler (DBH) roles at the bottom of the diagram in Figure 6. By providing the information needed in different creation activities, the DBH roles glue the service creation activities together. In the overview diagram in Figure 6, four DBH roles are assumed to store service related information whereas there is only one DBH role defined to store and maintain service constituents. It is as well possible to define just one DBH role for services or one DBH role to store both, services and service constituents. A fine grained structure with several data base handlers shows explicitly which information an activity requires, e.g. the identified service requirements are needed as input in the analysis, specification and testing activity whereas the service design is only input to the implementation activity. Which model is the most adequate can not be said in general. This depends on the model's purpose. The data base handlers are also essential to enable a reuse-based service creation process. Further aspects of storage of information in a service creation environment will be discussed in section 3.2.

The SCE role model may be characterised as follows. It

- is consistent with the service creation phase of the P103 service life cycle.
- does not put unnecessary constraints on the sequence in which the service creation phases may be run through.
- provides a general framework for service creation which does not impose unnecessary constraints on the service creation process, since the SCE model covers all individual models of the service creation activities without restricting the way they may be combined.
- reflects the reuse-driven approach towards service creation. The role models for the different activities are designed in such a way that reuse of existing results is integrated in the model whenever this makes sense.

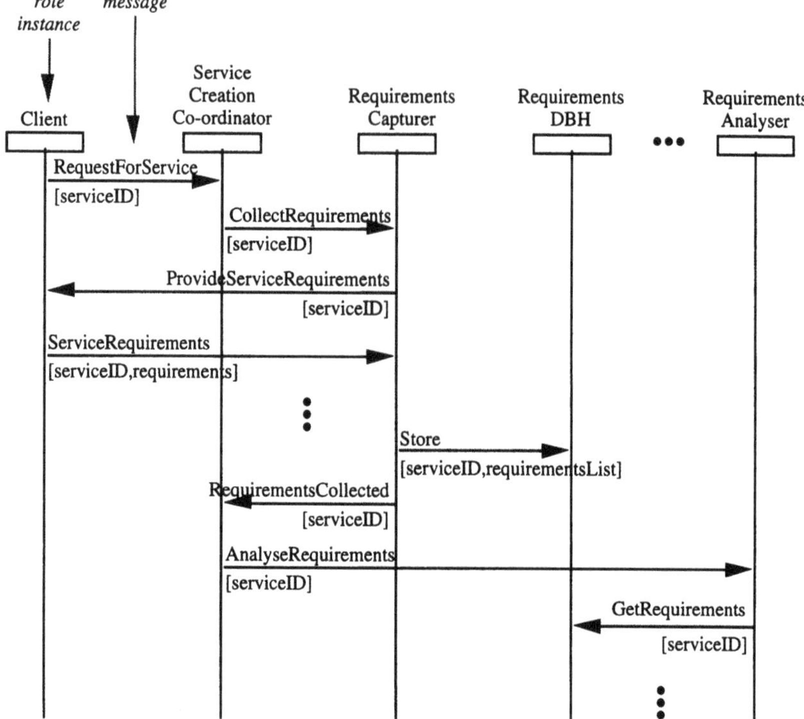

Figure 7 Example of an interaction scenario, represented as MSC.

The model presented so far captures the tasks to be performed during service creation in form of roles, and their relationships in form of contracts and interaction scenarios. Other important aspects like

- what information is important during service creation and how it may/should be organised (in order to support reuse),
- how to perform complex tasks like service specification and design, service interaction handling or reuse,

were also looked at. They are briefly discussed in the next two sections.

3.2 Reference model for storage

During service creation, a lot of information is generated and needed as input for different tasks or activities. Storage and maintenance of all this information so that it is available where it is required, and search and retrieval of certain information are essential for efficient and effective service creation. They become even more important in a reuse based creation process where a lot of information has to be maintained and very good search and retrieval facilities are needed. The problem of what information has to be stored and how to support search and retrieval is thus crucial for the definition of a service creation environment model. In P103 such a storage component (see SC Library in Figure 1) was therefore investigated more closely. A model describing what information should be maintained and in which structure was developed, always keeping reuse aspects in mind. The model is based on work carried out in the RACE project SCORE [SCORE93]. Figure 8 presents an overview of this model, the P103 component model. It is given in OMT notation [Rumb91].

The elements to be stored during service creation, here called components, may either be services or service constituents (reusable parts of IN services). For each component, three types of information are maintained:

- Administrative information: Here all the general information about a component is kept that is needed to administer an eventually big set of components. Examples of such information are the unique component identifier, the component owner or the (development) history, to mention only a few.
- Network related information: Here the information concerning network specific aspects of the component is stored, e.g. network requirements or target architecture.
- Role model: This is the core part of each component where its essence is described. Since in P103 services as well as service constituents are represented as role models, this core information about the component is structured according to the information kept in a role model, like area of concern, stimulus/response pairs, role diagram and so on. In addition, other information important during service creation like information required for service interaction handling (properties), information relevant for testing (test cases), or information about relationships between components (uses/used-in links) are stored here.

Note that the diagram in Figure 8 gives only the high-level information structure of a component. Some of the attributes are themselves defined as complex structures.

The component model defines the information structure of services and service constituents as required during the service creation process. It may thus be seen as one possible candidate for realising a data base handler within a service creation environment (see data base handler roles in Figure 6). In order to support reuse of components, search and retrieval functions play a very important role. The problem of efficient search and retrieval is a complex and difficult subject. Although the problem is not new, no general solution has been found yet. Within P103 a list of requirements was compiled that describe aspects relevant for efficient search and retrieval.

3.3 Important aspects: reuse and service interaction

Several sets of guidelines enhance the SCE model with respect to the support of reuse, service analysis, specification and design, service interaction handling and the incorporation of security aspects. In the following, reuse and guidelines supporting a reuse based service creation process as well as service interaction and how to handle this problem during service creation will be discussed.

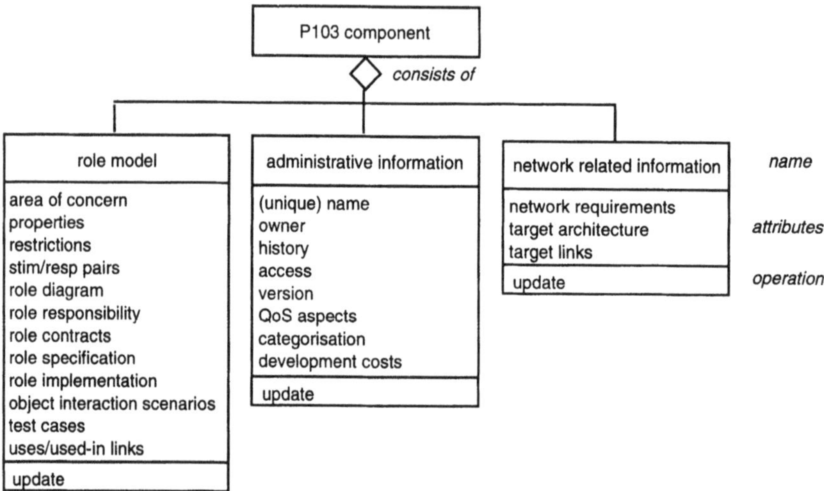

Figure 8 P103 component model.

3.3.1 Reuse

Reuse is one of the most popular buzzwords these days. Its appeal lies in the promise for (more) efficient software/service development. New developments can build on already existing components, thus saving development time and costs as well as reducing testing effort. Thus it is quite obvious that reuse of existing service components during the creation of new services is crucial for rapid service creation. Despite all these promises, one has to be aware that reuse is still an open issue in object-oriented research. Within P103, one step towards its concretisation was undertaken by developing guidelines to support reuse during service creation. These guidelines describe for each service creation activity what may be reused in this activity and in which way.

In general, the level of reusability (during service creation) is determined by several factors. In order to reuse a service component in different services, it needs to be independent of a specific service[3]. Reuse also depends on the way components are stored: information structure, search and retrieval strategies, etc. (see section 3.2). Another factor that influences the level of reusability, is the quality of the already existing components, their size and structure, with a trade-off between the latter two.

3.3.2 Service interaction

Service interaction is seen as one of the major obstacles to rapid service creation. The problem may arise that the operation of new services/features (unexpectedly) influences and alters the behaviour of the existing services. Figure 9 shows a well-known example where IN services CFU (Call Forwarding Unconditional) and SCR (Selective Call Rejection) interact.

---

[3] For a more general discussion of the notion of service independence, see [MuLiMi94].

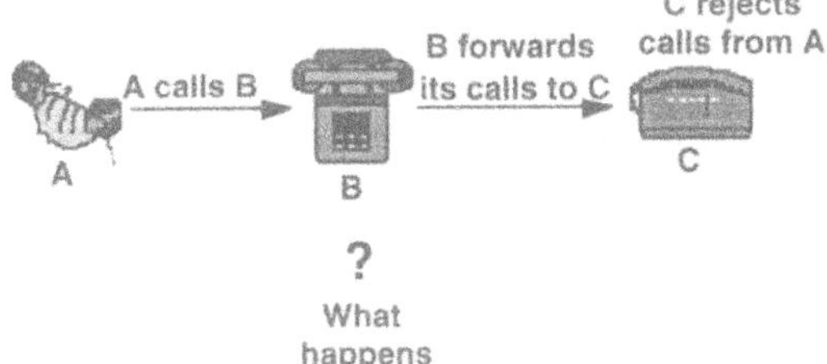

Figure 9 Example to illustrate the problem of service interaction.

In general there exist different ways how to handle the service interaction problem:

- Avoidance: Services (and networks) are developed in such a way that no interactions occur. However, this is quite unrealistic.
- Detection: (During service creation) potential service interactions are identified.
- Resolution: A solution to known interactions is provided.

Note that the problem of service interaction needs to be addressed during the complete service life cycle. Some interactions can be detected during service creation, whereas others will only be found out during service execution. Some interactions can be resolved reasonably during service creation, whereas others are better handled during service execution.

P103 has addressed service interaction during service creation only. Places in the P103 SCE model have been identified where service interaction should be handled. In addition to this, guidelines have been developed how to detect and resolve interactions. These guidelines were specifically developed in relationship with the P103 modelling technique. They are defined based on the concept of service properties.

4 SUMMARY AND CONCLUSIONS

A generic model of a service creation environment for future intelligent networks, as developed within EURESCOM project P103, "Evolution of the Intelligent Network", has been presented. Aim of the SCE model is to provide an universal framework for service creation. The model covers different aspects related to the service creation phase of the service life cycle. It integrates service interaction and security[4] issues in the process of service creation and incorporates a model for component storage. The P103 SCE model was elaborated using an object-oriented modelling technique developed within the project.

The SCE model is enhanced by several sets of guidelines that were not included here. Reuse guidelines state for each service creation activity what may be reused in this activity and which aspects have to be considered during reuse. Guidelines for handling service interaction in the different activities are defined based on the concept of service properties and in relationship with the P103 service modelling technique. Concepts and guidelines how to incorpo-

---

[4] Not covered in this paper; for more information, see [P103TR3].

rate security aspects into a new service right from the start are defined as well. In order to support the analysis, specification and design activities, the P103 modelling technique may be used.

By presenting parts of the P103 SCE work, important aspects of service creation like

- the different service creation activities,
- reuse of existing service components,
- storage of information,
- handling of service interaction

have been discussed. Others, like the integration of security issues or a smooth transition to service deployment/management, could not be included here.

In its overview diagram (see Figure 6), the P103 SCE model clearly indicates the major parts of service creation environments:

- functions that support the creation *process* (encapsulated in the *Service Creation Coordinator*,
- functions that maintain the creation *data* (data base handler roles), and
- tools that are used during the creation activities and work on the creation data (role models for creation activities).

In that sense, the P103 model is in line with the ECMA reference model for CASE environment frameworks (see [ECMA]), although both models were developed completely independent of each other.

The P103 SCE model may serve as a basis for

- the comparison of existing SCE tools,
- a requirements list to SCE tool developers,
- the specification of a company specific service creation environment (for a discussion and illustration of the derivation of a specific SCE, see [Cap95]),
- the development of a SCE reference model.

5 ACKNOWLEDGMENT

The P103 Service Creation Environment Model was elaborated in the EURESCOM project P103 “Evolution of the Intelligent Network”. Thanks are thus due to all partners who participated in the project, especially to those people who contributed to the development of the SCE model, namely Pierre Combes (France Telecom), Raymond Nilsen (Norwegian Telecom Research), Tapio Vaarnamo (Telecom Finland), Raúl Guitérrez and Marisa Felipe (both Telefónica de España).

6 ABBREVIATIONS

CFU Call Forwarding Unconditional
DBH Data Base Handler

MSC Message Sequence Chart
OMT Object Modelling Technique
OOram Object Oriented Role Analysis and Modelling
RM Role Model
SC Service Constituent
SCE Service Creation Environment
SCR Selective Call Rejection
SDL Specification and Description Language

7 REFERENCES

[Cap95] Capellmann, C.: A Generic Service Creation Environment. In Proceedings of TINA'95, Melbourne/Australia, February 1995.

[ECMA] Earl, A.: A Reference Model for Computer Assisted Software Engineering Environment Frameworks. In: Softwaretechnik-Trends, Band 11 Heft 2, Mai 1991. (also published as ECMA Technical Report TR/55).

[P103TR2] EURESCOM Project P103: Guidelines for Service Creation and Composition. Technical Report No 2, December 1994.

[P103TR3] EURESCOM Project P103: Service Creation Environment. Technical Report No 3, December 1994.

[Reen92] Reenskaug, T. et.al.: OORASS: Seamless support for the creation and maintenance of object oriented systems. Journal of Object-Oriented Programming, October 1992.

[Rumb91] Rumbaugh et.al.: Object-Oriented Modelling and Design. Prentice-Hall, 1991.

[SCORE93] The SCORE Component Model. Deliverable D104 Volume II, 1993.

[Vaar94] Vaarnamo, T. (ed). EURESCOM project P103, Deliverable No 6a. 12/1994.

[Z.100] ITU (CCITT) Recommendation Z.100, "SDL", 1992, Revised Recommendation.

[Z.120] ITU (CCITT) Recommendation Z.120, "Message Sequence Charts", 1992.

8 BIOGRAPHY

Carla Capellmann studied Computer Science at the University of Koblenz. After having finished her studies in 1991, she joined Deutsche Telekom AG, Technologiezentrum at Darmstadt. She is member of the research group "Functional Aspects of Networks" which deals with Intelligent Networks, Telecommunication Management Networks, and Formal Description Techniques. In EURESCOM project P103 "Evolution of the Intelligent Network", she worked on service creation and methodology aspects. Currently she is task leader within EURESCOM project P509 "Handling Service Interaction in the Service Life Cycle". She is also involved in EURESCOM project P508 "Evolution, Migration Paths and Interworking with TINA".

Mr. K. Hermansen holds a Ph.D. in physics from the University of Oslo. He has been involved in narrowband data and broadband projects at Telenor R&D. Currently he is working on services in Intelligent Networks.

Jørgen Nørgaard graduated with a MSc. in computer science from the Computer Science department at Aarhus University, Denmark. After joining Tele Danmark Research, he has been working at the software engineering department and participated in the RACE Cassiopeia project and in the EURESCOM P103 project in which he lead the task on activities for service creation and methodology work. At the computer science department at Aarhus University he co-designed and developed an open, distributed hypermedia system on object oriented database system and the X window system. At Nokia Telecommunications he worked with object oriented techniques and their application in telecommunications software, using the notation SDL-92 and OMT.

6

The Feature Interaction Problem in the IN: In search of a global solution

Nadir Belarbi
University of Versailles - PRiSM Laboratory
45, avenue des Etats Unis- 78000 Versailles - France
Tel : +33 1 39 25 43 32, Fax : +33 1 39 25 40 57
E-mail: NaBe@prism.uvsq.fr / belarbi@inf.enst.fr

Dominique Gaiti
Center for Telecommunications Research - Columbia University and UPMC - Casier 204
4, place Jussieu 75252 Paris Cedex 05 - France
Tel: +33 1 44 27 38 89, Fax: +33 1 44 27 48 37
E-mail: gaiti@ctr.columbia.edu

Abstract

Telecommunication operators are looking for new types of services to be introduced in multimedia networks. It is a good way to speed up the design of such networks and to raise the reliability and the efficiency of the actual telecommunication networks. During the last years, new types of services were introduced by the bell operating companies around the world. The toll free number is a key service that first permitted to realize that the introduction of new services in classical telecommunication networks is a difficult task. Ever more logical concepts and less and less physical attachment are needed to bring up more flexibility in the telecommunication network architecture to be able to take in charge new services. The Intelligent Network (IN) as a distributed system working in real time is particularly complex to design and to manage. The purpose of our article is to give a general overview of the feature interactions problem considered as one of the main problem in the IN.

1 INTRODUCTION

During the 1980s, the hardware components of a typical 15,000 lines PBX decreased by 45% (from 55% of hardware to 30%) while the software components increased of 1500% (from 5 packages to 80) [NT 89].

Today, it seems logical to say that telecommunication networks rely essentially on software. Therefore, problems of such networks are now, in major cases, software problems.

The difficulty for designing and for developing software of such complex distributed systems is an important obstacle in the development of new services in the Intelligent Network (IN). Telecommunication software must proceed in real time and has to be distributed. It is difficult to add new services to an existing system without modifying its operations. Each new service may interact with many existing services, causing an unexpected service behavior or a system failure. This problem is generally called a feature interaction in the area of telecommunication systems [Bowen 89].

Many papers have been published about the design process of a service (from waterfall models to the idea of growing software); many methodologies have been proposed (from the structure analysis and design to the object orientation), and many approaches have been suggested (from drawing pictures to formal methods with mathematical foundations). Tools have been provided for most of the above schemes [Linden 94]. None of these have turned out to be the Silver Bullet they promised to be, may be because most approaches tackle the accidental tasks that arise from software production and not the essence of the conceptual model [Brooks 87], [Harel 92]. It seems obvious to say that there is no global methodology to solve the feature interaction. The results of different research fields are needed to address all facets of the feature interaction problem. These fields include, at the minimum, formal description methods, software engineering, protocol engineering, distributed computing and distributed artificial intelligence [Cameron 93].

The consequence of this situation is that several methodologies try to address the interaction problem at different stages of the service feature life-cycle. We are going to introduce some of the main methodologies involved in the management of the interaction problem.

The remainder of this paper is organized as follows. The next section introduces the feature interaction problem in the IN and the main strategies used for its management. The third section gives first, an overview of the main concepts used by these strategies. Then, we present some of the main solutions corresponding to each strategy. Finally, some remarks conclude the paper. We will not investigate here all the existing methodologies and examine in detail their mechanisms. For more details see [Cameron 94], [Cameron 93], [IEEE 93a], [IEEE 93b], [IEEE 93c], [Bowen 89]. We rather try to focus on the main concepts they use to reach their goals in each kind of strategy.

2 THE FEATURE INTERACTION IN THE IN

2.1 An introduction

The feature interaction problem seems to be an inherent characteristic of the IN. Every time a new service is introduced in the IN, there is a high probability that this service process will meet problems within the new environment. The design of new services is a critical task because we cannot anticipate all the possible situations new services will meet during their life time in the IN. The problem is very similar to the problem of designing new software in computer science: a long life cycle is needed to validate the product.

To fulfill customer demands, an increasing proportion of value added services must be provided through the software. This adds significantly to the flexibility of the switching and transmission components but this increases the complexity of the resulting systems. Software design, installation, testing, configuration, and maintenance in a widespread heterogeneous environment are becoming increasingly difficult. With each new service and/or feature, systems become more complex. Those systems are connected in places where no such interconnections were foreseen. It is hard to understand the whole system and to be sure that additions and interconnections will have the desired effect and not another [Linden 94].

Generally three categories of approaches are used to address the problem of feature interactions [Cameron 93] :

- Avoidance;
- Detection;
- Resolution.

These three approaches are complementary. It seems impossible to avoid completely interaction problems even when a correct specification of the services is provided. For that reason, detection and resolution steps generally follow the avoidance step. Detection and resolution steps seem to be related. The detection of an interaction can for instance call a resolution procedure and so on. Many approaches are used in each category, and they can be divided into two main groups: *off-line* and *on-line* approaches [Bouma 94]. The first type of approach operates before the process of a service in the IN. The distinction between off-line and on-line approaches is not significant in the avoidance step, but has a strong meaning in the detection and the resolution steps [Bouma 94].

We prefer to talk about the management of an interaction when we use the concept of avoidance, detection and resolution: solving completely an interaction is an optimal situation that can rarely be completed without effects on the process of one or several services.

We notice that few methodologies, dealing with the IN interference problem, address only one category of the mentioned approaches. Detection methodologies for instance try to propose a resolution step for the detected interaction and that seems natural. Avoidance methodologies are generally not well adapted to the resolution techniques. However, in some cases avoidance methodologies can include concepts strongly associated with detection and resolution problems (the Meeting Organizer in [Cross 94]).

It is also very difficult in some cases to say if a methodology avoids or detects interactions. The general tendency is to have a methodology including both aspects: first a reliable framework based on the experience from the software life cycle and second, a mechanism that permits a controlled management of the service feature operations. It is important to see that some detection methodologies try also to avoid feature interactions, in supplying control mechanisms. The on-line and off-line concepts [Bouma 94] are also strongly correlated in the detection approaches: the simulated environment can be very close to the real IN environment and so can permit the direct use of off-line detection concepts. Thus, a solution can be obtain at the service interactions through the on-line situation. All these common points underline the relativity of the avoidance, detection and resolution categorization. In spite off all, such classification is needed for a best description of the feature interaction problem, during the life cycle of a service. These three different approaches can be combined in different ways for the elaboration of a particular solution, for the management of the interaction problem (Figure 1).

Figure 1 Functionalities of a complete system for the management of interaction problems.

An optimal system for the management of interaction problems can be composed by three components having three different functionalities (Figure 1). We suppose here that the service is at its final life cycle stage, running in an environment close to the IN or directly in the IN.

The service will check at each of its actions if there are no interaction problems. If an interaction problem is detected by the detection component, a resolution strategy is engaged by another specific component until a solution is found. Once this solution has been applied successfully to the resolution of the interaction, it is recorded by the resolution component and the detection component. This new detection information can also be used to avoid situations that could lead to an interaction in the IN.

2.2 A definition

A feature interaction problem is generally defined as an unexpected service behavior. The introduction of a new service may cause behavior changes for other services (policy interference). The new service may also behave itself in a different way it was supposed to do. The problem is that this new behavior in the service and their consequences are not wanted and planed. Operators and users can have specific views of feature interactions.

Let us provide with an example of a feature interaction from the user's point of view. We consider the emergency number (911 in the USA) and the three ways calling.

One characteristic of the 911 number is that when the 911 operator answers a call, it could be ended only by this operator. We suppose A is calling B, and during their conversation suddenly A is in an emergency. If B adds the 911 operator as a third party, the properties of the 911 number apply to B's line, but not to A's. Ideally, the properties of the "911" should be apply to A's as well; A can still hang up, and B cannot [Cameron 93].

The feature interaction problem from the operator's point of view can be different and varied. The problem may occur from an incompatibility with the user profile [Mierop 92], an unreachable goal [Faci 94], a logical model unworkable [Gammelgaard 94], an inconsistent system description [Blom 94], a state transition different from the corresponding state signature [Tsang 94] or an updated inconstancy [Cameron 91].

3 THE MANAGEMENT OF FEATURE INTERACTIONS IN THE IN

Several approaches are available to solve a feature interaction problem. All these approaches agree with the necessity to downsize the complexity of the system with the use of different architecture models. Modularity and decomposition of the systems in several blocks are key concepts of these methodologies. We remark that all these methodologies tend naturally to use object oriented models and their interesting properties [Mierop 92], [Erradi 92], [Arango 92].

The modularity of a structured system is a natural characteristic of an object oriented system This one may be easily extended and managed. The object model is based on several elements, as the abstraction, the encapsulation, the modularity, and the hierarchy. This provides the way to define a precise responsibility to each object of the system. All these characteristics downsize the complexity of the system and permit to express temporal and space constraints. Using the Case Driven Modeling (UCDM), coming from the object oriented software engineering, seems to be an efficient and helpful technique to design new services [Kimbler 94b].

Generally all the methodologies agree with the necessity to model the system with an architecture based on several levels of abstraction. These levels are related to each other through interfaces that exchange precise data. TINA and INCM are two examples of such architectures [TINA 92], [ITU-T Q120xx, 93]. The IN should take into account heterogeneous equipment that are distributed over its whole structure. For that, IN architecture can follow different existing models like ANSA [APM 91], ODP [ISO/IEC 10746] or CORBA [OMG 94].

Thus, it seems logical to see the IN as a distributed system, composed of several entities, that must work together to achieve several goals. The object is the fundamental entity that constructs the whole IN system. This component can be called application, entity or agent, but it is an object in its role and in its structure. These objects must communicate and exchange information to ensure the operation of the system. The intelligent agent approach, which is an extension of the object concept, tends to permit the modification of the internal structure of these objects and thus to adapt their behavior to the system operations by using the negotiation principle [Griffeth 94].

The use of formal specifications and languages to describe the structure of the modeled system seems to be a mandatory condition to get a correct and coherent description of the system. LOTOS [ISO 8807], SDL [ITU-T Z.100] and ESTELLE [ISO 9074] are some examples of such formal languages. LOTOS is strongly used because of its capability to define parallel and synchronizing constraints in the behavior of the modeled process.

The dynamic of the system (the service) is generally modeled by a sequence of different states which takes part of a Finite State Machine (FSM). Each state or a combination of states corresponds to a particular event of the system (service) functioning. The use of a formal framework as a Finite State Machine (FSM) [Hopcroft 79] permits a powerful management of the system with time constraints, using efficient mathematical techniques.

Avoidance and detection schemes can be associated to these mathematical techniques. Constraints [Halpern 90] or logical rules [Gammelgaard 94] [Combes 94] can control the operations of the system and can check the different features of the system. These techniques validate the state transitions and thus detect possible interactions. A similar approach consists to associate to each Service Feature a state transition sequence called a state signature [Tsang 94] or a Feature Profile [Mierop 92]. These approaches can be implemented using a centralized or a distributed knowledge database. A distributed knowledge base permits the association of a specific control entity to each feature: Feature Manager (FM) [Cain 92], [Tsang 94], Feature Interaction Manager (FIM) [Griffeth 94] or User Profile [Mierop 92].

In the resolution step we are free to redesign or to adapt the service to the system (after testing the new process). Redesigning a service is an efficient way to control its behavior and to avoid interaction problems and unspecified operations. But the redesign step is a long and a hard (off-line) technique which is generally experimented in a simulated environment.

Another solution consists to give to the service feature an intelligent and autonomous behavior that permits the negotiation concept [Griffeth 94]. Such intelligent services will be introduced in the IN to provide an automatic adaptation to the structures and the processes. A negotiation approach will require a certain delay to solve feature interaction problems [Kimbler 94b]. The problem is to know if this negotiation period will not affect the quality of service.

3.1 Avoidance step

The schemes to avoid service interaction problems may be provided directly by the platform supporting the service or by the service creation environment. That leads to a service implementation which is intrinsically independent on the interactions [Cameron 93]. Such approaches incorporate open distributed platforms in telecommunications systems to deal with interactions that are qualified of "intrinsic problems in distributed systems" [Cameron 94].

Experiences to validate the restructuring approach to feature interaction, indicate that the feature interaction needs to be viewed at different levels in the system, starting at the highest level with the customer or enterprise point of view and going through various levels [Herbert 93] [Cross 94]. A feature interaction at one level is not necessarily the same feature interaction at another level. Thus, an undetected interaction at one level may involve problems at different levels. The relationship between two levels may be complex and the detection of the interaction quite impossible to find.

[Linden 94] suggests that there are two ways to reduce the complexity of distributed computing systems and their design. First, it is possible to introduce constraints on the structure of the systems to define clearly the interfaces between the software components and the ways these components may interact. In that case components can be easily analyzed and replaced due to these well-defined interfaces. Second, it is possible to provide a framework to describe the system, the software components and the interactions between them in an independent way of the system design. Principles of separation and substitutability are used to reduce the feature interaction. Separation suggests that two entities (features) have to be completely physically separated. In that manner there is no way for two features to interact (in a positive or a negative manner). Substitutability suggests that software components can be replaced or extended by other components that are compatible.

A similar view is used by [Cross 94] where the entities added to the system are not features, but non-interacting applications. The features of a service may be modified through new components that have to be added in the communication units.

This new paradigm affords an organized range of parameters that includes a choice in the temporal connections between services, a choice in the route to transport the information and a choice in the participants. MORGAN is an implementation of this model with an object-oriented architecture that permits to organize meetings. In this architecture there is a collaboration between the different classes that represent the system. Additional objects include the Meeting Organizer which incorporates the functionality for organizing meetings and a knowledge base which contains the rules to schedule the meetings [Bendeich 93]. MORGAN solves feature interaction at its lowest level by using a matrix.

A knowledge-based representation is also used to create an unambiguous understanding of the requirements of the telephone features [Dankel 94].

Specifications of these requirements, written in an English-based language, are described in a natural language-based system into the knowledge-based representation (requirement capturing system : GATOR). A natural language is used to introduce the information on the interface. An interpreter updates the knowledge base to store this information. The goal of this conversion is to provide less ambiguous written mandatory documents and to update automatically existing requirements documents when new interactions are found.

Other approaches try to extend the existing signaling protocols to prevent interactions [Cameron 93]. Guidelines for a service creation and a service management also take part in those avoidance schemes.

3.2 Detection

All interaction problems cannot be solved just by applying a good avoidance methodology for the service. The detection step is necessary to detect new feature conflicts and to give to the resolution step some helpful information. The detection step can use both off-line and on-line techniques to check an inconsistent service operation in the IN. Off-line detection techniques come mainly from verification and validation tools, simulation schemes of the environment, and the use of checklists. The number of services and service features involved, the number of ways in which features can interact, as well as the size of a complete, detailed feature implementation description will usually make the complexity of the off-line detection of all possible interactions too high to be feasible. Therefore, it is necessary to use on-line detection techniques to check the service process in the IN. Such techniques analyze the behavior of the service at different stages. The instantiated parameters and invoked features make easier the analysis of the interactions. The next sections present some of the main concepts used by the detection techniques.

3.2.1 State-transition machines

State-transition machines and detection algorithms can take part of an off-line automated solution, for the detection of feature interactions. In [Braithwaite 94] the service features are arranged in layers that represent the machines Each machine can pass its unaltered input, represented by a token, to the next machine and thus, the token is propagated through the layers. A call is represented by a single-ended call model based on agents rather than on connections.

An agent is a component of the system that participates in the establishment of a call. To avoid feature interaction problems, a state-transition machine is created for the different types of calls that the various features can modify. Finite-state machines are created to model the set-up and the release of the call. Then, features may be added on top of these machines. In that case, a detection technique is based on the execution of all possible configurations of the component feature machines on all possible call models. Some interactions are detected as a conflict between the required information for a certain state and the real information found in that state. Thus interactions can be detected by testing reachable states of a composite machine. A backtracking algorithm is used to build the reachability graph of the aggregation of the two features.

A similar solution using state-transition machines is proposed by [Ohta 94] where the approach takes into account the network from the user point of view, which is considered as a black box. Service specifications are given by the description of the behavior of the terminal. Interactions are modeled as unwanted properties (illegal or lost transitions) of these descriptions and a service specification language based on state transition rules is used to ensure this function. Building blocks can be also used to model features and their operating context. [Lin 94] suggest that only three operating contexts are enough to treat all relevant cases of interactions: originating side of the call, terminating side, and two-parties call. Again finite-state machines are used to obtain a procedural description of the feature behavior.

A composite technique is presented by [Gammelgaard 94]. It use Finite State Machines and logic to define labeled transition systems over which network properties (properties in all states) can hold transition rules. If a state fulfills a (first-order) precondition, a certain transition is enabled and the target state fulfills a corresponding post-condition. Feature interaction problems appear in that case when certain logical models are not possible. Network properties and pre and post conditions of declarative transition rules are formulated in a simple logic that is a restriction of an ordinary first order logic. Both rules are taken simultaneously into account. There is a detection of an interaction if some requirements on the resulting state are not satisfied. The advantage of using such a technique is the possibility to carry out standard logical deductions and to support them by explicit network properties.

On the other hand [Inoue 92] and [Ohta 93] use non determinacy in the selection of transition rules to detect interactions; if two rules are both true in the global state and if a certain priority rule does not apply, there is an interaction. We can notice that the method of [Ohta 93] is well suited for the detection of interactions while the method of [Gammelgaard 94] deals with the interference [Mierop 92] or with the policy interaction [Gaarder 93], which are very closed concepts.

3.2.2 Logic

The concepts of constraints and knowledge goals [Halpern 90] are also used to detect feature interactions [Faci 94]. Constraint-oriented specifications are developed in LOTOS [ISO 8807]. A type of operator corresponds to each type of constraints. The designer defines a set of knowledge goals and verifies their reachability when the involved features in the service process are active. If a goal is unreachable, the designer concludes that a feature interaction (or design error) exits. The goal is materialized by the behavior section in LOTOS process specifications.

[Combes 94] propose a user view abstract model using the SDL language [ITU-T Z.100]. The methodology is based on concepts coming from the Service Plane level of the INCM architecture [ITU-T Q.1205, 93] and from the temporal logic. The feature requirements are expressed as properties through a language based on a modal or a temporal logic. Interactions are detected when the system is not able to satisfy these properties [Combes 93].

Temporal logic [Manna 92] and Finite State Machine models are also used by [Blom 94] for an off-line detection of feature interactions. This logic coming from Lamport's Temporal logic of Actions [Lamport 91] is based on first-order formulas (with event predicates and actions describing the state-space changes) using the linear temporal operator always. Interaction in this case is characterized by an inconsistent system description resulting from the interaction behavior of two features.

3.2.3 Design methodologies and creation environments

Another way to decompose the system and to downsize the complexity of the interaction problem is to group together all the features which have similar functionalities.

[Kimbler 94a] suggest to break up the services into their different features and to concentrate the analysis on possible interactions between features which belong to categories subject to interactions in different stages of their life-cycle. A spotting algorithm is used to recognize combinations of features subject to interactions. This algorithm uses information about the interactions among the feature categories and the interactions among the services.

Use Case Driven Analysis (UCDA) is also proposed to detect combinations subject to interactions of service features [Mills 87]. The Use Case Driven Analysis is a widely spread technique using an object oriented software engineering [Jacobson 92], [Rumbaugh 91]. An approach based on UCDA is given by [Kimbler 94b]. The method starts from an informal user-oriented description of a service by generating scenarios described as sequences of events and user interactions with the system. These scenarios are transformed into service usage models describing the dynamic behavior of the system from the user point of view.

The analysis aims to find different possible scenarios of the system operation. Several models are built during the analysis; the most important models are the User Case Model (UCM) and the Service Usage Model (SUM) [Kimbler 94a].

The User Case Model (UCM) describes Actors and User Cases. The Service Usage Model (SUM) contains dynamic models of services. A User Case is a specific scenario of the system operations described as a sequence of events and interactions between the user and the system. It is obvious that User Case Models will not cover all the possible combinations of services and features and all the exceptional situations that might happen during the call set-up and the release process. But those models look like a structure that helps the designer to analyze the functionalities of a complex system under different point of views.

[Mierop 92] also use an object oriented environment to assist the designer to model and to analyze the effects of the introduction of a new service and to solve service interactions (interferences). The concept of interaction is here quite different from its general meaning. A service interaction is not considered as a negative point in the service process. The interference concept tends to underline the unwanted aspect of an interaction in general.

An approach tries to avoid, to detect and to solve the service interaction (interference) in the Service Creation Environment (SCE). ROSA is an object oriented architecture composed of two separate architectural frameworks: the Service Specification Framework (SSF) and the Resource Specification Framework (RSF) [Oshinsanwo 92]. ROSA has developed an object model, the ROSA Object-Oriented Model (ROOM), which focuses on the description of services. This model is very similar to the ANSA object model [APM 91].

A service is viewed by a client as a set of operations (on encapsulated data). In this approach, an object is a group of interfaces (Service Provision Points) and the associated data. An object has multiple interfaces and the number of interfaces can change. An object can also be created dynamically. The approach relies on dynamic objects and on an interface creation to configure dynamically telecommunication services. Each object has a clearly defined responsibility in the system, which helps the designer to understand the implication when new objects are added or when some objects are changed. We notice that changing the internal behavior of an object will not generally change its role in the system.

The detection strategy is based on a special object called user profile which knows to which services a user is registered and under which conditions certain (supplementary) services can be activated.

3.2.4 Other concepts

On the other side, [Cameron 91] use the concept of the update-inconsistency as a condition to detect interactions. If, using the select-all policy, two features force some variables to be updated more than once in a cycle, then there is an interaction. [Cameron 91] use a single variable to record active features. This gives an update-inconsistency as soon as two features start up at the same event, even if the features are not in conflict.

[Tsang 94] suggest an on-line detection technique, based on the introduction of a Feature Manager (FM) that checks if an activated feature displays the correct behavior. The feature is required to behave as a deterministic Finite-State Machine could do. The FM should be able to perform a behavior analysis and an analysis of the resources modified by the feature at each step. The advantage is that we do not need a centralized knowledge. The FM as a part of the Feature Management System (FMS) represents a low-cost approach requiring no modifications on the IN architecture. When a feature is instantiated, a Feature Manager (FM) instantiation is also performed. This instantiation includes the storage of a copy of the sequences that monitor the features. These sequences are independent from the other FMs. The stored sequences stay in a central database and each FM monitors only one feature instantiation.

Each feature has a particular sequence of state transition called state signature, which represents the behavior of the feature. When the activated state sequence does not match with the sequence signature of the feature, an interaction problem has to be expected. It appears that using only state information is a disadvantage. The size and the number of sequences have to be limited. If not, the inclusion of data is expected to cause a state explosion.

We notice that off-line validation tools and simulation environments try to recreate the original and real environment of the IN as it is going to be tested with on-line approaches. Clearly these techniques cannot replace completely and efficiently the environment of the IN. Moreover, on-line detection approaches are limited to certain classes of interactions when a service is tested. It was shown that certain parts of these interaction situations can occur when very specific conditions are met. This makes that on-line detection techniques are not able to eliminate all possible causes of interactions. Then, it is necessary to solve these interactions when they occur.

3.3 Resolution

Once an interaction is detected, resolution techniques have to be engaged to eliminate this interaction or at least to minimize the effects on the environment. Off-line and on-line resolutions can be used in that case. Off-line approaches include the manual redesign or the re-implementation.

On line approaches try to find a way to solve an interaction when this interaction occurs or even when it has occurred, to try to minimize its consequences. Feature-interaction manager (FIM) in the IN, event-based resolution mechanisms and negotiation schemes are different approaches to solve the interaction.

[Cheng 94] presents a technique for the incremental service specification and for the interaction management concept to support the incremental specification and addition of services. This method is mainly based on the process composition specified in LOTOS [ISO 8807]. Each new service has to be specified independently using a common call model and/or an existing or a new Service/Service Feature (S/SF). Thus, the new service is developed by chaining Service/Service Feature behaviors together. The overall specification approach consists in a sequence of process behavior interconnected by links. The links establish a control and/or a data relationship (interaction) between two or more process behaviors.

A starting and a terminating process have to be defined when the different activities are chained. Constraints are applied during the process composition to manage and to solve interactions. LOTOS permits the introduction of such formal constraints on the parallel composition and on the synchronization of several processes.

The appropriate resolution of the interaction requires to know the intention of the subscribers when they activated the features. Typically, a particular implementation of a feature does not represent the ultimate intention of a user, but just a way to achieve the intention of the customer. There might exist alternative ways to achieve the intention of the customer. [Griffeth 94] assume that constraints can be expressed in terms of operations that a user or a provider is willing or unwilling to perform. To solve conflicts between the constraints of different users, a mechanism of negotiation is provided to determine which set of operations will be used to initiate or to modify a call.

An important consequence of the use of the negotiation principle is that the autonomy of different users and providers can be preserved [Griffeth 93a], [Griffeth 93b]. It appears that to solve the interaction, one must favor one feature over another [Griffeth 94], introducing this way, a priority concept.

A negotiation mechanism must provide the user agent a way of exploring all the possible alternatives to achieve his intention. After receiving a request to set-up a call which is unacceptable for the user, the user agent must recognize which intention might be behind the request and then to derive from that intention alternative ways (possibly acceptable) to achieve it. Even when a user is not informed explicitly about the intention of other users, he may be able to speculate about the intention, based on the information he has, i.e., the received request.

The system is composed of three concepts which use negotiation: the platform, the negotiation objects and the user interface. The mechanism of negotiation assumes that a call initiation and a modification are described by a collection of operations that determine the form of a call. Policy feature and technology features are both related to these operations. The technology feature determines which operations are available and the policy feature determines which operation a system user is willing to use. Object-oriented platforms help to address a technology feature which is an individual operation that the platform provides. Thus, the feature can be added or removed rapidly and easily from the system.

A policy feature is a constraint on a set of operations that a user or a provider is willing to perform to initiate or to modify a call. This set of operations is sent to the negotiating system, which returns different sets of operations to be executed. The negotiation mechanism guarantees that the concerned users and providers have authorized any set of operations it returns.

This mechanism works on behalf of entities, which have policies to select operations they want to perform. Agent objects represent the various entities of the system and try to carry out their policies. Negotiators help the agent to reach an agreement.

An agent-object is assigned to each entity in the network. The functions of an agent are to produce proposals, to initiate or to modify a call, to evaluate proposals coming from other agents, and to generate counter-proposals. The proposals have to specify the desired operations of the calls. If a proposal is not acceptable to one entity, the agent constructs a counter-proposal.

Three ways of negotiation exist [Griffeth 94] :

- Direct negotiation, where the agent negotiates directly with the other agents without the assistance of a mediator.

- Indirect negotiation, where a monitor called negotiator is used for the progress of the negotiation. The negotiator is a dedicated entity that recognizes which agent has to approve a proposal and routes proposals and counter-proposals to the appropriate agents.
- Arbitrated negotiation, where the agents do not need to generate and to evaluate proposals.

An arbitrator takes the complete script of each agent and has the responsibility to find a resolution of the conflict.

The use of a negotiation object permits to avoid some problems as:

- Non terminating sequence of proposals and counter-proposals;
- The introduction of new agents, and the facilities to communicate with them;
- Negotiation can store knowledge and use it for further conflicts (experience);
- Different mediators can be used for different situations providing benefits of specialization (smart or dumb negotiators).

The mediator is supposed to find an agreement that should be quite good for all parties.

The user interface allows the user and the service provider to access two kinds of functionality : to express constraints and to submit proposals. These constraints can be expressed in any logic languages. The basis of the mechanisms necessary for the negotiation is to provide a goals hierarchy, where the lowest level (basic) is characterized by operations on a given platform and where the higher level corresponds to the possible goals or intentions of the users. New goals can be defined by combining other goals. A goal is realized when the combination of the sub-goals leading to a final goal is achieved.

4 CONCLUDING REMARKS

A global solution for the interaction problem will probably use many concepts and solutions from different research fields. At a logical level, we trust that this solution will use object-oriented concepts, associated with a strong formalism to build a representative model of the IN system. The object methods for the analysis and the conception of object-oriented systems will probably also have an important role to play for the design of such telecommunication systems.

The dynamic operations of the system will probably be controlled by Finite State Machines associated with logical rules of transition. State signature, object profile or feature manager concepts may also constitute good strategies to detect feature interactions. Even if we take care of the two first stages, avoidance and detection, the resolution stage will always be necessary due to the limitation of simulation tools to approach the real environment and the operations performed in the IN. Negotiation techniques seem to be an efficient solution for the real time and for the autonomous resolution of interaction problems. The problem with these techniques is first, to know if the search of an interaction solution will not take too much time and second, if such an autonomous system will not perform an unexpected behavior.

5 REFERENCES

[APM 91] An Application Programmers Introduction to the Architecture, Release TR.017.00, APM Ltd, 1991.

[Arango 92] M. Arango et al., Touring Machine: A software Infrastructure to Support Multimedia Communications, Multimedia'92, 4th IEEE COMSAC Int'I, Wksp. on Multimedia Communications, Monterey, California, April 1-4 1992.

[Bendeich 93] J. Bendeich, I. Black, C. James, J. Ladmore and P. Sgangarella, Advanced Telecommunications services : MORGAN, User and Technical Manuals, 1993.

[Blom 94] J. Blom, B. Jonsson and L. Kempe, Using Temporal Logic for Modular Specification of Telephone Services, Feature Interactions in Telecommunications Systems, pp. 197-215, IOS Press, Amsterdam, May 1994.

[Bouma 94] W. Bouma and H. Velthuijsen, Introduction, Feature Interactions in Telecommunications Systems, pp. VI-XIV, IOS Press, Amsterdam, May 1994.

[Bowen 89] T.F. Bowen, F.S. Dworack, C.H. Chow, N.Griffeth, G.E. Herman and Y.J. Lin, The feature interaction problem in telecommunications system, SETS, 1989.

[Bowen 90] T.F. Bowen, C. Chow, F.S Dworak , N. Griffeth and Y. Lin, Views on the Feature Interaction problem, Manuscript, Morristown, Bellcore, 1990.

[Braithwaite 94] H. K. Braithwaite and M. J. Altee, Toward automated Detection of Feature Interactions, Feature Interactions in Telecommunications Systems, pp. 258-259, IOS Press, Amsterdam, May 1994.

[Brooks 87] F. Brooks, No Silver Bullet, Essence and Accidents of Software Engineering, IEEE Computer, vol 20(4), pp. 10-19, April 1987

[Cain 92] M. Cain, Managing Run-Time Interactions Between Call-Processing Features, IEEE Communications Magazine, pp. 44-50, February 1992.

[Cameron 91] E.J. Cameron and Y.J. Lin, A Real Time Transition Model for Analysing Behavioral compatibility of Telecommunication services, proceedings of the ACM SIGSOFT'91, Conference on software for Critical Systems, pp. 101-111, ACM Press, December 1991.

[Cameron 93] E. J. Cameron and H. Velthuijsen, Feature Interactions in Telecommunications Systems, IEEE Communication Magazine, August 1993.

[Cameron 94] E.J. Cameron, N.D. Griffeth, Y.J. Lin, M.E. Nilson, W.K. Schnure and H.Velthuijsen, A Feature Interaction benchmark for IN and beyond, Feature Interactions in Telecommunications Systems, pp. 1-23, IOS Press, Amsterdam, May 1994.

[Cheng 94] K. E. Cheng, Toward a Formal Model for Incremental service Specification and Interaction Management Support, Feature Interactions in Telecommunications Systems, pp. 152-166, IOS Press, Amsterdam, May, 1994.

[Combes 93] P. Combes, B. Renard,W. Bouma and H. Velthuijsen, Formalisation of properties for feature interaction detection, International Conference on Intelligence in Service and Networks, Paris, France, November 1993.

[Combes 94] P. Combes and S. Pickin, Formalisation of a user View of Network and Services for Feature Interaction Detection, Feature Interaction in Telecommunications Systems, pp. 120-135, IOS Press, Amsterdam, May, 1994.

[Cross 94] M. Cross and F.O'Brien, Restructuring the Problem of Feature Interaction : Has the approach been Validated ? Experience with an Advanced Telecommunication for Personal Mobility, Feature Interactions in Telecommunications Systems, pp. 249-257, IOS Press, Amsterdam, May 1994.

[Dankel 94] D. D. Dankel II, K. Nielsen, M. Schmalz, L. Muzzi, W. Walker and D. Rhodes, An Architecture for Defining Features and Exploring Interactions, Feature Interactions in Telecommunications Systems, pp. 258-259, Amsterdam, IOS Press, May 1994.

[Erradi 92] M.Erradi, F. Khendek, R. Dssouli and G.Bochmann, Dynamic extension of object-oriented distributed system specifications, International Workshop on Feature Interactions in Telecommunications Software Systems, pp.116-132, St. Petersburg, Florida, USA, December 3-4 1992.

[Faci 94] M. Faci. and L. Logrippo, Specifying Feature and Analysing Their Interactions in a LOTOS Environment, Feature Interaction in Telecommunications Systems, pp. 136-151, Amsterdam, IOS Press, May 1994.

[Gaarder 93] K. Gaarder and J. Audestad, Feature interaction policies and the undecidability of a general feature interaction problem, TINA'93, pp. II-189-II-200, 1993.

[Gammelgaard 94] A. Gammelgaard. and J. Kristensen, Interaction detection a logical approach, Feature Interaction in Telecommunications Systems, pp. 178-195, Amsterdam, IOS Press, May 1994.

[Griffeth 93a] N. D. Griffeth and H. Velthuijsen, Reasoning about goals to resolve conflicts, Proceedings International Conference on Intelligent Cooperating Information Systems (ICICIS-93), pp. 197-204, Rotterdam, IEEE Computer Society Press, May 12-14 1993.

[Griffeth 93b] N. D. Griffeth and H. Velthuijsen, Win/ win negotiation among autonomous agents, proceedings 12th International Workshop on Distributed Artificial Intelligence, pp. 187-202, Hidden Valley, PA, May 19-21 1993.

[Griffeth 94] N.D. Griffeth and H. Velthuijsen, The negotiation Agent Approach to Runtime Feature Interaction Resolution, Feature Interaction in Telecommunications Systems, pp. 217-235, Amsterdam, IOS Press, May 1994.

[Halpern 90] J.Y. Halpern and Y. Moses, Knowledge and common knowledge in a distributed environment, Journal of the ACM, 37(3), pp.549-587, July, 1990.

[Harel 92] D. Harel, Biting the Silver Bullet, IEEE Computer, Vol 25(1), pp. 8-20, January 1992.

[Herbert 93] A. Herbert, Personal Comment at the Feature Interaction Session, TINA'93, L'Aquila, Rome, September 27-30 1993.

[Hopcroft 79] J.E. Hopcroft and J.D. Ullman, Introduction to Automata Theory, Languages and Computation, Addison-Wesley, 1979.

[IEEE 93a] IEEE Communication Magazine, Special issue on IN, March 1993.

[IEEE 93b] IEEE Communication Magazine, Special issue on Feature Interaction, August 1993.

[IEEE 93c] IEEE Computer, Special issue on Feature Interaction, August 1993.

[Inoue 92] Y. Inoue, K. Takami and T. Ohta, Method for supporting detection and elimination of feature interaction in a telecommunication system, International Workshop on feature

Interaction in Telecommunications Software Systems, pp. 61-81, IEEE Communication Society, December 1992.

[ISO 8807] ISO, LOTOS, A Formal Description Technique Based on the temporal Ordering of Observational Behavior, Information Processing Systems, Open Systems Interconnection, International Standard IS 8807, ISO, 1989.

[ISO 9074] ISO, ESTELLE, A Formal Description Technique Based on Extended State Transition Model, Information Processing Systems - Open Systems Interconnection, ISO/TC 97/SC 21, ISO 9074, 1989.

[ISO/IEC 10746] ISO, ITU-TS, Recommendation X.9000: Basic reference Model of Open Distributed Processing, Draft International Standard, ISO/IEC 10746, March 1994.

[ITU-T Q.12xx 93] ITU-T, Recommendations Q.120x and Q.121x, 1993.

[ITU-T Q.1203, 93] ITU-T, IN Global Functional Plane Architecture, Recommendation text Q. 1203, 1993.

[ITU-T Q.1204, 94] ITU-T, IN Distributed Functional Plane Architecture, Recommendation text Q. 1204, 1994.

[ITU-T Q. 1205, 93] ITU-T, IN Physical Plane Architecture, Recommendation text Q. 1205, ITU-T, 1993.

[ITU-T Q. 1211, 92] ITU-T, Q.1211 Recommendation for Target Services and Services Features for IN CS-1, 1992.

[ITU-T Z.100] ITU-T, Specification and Description Language SDL, Recommendation Z.100, 1987.

[Jacobson 92] I. Jacobson et al., Object-Oriented Engineering, A Use Case Driven Approach, Addison-Wesley, 1992.

[Lin 94] F. J. Lin and Y.J. Lin, A building block Approach to Detecting and Resolving Feature Interactions, Feature Interactions in Telecommunications Systems, pp. 86-119, Amsterdam, IOS Press, May 1994.

[Kimbler 94a] K. Kimbler, E. Kuisch and J. Muller, Feature Interactions among Pan-European Services, Feature Interactions in Telecommunications Systems, pp. 73-85, Amsterdam, IOS Press, May 1994.

[Kimbler 94b] K. Kimbler and D. Söbrik, Use Case Driven Analysis of Feature Interactions, Feature Interactions in Telecommunications Systems, pp. 167-177, Amsterdam, IOS Press, May 1994.

[Lamport 91] L. Lamport, The temporal logic of actions, Technical report, DEC/SRC, 1991.

[Linden 94] R. van der Linden, Using an Architecture to Help Beat Feature Interaction, Feature Interactions in Telecommunications Systems, pp. 24-35, Amsterdam, IOS Press, May 1994.

[Manna 92] Z. Manna and A. Pnueli, The Temporal Logic of Reactive and Concurrent Systems, Springer Verlag, 1992.

[Mierop 92] J. Mierop, S. Tax and R. Janmaat, Service Interaction in an Object Oriented Environment, International Workshop on Feature Interactions in Telecommunications Software Systems, pp. 133-152, St. Petersburg, Florida, USA, December 3-4 1992.

[Mills 87] H. Mills et al., Cleanroom Software Engineering, IEEE Software, September 1987.

[NT 89] Northern Telecom, Northern Telecom Software, 5204.11/11-89, Issue 1, Northern Telecom, Research Triangle Park, NC, 1989.

[Ohta 93] T. Ohta, K. Takami and A. Takura, Acquisition of service specifications in two stages and detection/resolution of feature interactions, Tina' 93, pp. II: 173-187, September 1993.

[Ohta 94] T. Ohta and Y.Harada, Classification and Resolution of Service Interactions in Telecommunication Services, Feature Interactions in Telecommunications Systems, pp. 60-72, Amsterdam, IOS Press, May 1994.

[OMG 94] OMG, The Common Object Request Broker: Architecture and Specification, Revision 1.1, OMG Document number 91.12.1, 1991, Revision 1.2, OMG, 1994.

[Oshinsanwo 92] A. O. Oshinsanwo et al., The RACE Open Service Architecture Project, IBM Sys. J., vol 31, no 4, December 1992.

[Rumbaugh 91] J. Rumbaugh et al., Object-Oriented Modeling and Design, Prentice Hall, 1991.

[TINA 92] Proc. 3rd Wksp. on Telecommunication Network architecture (TINA), Narita, Japan, January 1992.

[Tsang 94] S. Tsang and E.H. Magill, Detecting Feature Interaction in the Intelligent Network, Feature Interaction in Telecommunications Systems, pp. 236-247, Amsterdam, IOS Press, May 1994.

6 BIOGRAPHY

Dominique Gaîti received the Ph.D. and the habilitation degrees in Computer Science from the University of Paris VI and Paris IX, France, in 1991 and 1995, respectively. She is currently an Associate Research Scientist at the Center for Telecommunications Research (CTR), Columbia University, New York, after two years spent as a Visiting Scientist in the same lab. She also works at the University of Paris VI, France, as a member of the Scientific Staff. Her research interests are in network management and control, intelligent networks and distributed artificial intelligence.

Nadir Belarbi was graduated in 1993 as a computer science engineer from the University of science and technology, Oran, Algeria. He obtained a Master degree (DEA) in computer science and telecommunications from the University of Paris V and the Ecole Nationale Superieure des Telecommunications (ENST), Paris, France. He currently prepares a Ph.D. degree from the University of Versailles, France. His research interests include management of the service interaction in Intelligent Networks (IN), Artificial Intelligence and distributed systems.

7
Formal Criteria for Feature Interactions in Telecommunications Systems

Jan Bredereke
University of Kaiserslautern
Postfach 3049, D-67653 Kaiserslautern, Germany
Phone: +49 631 205-3287, Fax: -2640
E-mail: bredereke@informatik.uni-kl.de,
URL: http://www.informatik.uni-kl.de/aggotz/bredereke

Abstract

The feature interaction problem in telecommunications systems increasingly obstructs the evolution of such systems. We develop formal detection criteria which render a necessary (but less than sufficient) condition for feature interactions. It can be checked mechanically and points out all potentially critical spots. These have to be analyzed manually. The resulting resolution decisions are incorporated formally. Some prototype tool support is already available. A prerequisite for formal criteria is a formal definition of the problem. Since the notions of feature and feature interaction are often used in a rather fuzzy way, we attempt a formal definition first and discuss which aspects can be included in a formalization (and therefore in a detection method). This paper describes on-going work.

1 INTRODUCTION

Telephone switching systems are a classical example for long-lived and perpetually evolving software in the telecommunications domain. The first software controlled switching exchanges essentially still provided the plain old telephone service (POTS). Step by step, new features have been added since then which were supposed to offer added value to the customer (e.g. by call forwarding) and/or to the service provider (e.g. by improved accounting)[1].

Up to now a large number of features has been developed (several hundred, [Bo+89]). Therefore the probability is high that augmenting such a system by one more feature will influence another feature, especially in a negative way. This is called a feature interaction. Both the no-

[1] Example: the features defined in the ITU-T recommendations for Intelligent Networks (IN) [ITU93b].

tion of feature and the notion of feature interaction are used quite fuzzy and informal most of the time. This does not make a solution of the problem easier. The formalization in this paper is intended as an attempt to improve this situation.

Since we do not know how to *avoid* feature interaction problems in telecommunications systems altogether, we need to *detect* and *resolve* them ([BoVe94]). Detection of feature interactions may be done in different ways, for example by simulation or by online behaviour analysis, but we will concentrate on an off-line *verification* approach here. Usually, a verification approach works like this (compare, e.g., the contributions in [BoVe94]): we write a high-level, property oriented description for each feature and for the basic system (e.g., in temporal logic), and we write a lower-level, constructive[2] description for them (e.g., in Estelle [ISO89]or SDL [CCI87). Then we define an "implements" relation and prove mathematically that the combined system built from the constructive description of the basic system and the added features still implements all high-level properties.

This approach surely is promising, and it needs to be studied further. But it also has some limitations and prerequisites. If the proof is done by logical reasoning, the work is painstaking even with tool support. Furthermore, one can only verify those properties that are explicitly specified. And in general, it is practically impossible to formalize *all* expectations at the system which the service provider or the customer may have.

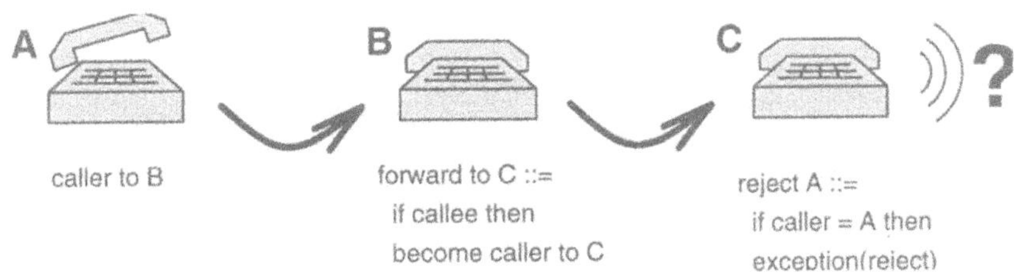

Figure 1 Feature interaction between call forwarding and terminating call screening.

Even worse, existing property descriptions may become "incomplete" or "inaccurate" by a later addition of a new feature: Imagine that you have expressed all relevant properties concerning a "caller" in a telephone system. Then, add a new call forwarding feature to this system. When a caller A calls a callee B, then B may forward this call to a user C (compare Figure 1). This way B may become a caller, too, in a sense. Who is the caller for C? Therefore, the notion of "caller" has become a little muddy, and our carefully written formulae on properties of a caller may not denote the meaning anymore which we wanted them to have.

In this paper, we propose a different verification approach which may help finding feature interactions that could go unnoticed by other verification approaches. We have proposed formal definitions for the terms "feature" and "feature interaction" in , which we sketch

2 "Constructive" means "executable" or that an implementation can be derived directly.

(restricted by space limitations) in Section 2. In Section 3, we describe detection criteria for feature interactions; in Section 4, we sketch a way to resolve the detected interactions; in Section 5 we discuss briefly how our generic formalism can be mapped onto real formal description techniques (FDTs); in Section 6 we present the tools developed up to now and outline the practical applications done so far; and the paper is concluded by a summary and an outlook.

2 DEFINITION OF "FEATURE INTERACTION"

2.1 What to Define and What to Leave Out

Before we tackle the feature interaction problem, we will give our definitions of the terms "feature" and "feature interaction" first. As has been pointed out by Cameron and Velthuijsen [CaVe93], it is crucial to define the problem first before trying to solve it. Our formal definitions surely do not cover all the aspects which are commonly associated with the informal terms, as no formal definition probably ever will be able to. But they state more clearly what the achievements are and which limitations have to be taken into account while they are used.

Cameron and Velthuijsen [CaVe93] describe two views on the problem. In the business view, "a feature is a tariffable unit", and "a feature interaction occurs when the behaviour of one feature is altered by the use of another". "A second kind of interaction occurs when the use of one feature should alter the behaviour of another, but does not." In the implementor's view, "a feature is any increment of functionality added to an existing system". "Just as in the business view, a feature interaction occurs when one feature's behaviour is altered by the use of a second." Other authors provide similar descriptions, compare e.g. Kimbler and Velthuijsen [KiVe95, page 2] and Aho and Griffeth [AhGr95].

Even if there does not exists a clear (formal) concept of what a feature and a feature interaction are, the notion of "*behaviour* of a feature" is surely a central one. It is a necessary precondition for a feature interaction that this behaviour is changed. In this paper, we disregard the case where an interaction occurs because a change in behaviour does *not* take place when a change in the intentions requires it to do so, since the notion of intended behaviour is still a topic of research ([CaVe93]). Maybe a property oriented approach (compare Section 1) will be able to handle it. In the remainder of this section, we will develop a formal definition which is based on the central notion of the change in behaviour.

When we have to decide if the behaviour is still the same as before, there are different aspects of the behaviour. First, there are the functional aspects. This concerns the sequences of possible states/events. But the non-functional aspects can become important, too. This concerns any real time or performance aspects. Also, there can be important properties on knowledge, e.g. security properties. Unfortunately, the formal treatment of non-functional aspects is much harder than that of functional aspects, and many areas are still a topic for research [BCN95]. Therefore, we restrict our notion of feature interactions to functional aspects. This excludes, for example, the treatment of a case where one feature is impaired because another feature causes an overload on some shared resource.

We will not define what a service is. Commonly, it is some collection of features, but only if this collection is somehow self-contained. Since we will concentrate on functional aspects anyway, we have no need to take the pains with designing a formal definition of a service; we will stick to the task of defining a feature and its interactions.

For the formalization of the functional aspects of the behaviour, we advocate the use of automata [HoUl79]. The *computations* (execution sequences) of an automaton are a good and formal way to capture this kind of behaviour. An automaton is a finite description that specifies a behaviour. There are many variants on the exact definitions; we will present a suitable one in the following.

2.2 Suitable System Architectures

One may choose to use a collection of concurrent communicating automata, or a single global automaton (comprising different local automata which, again, have some form of concurrency and communication). If we take the first alternative, we still have to define formally what concurrency and communication is. Since a modelling of concurrency by interleaving is sufficient for our purposes, this question will inevitably lead to an answer that equals the second alternative. Therefore, and since the semantics of the standardized formal description technique Estelle ([ISO89]) is defined in a very similar way, we take the second alternative. As soon as we want to apply our results to practical system specifications, we have to transfer the results to practical specification languages; a similarity in the underlying semantics will facilitate this.

The computations of an automaton can be formalized by a tree of reachable states or by a set of linear sequences of reachable states. Since we are not interested in terminating computations, we take the simpler second alternative.

In the effort of opening up existing telecommunications systems, abstract models have been developed, e.g. the intelligent network (IN) conceptual mode [ITU92, DuVi92] by the ITU-T. The basic call state machine (BCSM) in the IN conceptual model (INCM) is designed in a way that the control flow remains as much as possible in the BCSM. Among others, this results from the communication overhead involved in passing away the control. On the other hand, this system design results in a rather complex semantics of the *detection points* (DPs) where control may be given away. It turned out that such a complicated architecture obstructs the formal analysis of the system. Therefore, we use a simplified IN architecture. We do not use explicit detection points, we just specify a homogenous automaton, and extensions result in the addition of more transitions and states of the same kind as the old ones. A closer adaptation to the IN architecture (or *of* the IN architecture?) is a topic for future research.

2.3 Formalization by Automata

For space reasons, we cannot present our entire detailed formalization [Bre95]. Therefore, we will describe just the highlights and some important decisions which were taken during the formalization.

SG = ((waitForAck, 42 'abcdef'),
(connected, 41, 'hijklm'),
(idle, 0))

Figure 2 Example of a global state.

We start out by constructing a state space. It consists of a fixed set of system components which each have a fixed set of variables (Figure 2). Next, we define a simple global automaton, consisting of a state space, a set of "simple" state transitions and an initial state. Simple transitions have the disadvantage that their number can become infinite as soon as the domain of a variable is infinite, e.g., like for the natural numbers. Therefore, we allow to specify many simple transitions at the same time by a "transition". The number of these must be finite. The same concept to achieve a finite representation can be found in Estelle and SDL specifications, too. Next, we define a global automaton $A = (S^G, T, s_0)$ based on a simple global automaton, consisting of a state space S^G, a finite set of transitions T and an initial state s_0.

We define a computation of a global automaton to be a (possibly infinite) sequence of states which can be constructed using the set of simple transitions, starting from the initial state. Each global automaton defines a set of possible computations.

Now we come to the part of the formalism that will define what a feature is. For this, we describe how we add something to a global automaton. A prerequisite for the detection of interactions between features is to specify different features separately. Since we model the entire system by computations, it must be possible to associate each transition with a single feature[3]. For this, every transition must contain only functionality of a single feature. This implies that we have to use a *specific specification style* when we specify the system and its extensions. Two basic rules are that we *modify only on a coarse-grained level*, and that we *only add* to the specification. This means for the behaviour part that we never modify an existing transition to achieve the behaviour of a new feature, but we only add an entirely new transition. Also, we do not delete any obsolete transition (see Section 4 for how to get rid of it). The only way to extend a global automaton in our formalism is to add new possible computations.

We designed our formalism in such a way that it supports these stylistic rules for the state space part, too [Bre95]. One of the ideas is to use a state space extension function. It tells where new elements ("variables") are inserted into the system components, and where entire new system components ("modules"/"processes") are inserted into the global state space.

[3] For orthogonality, we consider the basic system as just another feature, too.

Based on this, we define what an extension of a global automaton is [Bre95]. The extension may comprise both the state space and the behaviour. The extended global automaton must have all the computations which the non-extended automaton has, modulo a state space extension, and it may have new computations.

Finally, we define a feature $f = (\phi^G, T^e, d^0)$ for a global automaton by a global state extension function ϕ^G, a set of new transitions T^e and the initial state d^0 for the new state parts. Therefore, if we add this feature to a global automaton, the resulting automaton will be an extension of the original one in the above sense. Figure 3 shows an original automaton which only contains transitions of a simple basic system B; in Figure 4, a (simple) feature f has been added. The state space has been extended by a new local component for the dialled number (via ϕ^G), the initial state has got a value for this number (via d^0), and there are new transitions (T^e).

Through its definition, a feature is an *increment* in the possible behaviour. This is a deliberate restriction. This way, we still have all the computations of the other features and can find deviations. (Compare the definition of a feature interaction below.) If the purpose of the new feature is to remove the other computation, then this is certainly an interaction, and it should be handled explicitly. In Section 4, we will discuss how this resolution may take place.

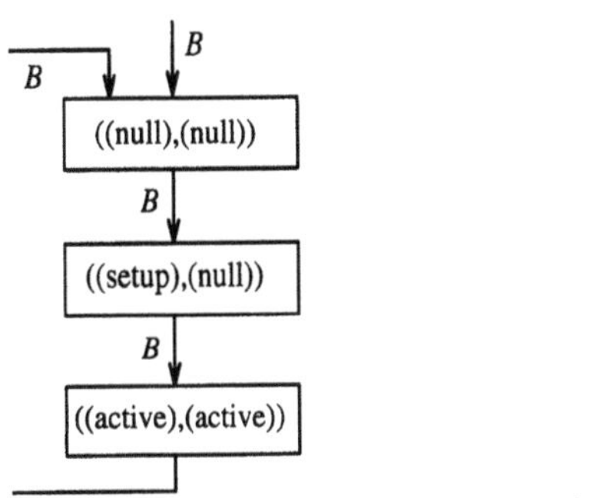

Figure 3 A global automaton with two local components.

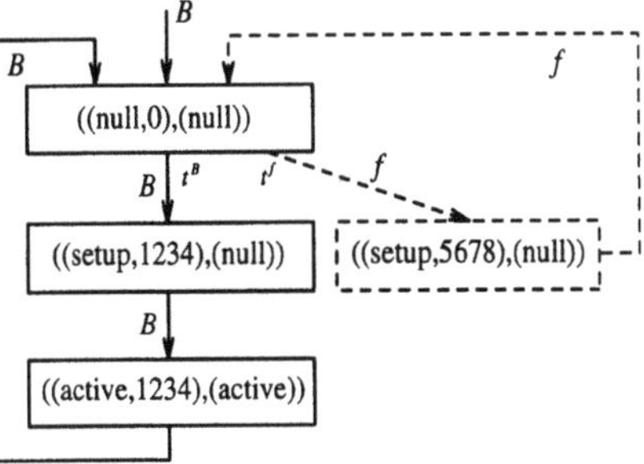

Figure 4 Adding the feature f of dialling numbers to the global automaton.

We require a global automaton to have at least one transition so that there is at least one feature name. By convention, the "basic system" of the global automaton which is always present carries the name B.

Now we consider different cases of feature interactions. If there is a computation c, whose steps out of simple transitions are partially associated with feature f, and partially associated with feature g, then this computation is only possible if both features are present in the global automaton (compare Figure 5). We cannot add both features independently to the automaton without influencing the behaviour of the other feature. Therefore, both features interact.

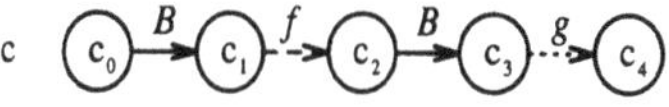

Figure 5 A computation c which is only possible if both features f and g are present.

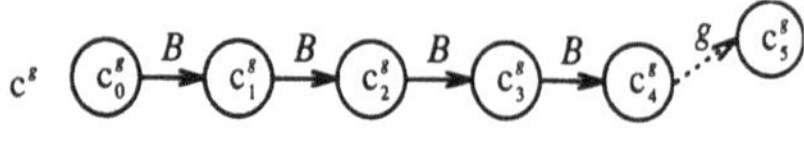

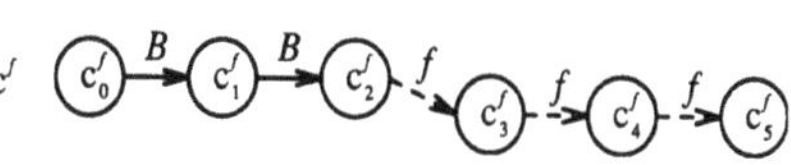

Figure 6 Feature f may divert the execution away from computation c^g of feature g.

The condition that a computation is associated with two different features (not equal to the basic system B) implies a feature interaction, but this is not true vice versa. There may be two different computations belonging to two different features, but they have a common prefix belonging to the basic system where it is not decided which one of the two will be taken eventually. If there is such a non-deterministic choice for the automaton then it may be possible that the original behaviour of the first feature does not take place anymore. If this choice is only due to the addition of the second feature, this denotes a feature interaction (see Figure 6).

A feature with some terminating computations may lead to such a diversion state even without a non-deterministic choice. Nevertheless, all of these cases are handled by the following definition:

Definition 1
A feature f in the global automaton A interacts with a feature g in A iff $f \neq g$ and there exists a computation c^f that comprises at least one simple transition from f and a computation c^g that comprises at least one simple transition from g and an index $i \in \mathbf{N}$ such that the prefixes of c^f and c^g up to the i-th state are equal, i.e., $\langle c_0^f,..., c_i^f \rangle = \langle c_0^g,..., c_i^g \rangle$, and the simple transition (c_i^f, c_{i+1}^f) belongs to feature f but neither to feature g nor to B.

Every useful feature which is added on top of the basic system will of course modify its behaviour. This is the intention behind adding the feature. Therefore, for practical purposes we do not count this feature interaction in the following definition.

Definition 2 A *feature interaction among the features of a global automaton A occurs* iff there exist two features f, g in A such that f interacts with g and where both are not the basic system B.

3 AUTOMATIC DETECTION OF POTENTIAL FEATURE INTERACTIONS

In Definition 2, we have defined what a feature interaction among the features of a global automaton is. Since computations are of infinite length in general, the criterion in Definition 1 cannot be used constructively (and efficiently). Therefore we have to find criteria for feature interactions which

1 can be computed efficiently,
2 are still necessarily fulfilled for feature interactions (wide enough), and
3 do include as few as possible potential interactions (specific as possible).

From Theorem 1 on (see below), we will present a sequence of criteria which all fulfill the first two items and which become increasingly better at the third item.

3.1 Necessary Conditions for Feature Interactions

Under the assumption of infinite computations[4] , there may be some feature interaction only if there is a reachable state s in which there is a non-deterministic choice between at least two transitions belonging to different features.

Proof: a feature interaction among the features of A implies that a feature f in A interacts with a feature g in A (Definition 2). So it remains for us to prove that this plus the premise that every computation is infinite implies the existence of the mentioned non-deterministic choice. Definition 1 provides us already with the state $s = c_i^f = c_i^g$ and the transition t^f. The only item still needed is a continuation transition t^g starting from s for the other computation. Since c^g cannot be finite, t^g must exist.

The condition of non-determinism is necessary for feature interactions, but (in general) less than sufficient because we did not count the interactions with the basic feature B. If a feature f interacts with B, then there may or may not be an interaction with another feature g, as shown in Figure 6. Therefore, if some feature f interacts with B, we are warned and have to check manually if there are transitions of g later in the affected computations. (We will give more specific criteria on this in the following subsections.)

Furthermore, the above criterion is still not efficiently applicable because of the condition that the state s must be reachable. This would imply a complete state space search. Therefore we weaken the condition further and drop that condition:

Theorem 1: Be A a global automaton such that its set of all possible computations contains only infinite computations. A feature interaction among the features of A may occur only if there exist a feature f in A, a transition t^f of f, a feature g in A with $g \neq f$, a transition t^g of g and a global state s such that both t^f and t^g are enabled (non-deterministically) in s.

Note that f or g may equal B in Theorem 1. Informally, Theorem 1 says that we can find any feature interaction if we compare all pairs of transitions from different features and check if their enabling sets overlap. If no such pair exists, then no feature interaction can occur. The proof is a straightforward extension of the above one.

A simple example for a situation as described in Theorem 1 may be found in Figure 4. The global state s is ((null), 0), (null)). Due to our very abstract modelling, you find in this state a non-deterministic choice between the normal setup transition t^B and the newly added transition t^f which aborts the setup in case of a wrong dialled number. This means that the new feature f of dialling

[4] The assumption that the basic system for itself is free of deadlocks can be justified by the applications in mind, which are reactive systems. A first feature interaction may invalidate the above assumption for a second, but we can always detect the first one and after a resolution also the second one.

numbers may possibly interfere with the basic system *B*. It was impossible that a call could fail in our very simple basic system.

3.2 Excluding Concurrent Transitions

The above criteria warn us of potential feature interactions. But even if the computations of a feature are affected by another feature, this may be harmless in some cases. E.g., think of two transitions belonging to different features which are enabled in the same states and where the resulting state when both have fired does not depend on the order of firing. (Compare Figure 7.)

Definition 3 *A potential feature interaction as determined by the criterion of Theorem 1 is harmless* iff the interacting transitions t_f and t_g are independent, i.e. iff firing one of them does not disable the other one and if both can fire then the firing order is irrelevant.

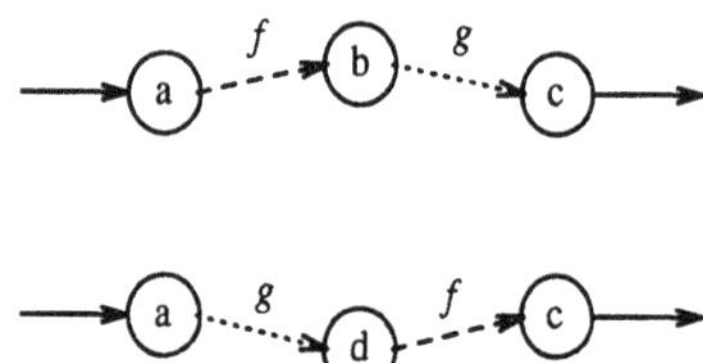

Figure 7 The firing order of these transitions is irrelevant.

This declares non-determinism as harmless if it, e.g., results from our modelling of concurrency of the local automata by interleaving. The condition of Definition 3 is usually hard to decide for two given transitions. Thus, we reformulated it based on the local state components modified by one transition and the local state components used by the other transition. As soon as we start to use a specific (constructive) formal description technique, these two notions are quite straightforwardly applied to specific language constructs. For example, an assignment to a local variable will lead to a modification of this local state component, and the use of the value of a local variable will be a use of this local state component.

The formal definition [Bre95] of the *output modification pattern* describes which local state components can possibly be modified by a transition. The *input pattern* describes which local state components are used at all by a transition. This comprises both the enabling conditions and the computation of the output. Output dependencies on the input in the fashion of the identity function are not counted. A transition t_2 has a *state space dependence* on a transition t_1 iff the intersection of the output modification pattern of t_1 and the input pattern of t_2 is not empty. Now we can reformulate our criterion on harmless potential feature interactions so that it should be easier to check for any specific formal description technique.

Theorem 2 *A feature interaction as determined by the criterion of Theorem 1 is harmless* if the interacting transitions t_f and t_g mutually have no state space dependence.

3.3 Excluding Interactions with the Basic System

Theorem 1 from Section 3.1 is a criterion that warns us of all possible feature interactions. As already mentioned there, it also warns us if a feature f interacts with the basic system *B*, because there may be situations like the one in Figure 6 where this influences another feature *g* indirectly. But since a feature which does not interact with the basic system at all is rather useless, we have to differentiate further cases when *f* interacts with *B*.

To this end, we use our knowledge of the typical structure of a telecommunications system (compare, e.g., [ITU93b, ITU93a]). Call processing is described there by one extended finite state automaton (EFSA) (if we take a global view) or by two EFSAs (in a distributed view). These typically have a *null state* where no call is established and which is reached again after the processing of one call has been completed. Usually, this is also the initial state. This allows us to reduce the analysis of the computations c^f and c^g in Definition 1 to the end of the prefix where the null state has been reached the next time.

In order to reflect the use of EFSA in our formal modelling, we adopt the convention that the first component of the local state tuple describes the *major state*. Any sensible specification of a system will not only have a finite, but also a small domain of values for each major state. This allows us to do an abstraction step and reduce a global automaton to a simpler one with the same number of local components, but where these local components only have a major state and no extended state. This abstraction step to major states and the limitation of the analysis up to the next null state allows us to do an exhaustive reachability analysis without the tractability problems that a state space explosion imposes.

We start a reachability analysis at the state s where a transition of a feature f shows non-determinism with a transition of the basic system *B* (compare Theorem 1), and we follow all transitions from this state which belong to *B*. If we cannot find any transition not belonging to *B* or *f* before we reach the null state again, it is impossible that the detected non-determinism signifies a feature interaction among the features of the global automaton (Definition 2). The sketched search algorithm is worked out formally in [Bre95]. If we can find such a transition, there may be a harmful feature interaction and we have to analyse this situation manually. The manual analysis is supported with the information where the non-determinism showed up, which was the offending computation (prefix) and which were the possibly interacting transitions, these in turn provide information on the features which are concerned.

3.4 Excluding Independently Inserted Features

We may use another criterion that reduces the class of potential feature interactions further which have to be analysed manually. For this, we partially take into account the extended state space, too. A common situation in a local call processing FSA is shown in Figure 8. Two short automata extensions/features are inserted into the mostly sequential control flow of the basic local automaton. Examples for such a feature may be "abbreviated dialling" which translates a single digit into a full directory number, or "reverse charging" which changes some charging control variables. The first extension can never disable the second because it returns control back to nearly the same point where it took it away. This return situation may be detected easily during our exploration of the major state space, and be treated accordingly.

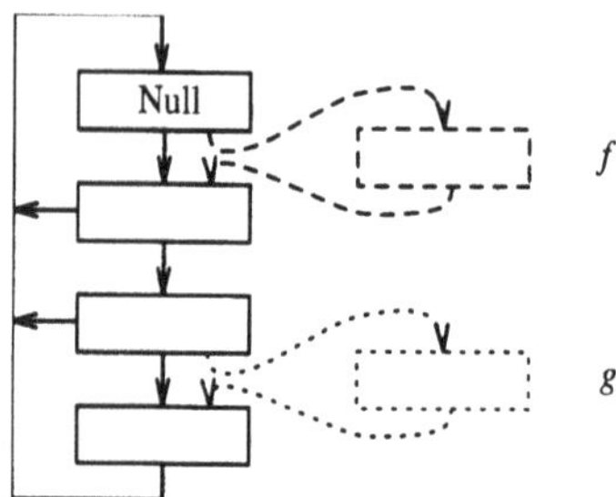

Figure 8 A typical situation with two new features.

We call a transition t *major state preserving* iff t either does not modify the major states at all or there is some transition t_B of the basic system such that t_B modifies the major states in exactly the same way as t. Note that this definition does *not* comprise the case depicted in Figure 8. There, more than one transition is inserted before control is returned to the basic system. Such patterns are only sensibly defined for a local automaton (see Section 3.5).

The only undesired thing that may happen now is that a first transition t_f modifies a part of the extended state space on which a second transition t_g depends. Here, we have to take into account indirect data dependencies, too. An example would be the case where the effects of a transition t_B associated with the basic system B depend on a state component which was modified by the first transition t_f, and where in turn a second transition t_g depends on these effects of t_B. To resolve this situation, we need to do some simple data flow analysis.

A sequence $\langle t_1,..., t_m \rangle$ of transitions contains an *indirect state space dependence on a transition* t_0 iff there exists a transition t_j, $1 \leq j \leq m$, which does neither belong to the same feature as t_0 nor to the basic feature B such that the intersection of the output modification pattern of t_0 (compare Section 3.2) and the input pattern of the concatenation of $t_1,..., t_j$ is not empty. As mentioned before, such state space dependencies are quite easy to determine for the language constructs of any specific (constructive) formal description technique.

Again, we start a reachability analysis at the state s where a transition t_0 of a feature f shows non-determinism with a transition from the basic system B (compare Theorem 1), and we explore the major state space up to the null state. If t_0 is major state preserving, and if we do not find any sequence of transitions which contains a state space dependence on t_0, then it is impossible that the detected non-determinism signifies a feature interaction among the features of the global automaton. The sketched search algorithm is worked out formally in [Bre95]. If we can find such a transition, there may be a feature interaction and we have to analyse this situation manually, again. Note that we did not require the sequence of transitions to be executable by the global automaton. Only the major states had to "fit". This makes the criterion weaker than it could be, but it saves us from exploring the full state space.

3.5 Analysis on a Local Level

Since we describe distributed and concurrent systems, the specified systems consist of local components (modules/processes) which operate with a certain degree of independence. (Except for descriptions on the global functional plane.) Many transitions have effects only on one local component. This is supported by most of the specific formal description techniques. Their syntax of transitions allows only local effects except for some special communication commands. This can be exploited by performing individual checks for feature interactions on a local level first. Due to the more restricted state space and the smaller degree of concurrency, this allows a deeper and more thorough automatic analysis without a state space explosion.

All detection criteria from the preceding subsections may as well be applied only to one local component. This can already discover those feature interactions which happen only locally. For example, an interaction may be discovered (and resolved) which results from an unspecified order of originating call screening and abbreviated dialling, in an appropriate local setting. We can determine if a transition has only local effects by the output modification pattern introduced in Section 3.2. In Sections 3.3 and 3.4, we proposed some (major state) data flow analysis after the detection of non-determinism between two transitions of different features. This analysis may be restricted to a single local component as long as only transitions with local effects are involved. If we reach a transition with non-local effects we have three options: we can study the consequences manually; we can apply the following analysis globally, as described in the preceding subsections; or we can continue with another local analysis. These secondary level local analyses start with any transition that depends on the state component in question, as determined by the input pattern defined in Section 3.2.

A local analysis has a further advantage: in Section 3.4 we discussed major state preserving transitions. On a global level, we could include only those transitions into the definition which either didn't modify the major states at all or for which existed a "similar" transition of the basic system. This did not comprise the case depicted in Figure 8. On a local level, we can sensibly extend the notion of major state preserving to entire groups of transitions. We replace the single transition of above by a sub-automaton consisting of several transitions and states, and we only require that the entry and exit of this sub-automaton exposes the properties described above.[5] .

4 RESOLUTION OF FEATURE INTERACTIONS

To resolve detected feature interactions, generally high-level design decisions on the policy of the service provider or on the needs of the users have to be made. And since the detected interactions are the typical cases nobody had thought of before, it is quite probable that a satisfactory answer cannot be deduced from any existing behavioural description (compare Section 1). Even if some solution can be deduced, we do not know if it also really satisfies the implicit and unexpressed requirements of the users and the provider.

But even if an automatic resolution is impossible in many cases, our approach to specify features and detect feature interactions allows for different methods to resolve the interactions. In the simple

[5] Some more details have to be considered, e.g., there must not be a deadlock, and the notions of input pattern and output modification pattern have to be extended suitably.

case, we have detected direct non-determinism between two transitions associated with different features. This is also the case when a new feature is "plugged" into the old system since, up to now, we do not have introduced a way to disable the old transition. We simply get this non-deterministic choice instead.

So, after the detection methods have been applied, and after we are sure that it is the right thing to override the old transition, we specify corresponding *precedences* between the transitions. Each transition of a feature is given the same precedence, therefore we only have to decide which feature takes precedence over which. This is done by setting up a precedence matrix for features. The matrix does not need to contain an entry for every pair of features, it is sufficient to enter the explicit design decisions we actually made.

Here is a sketch how the precedences are incorporated into our formalism. We define a global automaton with priorities $A^p = (S^G, T^p, s_0, p)$ for the hitherto existing global automaton $A = (S^G, T, s_0)$, where p is a partial order on the set of feature names. This also renders a partial order on the set of transitions T and it thereby allows us to construct the new global automaton A^p which is identical to A except that we remove any simple transition where all the corresponding transitions were overridden by other transitions with a higher priority.

Of course, there may be other cases where precedences are not sufficient. Then, we need to specify additional behaviour which resolves the conflict. We specify this behaviour formally like any other new feature, and we add it to the system iff both of the interacting features are added. This approach for the integration of the resolving behaviour allows us to investigate even interactions which may happen between this newly introduced behaviour and a third feature.

More details on the resolution of feature interactions may be found in our technical report [BrGo94a]. There, we apply our ideas to the formal description technique Estelle.

5 USING EXISTING FDTS

Up to now, we introduced a generic formalism for the specification of telecommunications systems, and we included only those language aspects which were necessary for our discussion. To make our ideas applicable, we need a full-fledged formal description technique which offers the appropriate expressive power that allows a system developer to specify real systems. Our ideas have to be adapted for such a FDT. But we designed our generic formalism in such a way that this adaptation is supported well. E.g., the FDT Estelle is rather rich in language constructs, but its semantic model ([ISO89]) is very similar to our global automaton. Therefore it is possible to find syntactic criteria in Estelle for possible feature interactions that correspond to our definitions in the generic formalism. We have already investigated the notion of non-determinism between Estelle transitions in depth [The95]. The richness of Estelle provided a rather large number of special conditions to consider, but this only justified our decision to investigate appropriate feature interaction criteria in a simpler formalism first and to apply them to real FDTs later.

An application to the FDT SDL should pose no fundamental problems as well since the semantic models of SDL and Estelle are very similar. Concerning the FDT LOTOS, our simple global automaton is basically already a labelled transition system which is the semantic model of LOTOS.

6 TOOLS AND CASE STUDIES

Several tools for the FDT Estelle are in use now. E.g., there is a tool that computes Estelle priority values from a precedence matrix and generates the appropriate Estelle code. Our tool CONFINE [ThBr95] detects non-determinism between Estelle transitions of different features; and it detects and automatically eliminates Estelle transitions which have become non-executable because they are completely overlapped by new, higher-prioritized Estelle transitions. The Pet/Dingo toolset [SiSt93] from NIST automatically generates executable code from an Estelle specification and animates the execution.

A first case study [BrGo94a][6] used a simple global service specification of a telephone switching system. Our detection tool Confine did not only find the already known feature interactions but also two interferences which escaped us while we specified this simple example. A manual analysis led to the conclusion that both cases are harmless if an implementation is sufficiently fast. Nevertheless, we achieved a deeper understanding of some underlying problems. Furthermore, Confine detected and removed automatically the inactive transitions which had become obsolete through the resolution procedure for the known feature interactions. Currently, we are working on a second case study which is based on a simplified version of the IN conceptual model [ITU92], specified in Estelle again, which takes into account the distribution aspect, too [Ill95, Jer96]. Concurrently, we work on incorporating more criteria into our detection tool [Bar96].

7 SUMMARY AND FUTURE WORK

The feature interaction problem in telecommunications systems currently obstructs the evolution of such systems more and more. We presented an approach for the offline detection of feature interactions which is different from other verification approaches. It does not rely on a high-level, property oriented description against which a lower-level, constructive description is checked. We observed that the behaviour of a feature is a central notion, and that feature interactions are indicated by a change in this behaviour. We formalized the notion of change in a definition of feature interactions. From this, we derived a necessary (but less than sufficient) condition for feature interactions. It can be checked mechanically and points out all potentially critical spots. These have to be analyzed manually. We proposed further criteria to exclude some cases where the interactions are harmless, and we made use of our knowledge of typical telecommunications systems to define criteria for more cases in which there cannot be any negative influence even if the basic criterion indicates a possible problem. This led to a restricted reachability analysis.

Since the notions of feature and feature interaction, as they are widely used, had been too fuzzy to have hope for a resolution of the problem, we had to start out with an attempt of a formal definition and discussed which aspects could be included in a formalization (and therefore in a detection method). E.g., we restricted ourselves to the functional aspects of a system. This allowed us to use the computations (possible execution sequences) of a global automaton to catch the notion of behaviour formally. We formalized a feature as an increment in the possible behaviour of a global automaton. A specific specification style was a prerequisite to construct an unambiguous association of the state transitions to the different features.

---

6 An overview of this study is more widely available [BrGo94b].

The result of our analysis is a list of pairs of transitions from different features which may possibly interact. This list may be condensed to a list of pairs of features which may possibly interact. Such a condensed list is of value for the management of detected possible feature interactions: for each entry, we need to find an expert who knows both features and can determine how the interaction is resolved best. This is an improvement to the current situation. Up to now we need an expert who knows all features in the system, if we want to add one new feature. And the larger the system becomes, the harder it is to find such an overall expert.

Next, we sketched an approach for the resolution of detected feature interactions, which we have detailed in [BrGo94a]. We put up a precedence matrix for features and thereby specify a partial order on them. This allows us to disable unwanted transitions. There is also the possibility to specify new behaviour to resolve an interaction. Furthermore, we discussed how our generic formalism can be mapped onto real FDTs.

Some of the detection criteria are already supported by an automated tool. A first case study based on the FDT Estelle used a simple global service specification of a telephone switching system. Our detection tool did also find two (relatively harmless) interferences which escaped us while we specified this simple example. Currently, we are working on a second case study which takes into account the distribution aspect, too.

This paper describes on-going work. We are working on even more detailed detection criteria. They would reduce the manual work of, e.g., dismissing certain classes of harmless interactions. Parallely, we work on incorporating more criteria into our detection tool, and we will continue to perform case studies. The application of our method to other FDTs than Estelle (e.g., SDL) should be straightforward.

8 REFERENCES

[AhGr95] Aho, A. V. and Griffeth, N. D., Feature interactions in the global information infrastructure. In *Proceedings of the 1995 Conference on Foundations of Software Engineering* (1995).

[Bar96] Barthel, D., Implementation of criteria for the detection of feature interactions. Master thesis (in German), Univ. of Kaiserslautern, Dept. of Comp. Sci. (1996). To appear.

[BCN95] Bucci, G., Campanai, M. and Nesi, P., Tools for specifying real-time systems. Real-Time Systems Journal **8**, 117–172 (1995).

[Bo+89] Bowen, T. F. et al., The feature interaction problem in telecommunication systems. In *Seventh IEEE International Conference on Software Engineering for Telecommunication Systems*, (July 1989).

[BoVe94] Bouma, L. G. and Velthuijsen, H., (editors), Feature Interactions in Telecommunications Systems, IOS Press, Amsterdam (1994).

[Bre95] Bredereke, J., Automata-theoretic criteria for feature interactions in telecommunications systems, Tech. Rep. 273/95, Univ. of Kaiserslautern, Dept. of Comp. Sci. (1995).

[BrGo94a] Bredereke, J. and Gotzhein, R., A case study on specification, detection and resolution of IN feature interactions with Estelle. Tech. Rep. 245/94, Univ. of Kaiserslautern, Dept. of Comp. Sci. 55 (May 1994).

[BrGo94b] Bredereke, J. and Gotzhein, R., Specification, detection and resolution of IN feature interactions with Estelle. In Hogrefe, D. and Leue, S. (editors), *Seventh International Conference* on *Formal Description Techniques for Distributed Systems and Communication Protocols — FORTE '94*, Proceedings, pp. 366–368, Berne, Switzerland (4–7. Oct. 1994).

[CaVe93] Cameron, E. J. and Velthuijsen, H., Feature interactions in telecommunications systems. IEEE Commun. Mag. **31**(8), 18–23 (Aug 1993).

[CCI87] CCITT SG X, Contribution Com X-R15-E., Recommendation Z.100: Specification and Description Language SDL (1987).

[DuVi92] Duran, J. M. and Visser, J., International standards for Intelligent Networks. IEEE Commun. Mag. **30**(2), 34–42 (Feb. 1992).

[GoBr95] Gotzhein, R. and Bredereke, J. (editors), GI/ITG Workshop on Formal Description Techniques for Distributed Systems, Univ. of Kaiserslautern, Dept. of Comp. Sci. (22–23 June 1995). URL http://www.informatik.uni-kl.de/aggotz/fachgespraech95.

[HoUl79] Hopcroft, J. E. and Ullman, J. D., Introduction to Automata Theory, Languages and Computation. Addison-Wesley (1979).

[Ill95] Illerich, J., Design of a telephone switching system in Estelle. Projektarbeit (in German), Univ. of Kaiserslautern, Dept. of Comp. Sci. (Oct. 1995).

[ISO89] ISO/TC 97/SC 21, ISO 9074. Information Processing Systems — Open Systems Interconnection — Estelle: A Formal Description Technique Based on an Extended State Transition Model (1989).

[ITU92] ITU-T, Recommendation Q.1201., Principles of Intelligent Network Architecture (Oct. 1992).

[ITU93a] ITU-T, Recommendation Q.931., DSS 1 — ISDN User-Network Interface for Basic Call Control (Mar. 1993).

[ITU93b] ITU-T., Q12xx-Series Intelligent Network Recommendations (1993).

[Jer96] Jerusalem, D., Extension of a telephone switching system in Estelle. Projektarbeit (in German), Univ. of Kaiserslautern, Dept. of Comp. Sci. (1996). To appear.

[KiVe95] Kimbler, K. and Velthuijsen, H., Feature interaction benchmark. Discussion paper for the panel on benchmarking at FIW'95, Kyoto, Japan (Oct. 1995).

[SiSt93] Sijelmassi, R. and Strausser, B., The PET and DINGO tools for deriving distributed implementations from Estelle., Comp. Networks and ISDN Syst. **25**(7), 841–851 (Feb. 1993).

[ThBr95] Thees, J. and Bredereke, J., A tool for the analysis of feature interactions in IN. In *Gotzhein and Bredereke* [GoBr95], pp. 199–208. URL http://www.informatik.uni-kl.de/aggotz/fachgespraech95/thees.ps.gz (in German).

[The95] Thees, J., Design and implementation of a tool for the analysis of feature interactions in Estelle specifications. Master thesis (in German), Univ. of Kaiserslautern, Dept. of Comp. Sci. (Apr 1995). URL http://www.informatik.uni-kl.de/aggotz/thees/diplomarbeit.ps.gz.

9 BIOGRAPHY

Jan Bredereke received his Diploma in computer science from the University of Hamburg, Germany, in 1992. From 1992-93, he became researcher there, in a project on Estelle for high-speed networks. Since 1994 he is a research assistant of the computer science faculty at the University of Kaiserslautern, Germany. He is currently interested in the communication system design with formal description techniques, especially in the feature interaction problem in Intelligent Networks and in the influence of the specification style, e.g. on the efficiency of automatically generated implementations. He is a member of the Gesellschaft für Informatik.

8
Open Service Node For Intelligent Networks

Pasi Kemppainen
Systems Software Partners Ltd.,
Laserkatu 6, 53850 Lappeenranta, Finland,
Tel. +358 53 624 3272, Fax. +358 53 402 0949
E-mail: Pasi.Kemppainen@ssp.kareltek.fi

Olli Martikainen
Telecom Finland Ltd.
P.O.BOX 106, 00511 Helsinki, Finland,
Tel. +358 2040 3503, Fax. +358 2040 3251
E-mail: Olli.Martikainen@tele.telebox.fi

Abstract

Intelligent Networks (IN) is a concept that supports fast and flexible creation of new services in fixed and mobile networks. In addition to generic IN services provided by telecom operators there is a need for customer oriented smaller scale services, which can be integrated to the special needs of the customer or service provider. In this paper a lightweight IN service control environment called the Open Service Node (OSN) is introduced. The OSN environment is designed for third party service creation by service providers and it can be applied in the control of fixed, mobile and broadband networks. The OSN has been developed at Telecom Finland since 1991 and it has been used in pilot projects since 1993. Systems Software Partners Ltd. is a software house specialized in IN service creation. SSP has co-developed the OSN with Telecom Finland and is currently productizing it for the international markets.

1 INTRODUCTION

The Open Service Node (OSN) offers an open and cost-efficient solution for creating service applications on top of Intelligent Network systems. It decreases IN service time-to-market through its flexible approach to service delivery from laboratory prototyping into production network. Because the OSN system is designed to be cost efficient, it is based on open standards (*IX operating system, SS#7 protocol, TCP/IP protocols and commercial SQL database system e.g. Sybase as SDP) and open application programming interfaces for service creation.

The OSN system is based on open modular software architecture. This allows for instance the development of new interfaces to different network modules (e.g. ATM switches) and introduction of new transaction processing systems (e.g. Tuxedo) also for third parties. Furthermore, the system modularity gives the flexibility needed to implement future requirements for multiple supported INAPs, SDPs, and others. Because the OSN system is developed with C-language based protocol framework, it is highly portable to different operating systems. It is first available for HP-UX and Linux operating systems.

Service development with the OSN is at the moment done using a service script language, which consists of extended finite state automaton language. The automaton language contains the possibility to embed C-program code within the automaton script. These embedded C-routines contain, for example, methodology for logging transactions and interfacing the OSN to the Service Data Point (SDP).

This paper describes mainly the functionality of the OSN environment in the development phase 3Q/1995.

2 THE OSN SYSTEM ARCHITECTURE

The OSN system is designed to be a complete environment for developing, testing and running reliable IN service applications in the network. The architecture is shown in Figure 1. The doubled system architecture offers high availability for running IN services.

The OSN environment is designed to implement several different network configurations. The OSNs switch interface is adaptive to different switches, because the services in the OSN have built-in INAP interfaces and the services can be implemented with one or more INAP interfaces. This enables the integrity of the service management in heterogeneous networks.

Furthermore, the OSN processes are reconfigurable for new functions, so that the needed parts of the OSN SS#7 process can be replaced e.g. with TCP/IP, ATM or GSM stacks to meet the network requirements. Also, the processes are distributable to different hosts, so that the service databases and services can be located in dedicated hosts, if needed.

The OSN system is available as

- laboratory version
- single service piloting version on the network
- doubled production network version

The OSN laboratory version runs on freeware Linux operating system and requires no expensive commercial SQL database. Therefore it is ideal for developing and testing new IN services and applications. Also, because of the OSNs open concept, the OSN is suitable for academic usage such as protocol engineering.

The single piloting version is based on HP9000 platform and is suitable for piloting services for example in real SS#7 network. It includes Sybase database system for fast and reliable transactions. The doubled production version (see Figure 1), which is based on HP9000 platform (with multiprocessor platform option), is a high availability version of the OSN for production network.

The estimated overall performance of the OSN is expected to be 20-30 tps, depending on traffic profile.

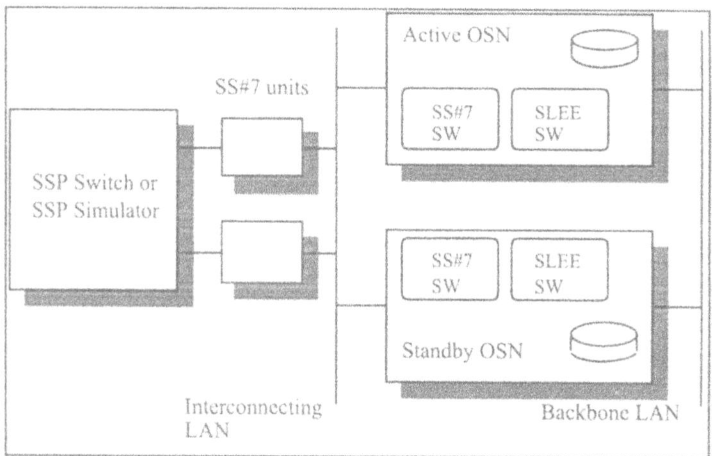

Figure 1 The OSN physical architecture.

3 THE OSN SOFTWARE ARCHITECTURE

The OSN software is based on CVOPS (C-language based Virtual Operating System) software architecture [1]. The CVOPS protocol framework is a portable protocol development and run-time environment which has been developed at the Telecommunications Laboratory of the Technical Research Centre of Finland (VTT). The main reason for its selection as a framework is its portability and design for protocol stack implementations.

The OSN software architecture is designed to be modular and flexible (see Figure 2). The main modules in the OSN are (the modules may include several processes):

- SS#7 stack
- SLEE
- Scheduler
- Database query and update
- SQL database
- Service Management System server

- Watchdog
- OSN system management

The OSN SS#7 stack consists of front-end hardware and back-end software as shown in Figure 2. The front-end hardware consists of MTP2 Interface Units (MIUs), which are based on PC servers (see Figure 1). The interconnection to back-end is done with TCP/IP LAN and communication between systems is implemented through TCP/IP sockets.

The MIUs include signaling data link (MTP level 1) and signaling link (MTP level 2) functions. MTP level 1 protocol offers a connection to 2 Mbit/s PCM interface. MTP 2 level protocol is implemented according to CCITT blue book. The implemented architecture enables SS#7 change-over procedures over separate links (MIUs), thus ensuring the availability of the services in the OSN. The SS#7 back-end can use several links to increase throughput in SS#7 network, and to give protection of link failures.

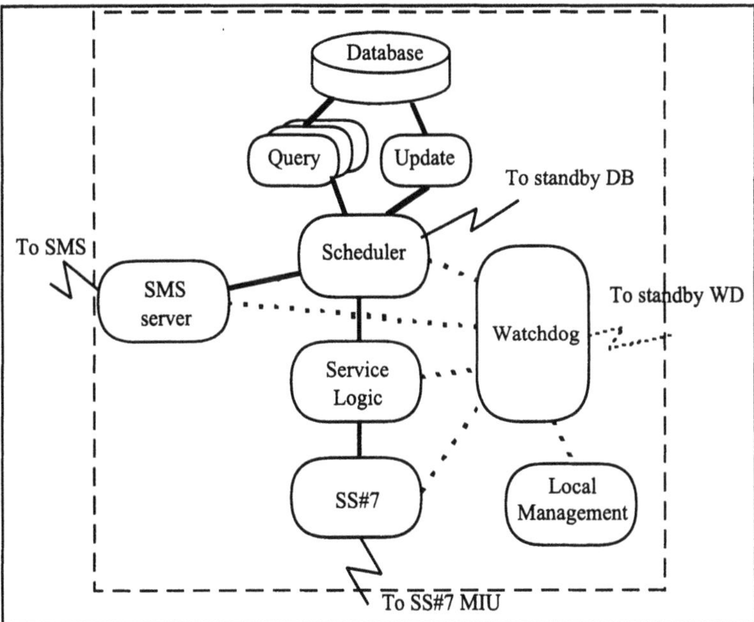

Figure 2 The OSN process architecture.

The OSN SS#7 back-end consists of processes for MTP level 3 and SCCP and their management (see Figure 2). TCAP and INAP are implemented within the service processes, Service Logic Programs (SLPs). Current implemented INAPs are Nokia's AIN rel. 0.1 FINAP and

ETSI Core CS-1. The modular architecture of SS#7 and services gives the option to replace parts or all of current SS#7 processes with different parts, if needed [2, 5].

The OSN SLEE (Service Logic Execution Environment) is based on CVOPS protocol framework. SLEE includes service processes, their management and data.

The interface between services and database is CCITT X.700 CMIP-compliant object oriented XI-interface. XI offers a programmer high abstraction of handling service parameters in database with GET, SET, CREATE, DELETE, ACTION - operations. Furthermore, the interface to operator's OSS (Operations Support System) is provided through OSN Service Management System (SMS) server XI-interface. With the XI, the database interface is possible to adapt for other database-systems, too (e.g. Oracle). The XI-interface has been developed using the previous work reported in [3].

The managed objects, methods and attributes in the OSN services' data are described in XI MIB (Management Information Base) or XIB. The execution of XI-methods are done in database query and update processes (see Figure 2). These query and update processes are used for the data access by Scheduler process.

Scheduler process is responsible for scheduling incoming database operation requests from service and system management processes in the OSN. The request operations are divided as database query requests and database update request. Scheduler offers definable priorisation for the request operations. For example, higher priorisation can be defined for services' requests than for service management requests, which are usually either large queries or batch updates to the database. Furthermore, in the doubled OSN, Scheduler controls databases' consistency. It writes data updates to both active and stand-by OSNs databases and in the system switch-over it controls that the information is equal in both databases.

Service Management System server process is a open interface to operator's OSSes. It provides operator with CMIP compliant XI-interface to service databases, provisioning of service data and access to service business functions such as billing.

The OSN is managed internally by the Watchdog process. The Watchdog starts other processes in the OSN at system reboot. The Watchdog expects a heartbeat signal from managed processes and checks that processes are alive. In error situations the Watchdog can restart failing processes or perform controlled system switch-over.

The operator management interface to the OSN processes is given through the Watchdog to make it easy for the operator to manage all processes and to reduce the number of connections between processes.

The OSN System Management is a menu driven management program in remote workstation and/or in local OSN system. System management includes e.g.

- audit processes and system usage
- log transactions and set logging levels
- configure parameters in OSN processes
- add, remove, configure services
- view service and system statistics
- perform system switch-overs and database consistency warm-ups
- perform OSN setup and shutdown

- setup processes in OSN

OSN Management also includes Logger process for logging OSN systems' transactions in remote management workstation. It alarms operator if erroneous or fraud actions has been detected in the OSN system.

4 SERVICE CREATION AND MANAGEMENT IN OSN

The service creation process [4] can be supported with the OSN environment tools. In the OSN Service Creation Environment (SCE) the Service Logic Programs (SLPs) are written with CVOPS protocol framework scripting language. The scripting language is a high level, C-language like protocol EFSA (Extended Finite State Automaton) language which defines protocol states, their inputs and actions taken on inputs (see dummy example in the Appendix). The mapping between the OSN EFSA language and SDL graphical representation is seen in the Appendix. The general actions and processings can be defined as macros for higher abstraction e.g. CollectDigits, AnalyzeDigits, PromptAndCollectUserInformation and so on.

The scripting language can be easily extended with user-written C-functions for new functions and service specific actions. The automaton language is compiled into C-language using the Service Creation Environment tools. The C-code will in turn be compiled into an executable program using an ANSI compatible C-compiler.

The service creation in the OSN can be divided in four layers (see Figure 3):

- Graphical user interface for service creation (not implemented yet) which uses SDL graphical representation (see appendix) and produces EFSA language for actual service compilation (see appendix)
- High level service description language and macros which uses OSN API functions such as INAP-message coding/decoding
- OSN system level functions needed in service creation (e.g. database management, functionality of macros and new INAP interfaces)
- Operating Systems functions needed in services (e.g. string operations, socket communications and handling of exceptions and interrupts). These functions are common to all services and are used as already implemented library functions.

The OSN service creation can be started from laboratory prototyping with the switch simulation program. The tested and proofed service files can be then moved to the actual pilot and production network without any modifications to the files. This innovative approach to service creation decreases services' time-to-market and thus enhances operator's competitive advantage.

Another important feature of the service creation in the OSN is the available toolset for developing services. Toolset includes e.g. following programs:

- A SSP switch simulation tool for emulating switch operations. It can be connected to the OSN either through TCP/IP connection or SS#7 connection
- A telephone simulation tool which provides the user an easy and fast method to ad hoc test services. Several telephone simulators can be attached to SSP simulator and they offer both A- and B-subscriber facilities
- A traffic generator that can be used to create IN user traffic and to test service memory handling. The traffic can be defined with script language to match wanted traffic profile

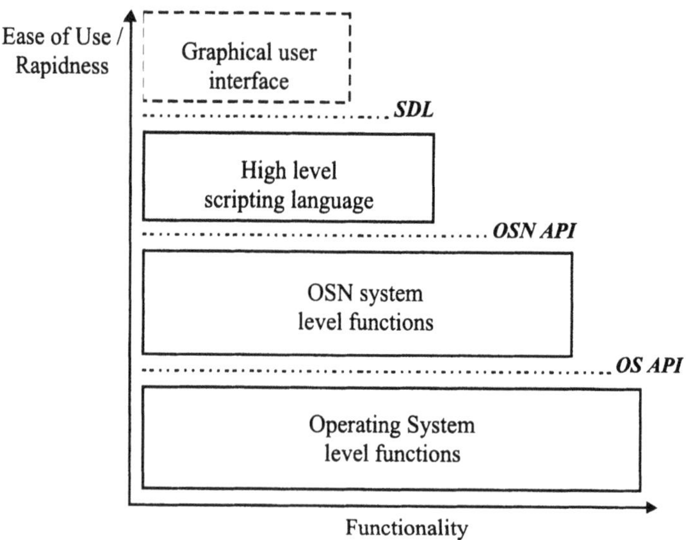

Figure 3 Layering for the OSN service creation.

- A third party commercial SQL database engine and tools for database management (Sybase 10)
- A third party commercial 4GL cross-platform UI development environment for IN service management software implementation (Uniface 4.0)
- Detailed tracing and debugging facilities

The service business functions such as billing information gathering, service data setup and service data altering are deployed through Service Management System server process. The XI interface is used for accessing these functions. The service management functions such as

- deployment of a new service
- removal of an existing service
- modifications to an existing service

are done by using menu driven OSN system management program. Furthermore, the OSN system management manages the service logging, auditing and statistics.

5 FUTURE DEVELOPMENT

It is obvious that the one of the main future development aspects of the OSN is in graphical user interfaces (GUIs) both in service creation and in system management. The OSN SCE GUI will resemble SDL-type of graphical definition of state automatons (see dummy example in appendix). The programmer will be provided with graphical tools for service testing with a phone interface, a mass traffic generator and switch simulators. The system management GUI will include within graphical interface also statistics and clarified reports of the OSN system usage. The implemented interfaces will be X-windows and Windows NT.

Furthermore, the enhancement of the overall performance of the OSN will be examined and tested. These enhancements include e.g. more efficient disk and memory usage of the database system, multiprocessor UNIX and dedicated hosts for services and database systems.

There is also work going on to apply the OVOPS (Object Virtual Operations System) environment for service application development [6]. Another future development area will be the broadband intelligence for ATM-networks, which has been discussed in detail in papers [7, 8, 9].

Because the OSN is very suitable for different network configurations, it is already used in various pilot projects in Telecom Finland and Technical Research Center of Finland (VTT). For example, the OSN system will be available in the EU ACTS Mobile National Host in Otaniemi, Espoo to provide mobility management in the UMTS network trials.

6 CONCLUSIONS

In this paper we have introduced a lightweight service control system called the Open Service Node (OSN) which can be applied in the control of fixed, mobile and broadband networks. Because the OSN system has been designed for differentiated services, the main objectives in the development have been

- cost effective IN service controlling system
- adaptation to different network and system configurations
- open interfaces for third party developers
- flexible and modular software architecture
- scalable and portable system
- reliable high availability doubled system
- TMN-compliant OSS interface.

7 ACRONYMS

4GL	4:th Generation Language
ACTS	Advanced Communications Technologies and Services
AIN	Advanced Intelligent Network
ATM	Asynchronous Transfer Mode
API	Application Program Interface
CMIP	Common Management Information Protocol
CMIS	Common Management Information Service
CVOPS	C-language based Virtual Operating System
CS	Capability Set
DECT	Digital European Cordless Telephone
EFSA	Extended Finite State Automaton
GSM	Global System for Mobile communications
GUI	Graphical User Interface
IN	Intelligent Network
INAP	Intelligent Network Application Part
IP	Intelligent Peripheral
LAN	Local Area Network
MIB	Management Information Base
MIU	MTP Interface Unit
MTP	Message Transfer Part
OSN	Open Service Node
OSS	Operations Support System
PCM	Pulse Code Modulation
SCCP	Signaling Connection Control Point
SCE	Service Creation Environment
SCP	Service Control Point
SLEE	Service Logic Execution Environment
SLP	Service Logic Program
SMS	Service Management System
SN	Service Node
SQL	Structured Query Language
SS#7	Signalling System Number 7
SSP	Service Switching Point
TCAP	Transaction Capabilities Application Part
TCP/IP	Transmission Control Protocol/ Internet Protocol
UMTS	Universal Mobile Telecommunications System
XI	X Interface

8 REFERENCES

[1] J. Malka, E. Ojanperä, "Reference Manual for CVOPS 4.0", Technical Research Centre of Finland, Telecommunications Laboratory, 1992, 1 - 107.

[2] O. Martikainen, V. Naoumov and K. Samouylov, Signalling System No. 7 Portable Software Implementation, St. Petersburg Regional International Teletraffic Seminar, 15-20.6.1993, 344 - 348.

[3] J. Airaksinen, O. Martikainen, J. Sonninen and H. Töhönen, UPT Service Management, International Workshop on Intelligent Networks, Lappeenranta 10.8.1993, 1 - 14.
[4] T. Karttunen and O. Martikainen, Open Intelligent Network Application Development, Proceedings of the International Symposium on Network Information Processing Systems, Sofia 12-14.10.1993, 77-90.
[5] O. Martikainen, V. Naoumov and K. Samouylov, Portable Intelligent Network Software Implementation, Proceedings of the International Symposium on Network Information Processing Systems, Sofia 12-14.10.1993, 91-96.
[6] O. Martikainen, P. Puro and J. Sonninen, The OVOPS Environment for IN Applications, IN '94 Workshop, Heidelberg, Germany, 24 - 26.5.1994, 1 - 14.
[7] K. Molin, O. Martikainen, Broadband Intelligent Network Project, Workshop on Intelligent Networks, 8.-9.8.1994, Lappeenranta, 253 - 263.
[8] O. Martikainen, T. Karttunen, V. Naoumov, K. Samuylov, Comparison of Broadband Intelligent Network Architectures, Workshop on Intelligent Networks, 8.-9.8.1994, Lappeenranta, 265 - 283.
[9] O. Martikainen, J. Lipiäinen, K. Molin, Tutorial on Intelligent Networks, Workshop on Intelligent Networks, 8.-9.8.1994, Lappeenranta,1-88.

9 BIOGRAPHY

Pasi Kemppainen received his M.Sc. in telecommunications in 1995. He has worked in the area of data communications and telecommunications since 1991 concerning EDI, X.400, telecommunications and internet technologies in various companies and also as a free-lancer. Currently he is Development Manager in Systems Software Partners (SSP) Ltd. and is specialized in software technologies for Intelligent Network service control and creation. Also, he works as a part-time assistant in Helsinki University of Technology for telecommunications cources.

Olli Martikainen, Ph.D., from Helsinki University and M.Sc. from Helsinki University of Technology. He has been doing research in several positions in Helsinki University of Technology (1976 - 1982), Oxford University (1980 - 1981) and Technical Research Centre of Finland (1982 - 1985). Later he has been R&D manager at Nokia Electronics (1985 - 1986), department manager at Nokia Research Centre (1986 - 1988), professor and head of Datacommunications Institute at Lappeenranta University of Technology (1988 - 1989) and professor at Technical Research Centre (1989 - 1991). Since 1991 he has been research director at Telecom Finland. Currently he is vice president, R&D, at Telecom Finland Ltd., and professor at Helsinki University of Technology. His main areas of interest are telecommunication software methods and tools, network architectures, performance analysis and new industrial and economic structures in telecommunications.

APPENDIX

From SDL-specification to OSN-service

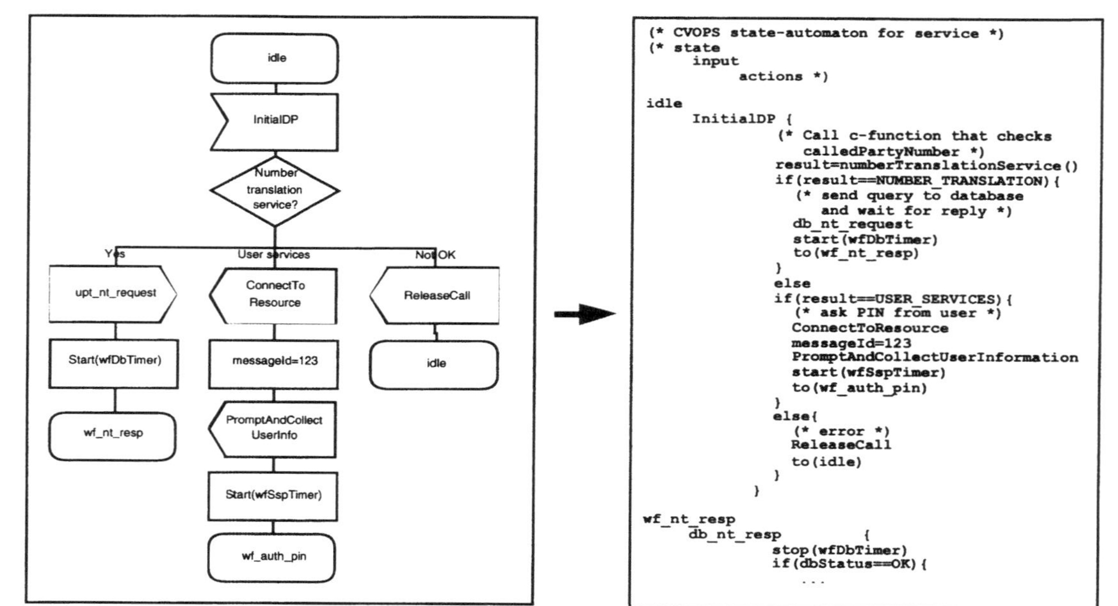

9
Advanced Concepts in an IN-Switching Platform
Standards and IN

Jørgen Dyst
L. M. Ericsson
Sluseholmen 8, 1790 København V, Denmark
Tel: +45 33 88 3 325, Fax: +45 33 88 31 28
E-mail: lmdjd@lmd.ericsson.se

1 INTRODUCTION

In this document it is assumed that the network is an IN-structured network. The description is based on standardized IN, e.g. the ITU-T IN- recommendations like Q.1214 for the CS-1 Distributed Functional Plane Architecture and ETSI Core INAP [1, 2]. The described Open Switching concept is based on the main ideas described in [3].

The focus is on advanced concepts in an IN switching platform, basic switching functionality and associated network aspects related to the application of IN technology within the N-ISDN network.

The following items are discussed:

- Open Switching IN-Platform
- Open Networking
- IN-Services in a Network Perspective

2 OPEN SWITCHING IN-PLATFORM

The Intelligent Network architecture is built above the basic Call Control Function (CCF), adding the Service Switching Function (SSF) to enable the recognition of calls that require IN support, linking call and service control to another functional entity, the Service Control Function (SCF).

An Open Switching architecture is to separate the basic switching functionality of IN calls from the Service Control Function. This separation enables third party call and service control.

A service switching architecture that clearly describes the relationship between basic call handling and services is essential for this external call and service control.

This control relationship is to be symmetric as IN services may be requested during call handling (i.e. from CCF), but a service (i.e SCF) may also initiate a new call (e.g. a wake-up call).

The basic switching functionality integrates the aspects of user & network access, connection handling (transmission) and call handling made up of the IN functional entities CCF and SSF - and optional SRF (Specialised Resource Function).

Message interfaces with defined procedure handling are used between the call handling and the service applications. Between SSF and SCF these message interfaces are standardized IN Application Protocols (INAPs). Open Switching allows to support different call and service control interfaces.

A protocol manager in SSF have to translate from standard protocols such as the ETSI Core INAP* to an internal protocol - a so called 'open switching' protocol - that establishes the connection to core SSF internal parts like DP Processing/Call Control Server. The internal protocol in SSF is to offer one uniform interface to the lower layer by integrating the different INAP protocols supported by the Service Control Functions. As a first step an Open Switching protocol may support e.g. only the ETSI CS-1 Core INAP protocol. However other INAP protocols for CS-2 and CS-3 will follow - with possible INAP protocol subsets (e.g. an INAP CS-1+) inbetween in support of different market demands.

The CCF builds-up, maintains and releases connections. The CCF functions have been modelled through the definition of the Basic Call State Model (BCSM) which represents the various phases of a basic call, on both the originating and terminating ends (i.e. O_BCSM and T_BCSM). The purpose of this model is to allow the implementation of services on the basic call structure and to identify call check points (Detection Points) that can be used to recognize the Intelligent Network services. These points are known as Trigger Detection Points and Event Detection Points.

The CCF contains mechanisms to detect IN requests demanding the IN (SSF) invocation on the basic call. The CCF includes the BCSM (Basic Call State Model) which provides a high-level model of CCF activities required to establish and maintain communication paths for users. The BCSM identifies the DPs at which transfer of control to SCF can occur to allow IN service logic instances in SCF to interact with basic call and connection control capabilities. It also contains mechanisms to detect service requests and to process incoming control messages from an IN service in SCF. These are extensions to normal existing basic call handling.

The SSF is the core part of an Open Switching architecture; it is to offer a logical view on call handling and connection handling that abstracts from switch system specific details. SSF also provides the open message interface by means of INAP to the SCFs.

Furthermore, SSF mechanisms for call event handling, charging event handling and SSF's role as a server for call control and connection control are needed for Open Switching.

The service control part can be subdivided into three main parts:

- Service Control Function (SCF)
- Service Creation and Service Management function (SCE, SMS)
- Service Data Function (SDF).

---

* The CS-1 INAP protocol, defined within ITU-T, was reviewed and refined within ETSI with the aim of eliminating redundancies, superfluous operations and parameters in an European compatible version defined as the ETSI 'core INAP'.

A complete IN-platform also comprises the above products which are not further described here.

3 CONCEPTS FOR OPEN SWITCHING

The separation of call handling into PICs (Points In Call) and DPs (Detection Points) as defined in the BCSM is the basis for the call state model that enables the SSF to abstract from switch-internal details. The SCF needs an abstract description of what happens during the PICs and the information which DPs are supported.

An explicit description of the basic switching functionality is needed for external call control. For example, IN service designers need a precise description of the set of supported DPs and the call manipulation INAP operations and procedures.

3.1 Implementation aspects of Open Switching Architecture

Open switching functionality can be built into the call handling structure of existing switching systems without needing to redesign or redevelop the existing software.

A solution is to implement 'hooks' between the call processing software for the A-side and the B-side. These 'hooks' are here called a Service Switching Function Handler (SSFH).
Figure 1 shows an IN basic two-party call with SSFH represented by an IN Call Segment. The SSFH may for example have been invoked due to a number based trigger or a line based trigger on behalf of the calling party or the called party, implying the use of an O_BCSM respective an T_BCSM instance on the two-party call.

The Basic Call State Models (BCSMs) for the originating and the terminating ends, i.e. the 'half-call' models are build by the SSFH based on the messages that are exchanged between the A-side and the B-side.

These Basic Call State Models (O/T_BCSM) support the required detection point mechanisms to enable call manipulation by IN service logic instances.

The operation mode of a SSFH can be illustrated by an example as follows:
If a DP in the O_BCSM is reached upon receipt of a message from the A-side, then the message is buffered and an event report is generated and sent to DP Processing in SSF. The event report informs the SSF about the DP currently reached. The SSF responds either by sending a message to let call handling continue or, if an IN service is triggered and invoked in SCF to take over call control, a directive toward the basic call is awaited from the SCF.

In the latter case an IN service logic instance is activated in the SCF, and may request to influence call handling. In support of this the SSF may modify parameters of the buffered message before that message is sent to the B-side - or it may even generate new parameters and/or messages for the B-side or even for the A-side call handlers that are sent instead of or in addition to the original message.

If no IN service logic instance in SCF is activated or if an service logic instance do not want to interact, then the stored call handling message is forwarded to the intended receiver - in this example the B-side.

The SSFH is to provide basic call control capabilities independent of the line or trunk type. The existing call handling software (A-side and B-side) provides interworking toward the specific line or trunk signalling systems applied. The messages sent by SSFH on the switch internal information channel between A-side and B-side may be based on ISUP.

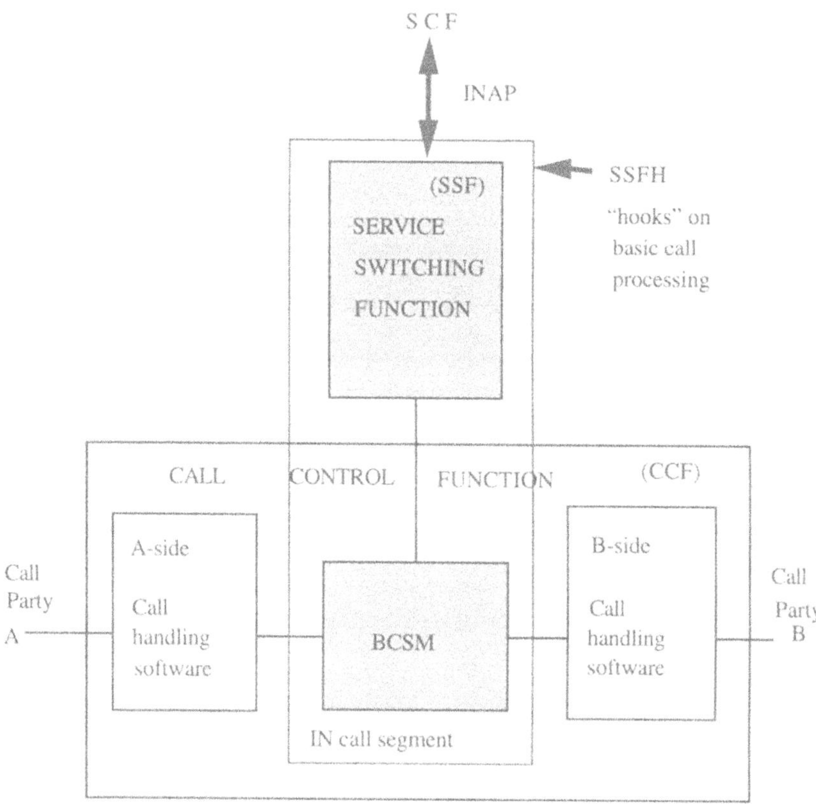

Figure 1 IN Open Switching Architecture applied on a two-party call

The integration of a SSFH with basic switching functionality allowing messages from the A-side to the B-side and vice versa to be intercepted by the SSF, checked, modified and again placed on the switch internal information channel, provides a flexible mechanism in support of call control (e.g. signalling manipulations) , but it also puts strong real-time requirements on efficient coding of especially this part of the implementation. However, the SSFH is only invoked on calls for which IN-services may be triggered- and the overhead in processing capacity occurs basically only where armed DPs are encountered.

4 DP PROCESSING PRINCIPLES

In the following a description is given to illustrate the principles of operation in the SSF in relation to DP processing, i.e. how an SSF may interact with the CCF (BCSM) and the SCF (via INAP).

4.1 Hand over IN basic call control

The point (DP) in a BCSM that may hand over the control and/or send event reports to service logic instances in SCF (s) is defined as a check-point where a trigger analyser and/or event handler within the DP Processing in the SSF may be called.

Within DP Processing an trigger analyser in the SSF decides whether a new IN-service is to be invoked, when a DP is armed as Trigger DP (TDP).

The trigger analyser transfers the control from the BCSM to the service logic instance in SCF only if it is possible to execute a service. Besides the encountering of an armed TDP other trigger criteria may have to be fulfilled to allow IN service invocation.

There may be several IN services invoked simultaneously from the two-party call segment in a IN basic call, but only one IN service will be allowed to have a control relationship (i.e. manipulate basic call processing). All other services will be in a monitor relationship and only be allowed to receive notification events.

Within DP Processing an event handler handles DPs armed as Event DPs (EDP-R/Ns), i.e. it distributes event reports to already invoked service logic instances in SCFs. The DP processing rules described in ITU-T recommendation Q.1214 applies [1].

Return to IN basic call control:

The following call control message types may be sent to BCSM call handling:

- Continue
  call handling continues at the same place where it was interrupted
- Continue with new parameters
  call handling continues with modified and/or new parameters or even new messages may be generated (for example for text sending)
- Proceed at PIC-n
  a 'go to' order to force call handling to proceed at PICn

A Call Control Server in SSF performs the state transitions on the BCSM instance used in the IN basic call.

4.2 Trigger Table Concept

The concept trigger-table as used in IN is generalised for the use as a basis for the triggering in the IN service switching platform on which also other switch based services are possible. This implies that especially the subscriber category handling and digit analysis handling in the basic call process becomes part of the trigger table concept.

The IN trigger process includes both the CCF and the SSF entities and may be considered as a two-step call related process:

- first step is to invoke the SSFH (O_BCSM or T_BCSM instance) in the basic call with armed TDPs, i.e. a prelimary trigger process.
- second step is the monitoring of the basic call process, i.e. the BCSM and on encountering of an DP armed as TDP invocation of the service logic in a SCF, if allowed.

The invocation of SSFH from CCF on a basic call may be defined on demand in the Digit Analysis to accomplish office (number) based triggering and in Subscriber Analysis data to accomplish line based triggering.

5 OPEN NETWORKING

5.1 IN Technology impacts the existing Network Architecture

The usage of IN technology provides added "network intelligence", but IN also brings specific IN-related problems. It may prevent some existing switch based services from being provided. These problems need to be solved in a open networking environment. In particular, signalling systems need to be enhanced!

The need for enhancements of existing signalling system capabilities can be illustrated with the usage of the existing switch based ISDN 'COnnected Line Presentation' service (COLP) together with IN technology.

A service example is where the calling party is using the IN credit card calling service and that in-band user interaction with SCF has occurred. This implies that a speech path has been set up from the calling party to the SSP e.g. on transit exchange level. With existing signalling systems it may imply that an 'early' answer message (ANM) is generated and sent from the SSP to the local exchange in order to establish the connection to SRF as needed for user interaction.

'Early' ANM sending and its consequences

An 'early' answer message is generated independent of the terminating access. It may be sent e.g. when an IP (SRF) is to be connected by SSP. The sending of an 'early' answer (ANM) message impacts the basic call and supplementary services which may be applied on the call.

After an ANM message has been sent backward to the calling user (corresponding to the answer of the called party for the first call), it is not possible to send back information received from the called user in the ACM and ANM or CON for the second and further calls.

Therefore information concerning the called party are lost for the originating local exchange. Moreover, it prevents some ISDN supplementary services from being offered to the calling user such as Connected Line Presentation (COLP) and User-to User signalling Services (UUS) during call setup.

This 'early ANM' method has impact on basic call and supplementary services in fixed/mobile networks. Existing signalling systems needs to be upgraded to avoid the consequences on 'early ANM' sending.

5.2 IN basic call requirement

An IN basic call should support the following versatile IN basic call handling concept: The IN basic call signalling capabilities (end to end message sending) should only be limited on demand. A limitation may take effect at the execution of an IN service due to imposed signalling system constraints. However, it should still be allowed to adopt improvements/enhancements in existing signalling systems, i.e. to remove limitations when signalling system upgrades will enable it.

There should in general be no restriction in the allowed service control level from SCF. The interactions with the IN basic call for the execution of an IN based service are defined in SCF on a per service basis or on a per service feature basis. However, the result of some specific interactions when applied on a call may be a restricted call signalling capability level.

5.3 Proposed near-term signalling system improvements

For user interaction (playing announcement and collecting in-band information), the transmission path has to be through connected in both directions from the calling user to the SSP. It may also be needed to stop network protection timers (e.g. T9). With existing ISUP procedures, the sending of an early ANM is necessary for both aspects.

A solution to avoid sending of early ANM by a SSP when this message is only needed to ensure through connection of the transmission path and/or to stop T9 'Awaiting answer' is being worked out in standardization bodies.

In this solution a new parameter is added in the backward direction in the ACM/CPG message to order the originating local exchange to through connect and stop T9 'awaiting answer'. Another parameter sent in the IAM message in the forward direction is to inform the SSP that the originating local exchange (OLE) has the capability of performing the through-connection and/or of stopping T9 on receipt of the backward parameter. However, the originating exchange is allowed to start a new timer Tuid, when it receives the order to stop or not to start timer T9. Thus, this new timer Tuid enables to guard the connection in the originating local exchange during the inband user interaction.

Hereby a mechanism is provided which enables to avoid the sending of an ANM when it is only needed for through connection and/or stopping T9 purposes. This mechanism is only applicable if the originating local exchange has the capability to perform it. When it has not, the SSP falls back on the sending of 'early' ANM. This solution is therefore compatible with existing exchanges not supporting this new mechanism.

However the problem with ANM sending remains related to follow-on calls (second and further calls) in IN, i.e. the problem how to carry information received in an ANM/CON message, to preceding exchange, when ANM has already been sent once.

If the intention is to solve the problem of all limitations, included for supplementary services (such as COLP), it implies a major impact on existing signalling systems.

The examples given is to illustrate the networking problems imposed with the introduction of IN CS-1. However, it is anticipated that IN technology also in forthcoming capability sets will put new requirements on the existing signalling systems. From IN CS-2 can be mentioned as an example OCCRUI (Out-Channel Call Related User Interaction) which denotes a proposed mechanism to be used to provide new IN services requiring the transparent transfer of information between a User and a Service (SCF) on a call. This will affect signalling protocols (DSS1, ISUP and INAP).

5.4 Main aspects

Open Networking demands that existing Signalling Systems needs to be upgraded to cope with the INAP capabilities:

- Align Signalling System capabilities (ISUP and DSS1) with the INAP capabilities.
- Impose network operators to implement this aligned ISUP and DSS1.

The associated call handling procedures in the SSP as well as in local exchange have to be defined, where affected by IN.

The impact on existing switching systems will depend on the level of information (for basic call and supplementary services) that is decided to be transferred for IN calls, i.e. is all known limitations to be resolved?

6 IN-SERVICES IN A NETWORK PERSPECTIVE

Third parties should be able to develop their services independently from other third parties as to foster competition.

The objective here is only to describe a few technical aspects in relation to a support of such competitive demands.

6.1 Types of services in a network

Some types of services it may not be desirable to be provided by third parties due to their nature. Services such as user to user signalling (UUS) are so 'entangled' with the network signalling procedures and capabilities that it is simply not possible for a third party to provide them. Furthermore, some services implying difficult service interaction issues should better not be provided by third parties when raising service-interaction problems. Unresolved service interaction problems may endanger network integrity.

Roughly we can define two types of services:

- Type A services using the existing network transport capability for providing information to the user.
- Type B services that are related to network internal capabilities (e.g. routing, charging...)

The first type of service does not raise any problems, i.e. could be opened for competition. The second type of services is not that easy to open for free competition because two services of this type may interact and not work if introduced independently at the same time

In ITU-T Recommendation Q.1211 this corresponds to Type A and Type B services, where CS-1 standards is targeted to support Type A services only.

The possibility to support multiple IN services at the same time is a problem. IN CS1 assumes single ended, single point of control technology at any time of the call.

CS1 and CS2 allows only one SCF to interact with one call segment association in the SSF. Multiple-point -of-control capability is considered as too complex and not implemental with present switching technology. For instance, how to resolve contradictory requests applied to the same call segment from two different SCFs [4].

The 'single point of control' technology implies in order to provide two services at the same time, it is necessary that both services are implemented in the same service logic program instance (SLPI) in the SCF. The SCF is then responsible for handling service interference. This approach may be preferable for type B services which otherwise may have serious service interaction problems.

However, it is likely that with IN services widely deployed, users will want to access multiple IN services simultaneously. Many IN-services could be activated on a single call; each providing specialized complex services and each operating seamless on the call.

6.2 Multiple SCF control relationships

The IN-CS2 (predicted release early 1997) is to include functions capable of supporting Call Party Handlingl (e.g. conference call) and enable triggering (e.g. midcall) in different phases of the call. Furthermore a support for global services like international freephone services etc. as well as a first limited set of capabilities to handle service interference between IN services in the network are included.

With increasing IN services' deployment, the handling of IN-IN service interworking becomes more and more important. IN services which are subject to office based trigger criteria (e.g. specific dialled digit strings) play a major role in many networks.

A support for independent service execution is needed. This is to allow for example two different IN services to be invoked during the same call, and if these services are provided by different service providers independent execution of both services is required.

An example where more points of control is applicable may be where a calling user via a Credit Card Call service wants to make a call to a premium rate service, both deployed as IN services.

In order to allow the SCF(s) to have multiple control relationships within a single call, a call segment model as used by Ericsson is presented that extends the model used for a basic two party call. Hereby a two-party call consists of a 'chain' of call segment associations, starting with an originating basic call segment and terminated by an terminating basic call segment, each representing the existing call handling software (A-side and B-side). In between a number of call segment associations may be invoked, here called IN call segments.

This IN Call Segment model allows more applications to be active on a single call at the same time. A call can be characterized as an originating or terminating type (O_BCSM or T_BCSM) dependent on how the IN call segment is invoked (IN trigger).

A separate control relationship must be established for each service logic program that can be invoked on the IN basic call. Each IN service is triggered in the IN switching system independently. This is enabled by the call segment model representing the basic two-party call as consisting of a number of independent call segment associations.

While processing a call the IN switching system detects an armed TDP-R and invokes the IN service logic program instance (SLPI). The IN service logic can initiate a persistent control relationship (EDP-Rs armed). During subsequent call processing the IN switching system (i.e. CCF) can detect another trigger event outside the call context of a previous invoked IN service. A subsequent call segment is 'linked' in the basic call to handle the new IN service logic instance in a SCF, which also may initiate a persistent control relationship - independently of any previous invoked IN service logic instances.

An example with two IN services invoked is shown in figure 2. Both services may have been invoked e.g. due to an office (number) based trigger whereby two independent BCSM instances are invoked on the call, each within a separate IN call segment as to allow independent IN service control. Both of the services (SLPI 1 and SLPI 2) may have a persistent control relationship with SSF as an SSFH instance is invoked for each of the control relationships.

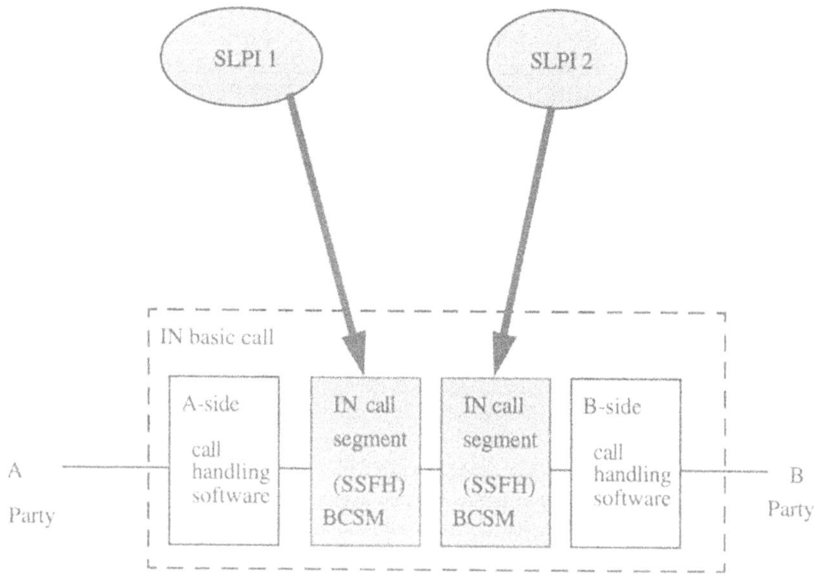

Figure 2 SCF control relationships, IN basic call, IN call segment model

The control capabilities allows for IN to IN service interworking and is for IN CS-1 applicable only to services (Type A) that do not have feature interference problems with each other. This is to avoid adverse interference between IN services simultaneously active on the same call.

For IN CS-2 service compatibility checks is to be added to the trigger mechanisms in the SSF for IN to IN service interference handling. On a call this enables to reject the invocation (triggering) of a new IN-service dependent on which IN-services may previously have been invoked.

In order to allow the same feature interference treatment independent of whether triggering occurs in the same SSP or in another SSP, the ISUP is required to be capable to convey this service/feature compatibility information.

6.3 DP processing rules

Event Detection Point processing requested by any of the involved IN service logic instances is performed as if triggering occurred in different BCSMs, which are separated by an ISDN network interface. This is secured by the invocation of an BCSM instance in a separate IN call segment for each subsequent IN call triggering. It is not visible for the involved IN services if triggering occurs in physically different SSP nodes or within the same SSP node.

The chain arrangement defines the priority of event handling. That is, originating events propagate down the call chain from the A-party towards the B-party and terminating events propagate down the call chain from the B-party towards the A-party.

7 MAIN ASPECTS

- IN demands distribution of services

  It is likely that with IN services widely deployed, users will want to access multiple IN services simultaneously. Moreover these services may be deployed in a multi-vendor environment. Therefore a demand beyond the CS-1 and CS-2 `single-point of control' exist to enable multiple services to be active on an IN call.
- Network topology should be transparent to IN service logic

  The IN services works independently of each other and need not to know how SSF functionality is distributed (e.g. between different SSPs or in a single SSP). Each IN service logic instance (SCF) invoked on the call has via its Call Segment Association own independent call models (e.g. Connection View states and BCSM).
- The Call Segment Model provides multiple IN control relationships

  The Call Segment Model provides a mechanism for handling the demanded distribution of services in a IN structured network. The model supports that event reports and sending of call control directives can be managed by the IN switching platform independently for each active IN service invoked on the call, enabling multiple services, i.e. multiple control relationships to be invoked simultaneously on the same IN call. This allows for a multi-vendor environment with distributed services (multiple SCPs).

8 ADDENDUM

The advanced concepts described herein before are based on the implementation of an IN CS-1 switching platform within Ericsson AXE10 switching systems.

9 REFERENCES

[1] Revised ITU-T Recommendation Q.1214, Distributed Functional Plane for Intelligent Networks CS-1 ITU-T Recommendation Q.1214 (May 1995)

[2 Intelligent Network (IN); IN Capability set 1 (CS1), Core Intelligent Network Application Protocol (INAP), Part 1: Protocol Specification. ETS: 300 374-1, (September 1994).

[3] Open Switching - Extending Control Architectures to Facilitate Applications, Martin Elixman, Bart L. de Greef, Armand M.M. Lelkens, Karl W. Neunast, Hermann Tjabben, Philips, ISS `95 Berlin (April 1995).
[4] Open Networking: is it technically possible, Roberto Kung, France Telecom/CNET-PAA-CER, ISS'95 Berlin (April 1995).

10 BIOGRAPHY

Jørgen Dyst was promoted IN Expert on April 1, 1994. He participates in IN standardization work in ITU-T and ETSI and works in the field of IN system management. Before assuming his present position, he served as IN Senior Specialist working with specification and analysis of new IN functionality and as operative IN product manager. He has been contributing to development on the Ericsson AXE10 advanced IN switching systems since the start. He has a background in the fields of switching and call control. Mr. Dyst attended Copenhagen Teknikum college of engineering and received a BS degree in Electrical Engineering in 1979.

10
Nokia SCE - An Architecture for a Lightweight SCE

Mikko Kolehmainen
Nokia Research Center
P.O.Box 45, FIN-00211 Helsinki, Finland
E-mail: Mikko.Kolehmainen@research.nokia.com

Abstract

This paper describes the architecture of a lightweight Service Creation Environment (SCE) that is developed at Nokia Research Center. The discussion focuses on an example of the future service creation architecture and on solutions that are considered to be worth further research. The aim of this article is to point out those architectural core capabilities that are essential for an effective Intelligent Network (IN) service creation environment. The core features of the architecture include a method to define service logic programs for several Service Control Point (SCP) platforms, a strong support for SIB based service creation approach, a possibility to define service logic programs with textual or graphical editors, a concept of using intermediate format of service definition language as a source format, and finally a possibility to use a network simulator via Core INAP (Intelligent Network Application Protocol) protocol.

1 INTRODUCTION

Nokia has a strategic alliance with Hewlett-Packard (HP) on IN products and platforms. As a result of this alliance Nokia delivers to its customers among other things HP based tools for service creation. The HP originated SCE has proven to be an effective tool for implementing new IN services. It provides capabilities for service creation for both fixed and mobile IN applications. Even though HP SCE is a powerful tool in the hands of a capable software engineer, it demands expertise beyond that of a person with a non-technical background on software engineering. This is mostly due to the relatively high expression power of the Service Logic Execution Language (SLEL) used in HP SCE. Bearing in mind that one of the basic purposes of IN service creation was the ability to produce services in a relatively easy manner, without deep knowledge of SCP programming, it is considered that there is a need for a lightweight service creation tool. Furthermore, since there is always a certain degree of trade-off between the ease of use and the capabilities of the language even regardless of the application area, the problem in the concept of SCE architecture is to determine the correct level of expression power of the service definition language.

Nokia Research Center has initiated a project to produce a prototype of an easy-to-use service creation environment. The architecture of the prototype SCE should be scaleable, fulfill the requirement of producing services to different platforms and conform to the ideas presented in IN recommendations. The architecture under construction has been named NOKIA SCE. The illustration below describes how the NOKIA SCE architecture can strengthen the current service creation approach based on HP SCE. When considering the service creation product line, it is important to realize that an SCE implemented according to the NOKIA SCE architecture is a lightweight SCE that is targeted towards the needs of second or third operators.

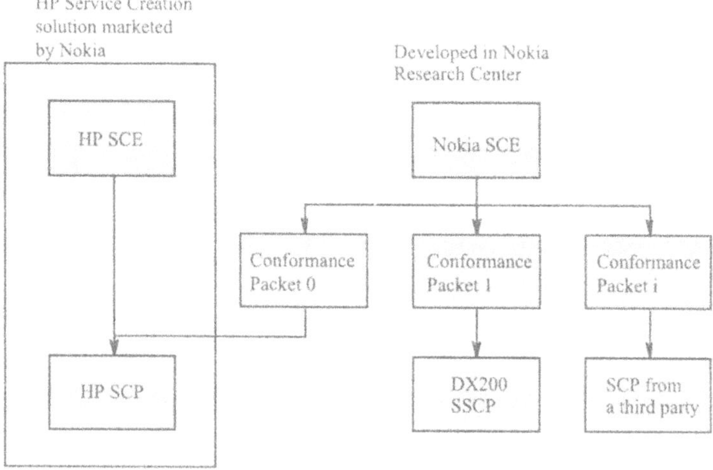

Figure 1 The goal of making NOKIA SCE a multitarget environment is achieved with the aid of conformance packets.

The idea is that services defined with NOKIA SCE could be ported to different SCP platforms. The NOKIA SCE provides the basic set of service creation functionalities that are expected to be present in most of the third party SCP platforms. The platform specific capabilities are added by means of a conformance package; each package is designed for a specific platform according to common guidelines.

2 THE GOALS OF NOKIA SCE ARCHITECTURE

The goals of the NOKIA SCE project can be divided into two classes: internal research work and a basis for future productization of a lightweight SCE. Both of these classes of goals are of equal importance and must be met.

In respect to research goals there are two issues that should be mentioned. First, there is a need for a tool that can be used for rapid service creation for Nokia's internal network simulation tool. Second, an SCE is needed for service logic definition for intelligent broadband network demonstrators.

The goal of initiating a product is to design and implement a prototype of an SCE that can be used as a starting point for a commercial lightweight SCE. That SCE will extend Nokia's IN service creation product line from the lower end. In the figure 2. there is an illustration that shows how the objective classes relate to each other and to the NOKIA SCE project.

The mission of Nokia Research Center is to concentrate on research work and to find out new solutions and technologies to support operation of Nokia business units which will make the ultimate decision on productization of individual architectures and concepts. Thus the design process of the NOKIA SCE architecture has been concentrating mostly on the research issues and ideas for features that extend the usability of the NOKIA SCE concept. Therefore the act of productization itself may appear somewhat distinct from the approach and decisions taken during initial design and prototyping phase.

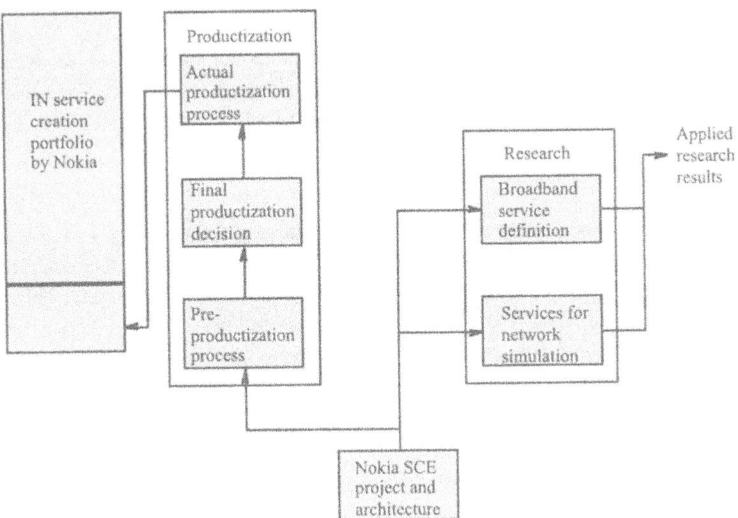

Figure 2 Objective classes of NOKIA SCE project and architecture design process.

3 THE OVERALL NOKIA SCE ARCHITECTURE

3.1 The set of NOKIA SCE units

The NOKIA SCE architecture consists of four basic elements, named *NOKIA SCE units*. The most visible NOKIA SCE unit is the graphical *NOKIA SCE Service Editor*, which acts as the main user interface towards the service designer. The purpose of the *NOKIA SCE Core* is to provide the functionalities related to service animation, validation, and Service Logic Program (SLP) database access for editing needed by the graphical editor. NOKIA SCE Core is also responsible for providing the tools required for processing the NOKIA SCE service logic language. The name of the language is *NoLo* - ***Nokia Logics***. NoLo is a scripting language that consists of an extendible set of SIB calls. Thus one SIB is considered to be one NoLo command. Besides tasks listed above, NOKIA SCE Core takes care of the communication with Inesim tool (*Inesim* - ***Intelligent Network simulation***) for SLP testing and demonstration purposes.

The third NOKIA SCE unit is the *NOKIA SCE mini-SMS* (Service Management System) that is used for service animation management and service data management. The fourth unit is one of the distinctive features of the NOKIA SCE concept: *NOKIA SCE conformance unit*. The unit consists of conformance packages that ensure the portability of SLPs created with NOKIA SCE. In the figure below there is an illustration that depicts the overall NOKIA SCE architecture.

In the following sections there is a short description of each NOKIA SCE unit. From end user's point of view the division into NOKIA SCE units is not visible. Also, the mapping between NOKIA SCE units and host operating system processes is not direct; in a workstation one process may be responsible for more than one NOKIA SCE unit.

3.2 The NOKIA SCE Service Editor

The NOKIA SCE Service Editor is used as a tool for service definition and as a user interface for service animation and testing. In the current prototype the main functionality of the NOKIA SCE Service Editor is to provide adequate tools for service creation. Therefore the set of tasks assigned to NOKIA SCE Service Editor is as follows:

- To provide the basic SLP graph editing functionalities.
- To ensure the linguistic equivalence between textual and graphical NoLo representation.
- To provide a dynamically configurable set of SIB icons and Service Support Data (SSD) definition routines.
- To provide a graphic user interface for service animation, testing and debugging.

The set of interfaces to the NOKIA SCE Service Editor bases on the tasks specified above. In the figure 4. there is an illustration that describes those interfaces.

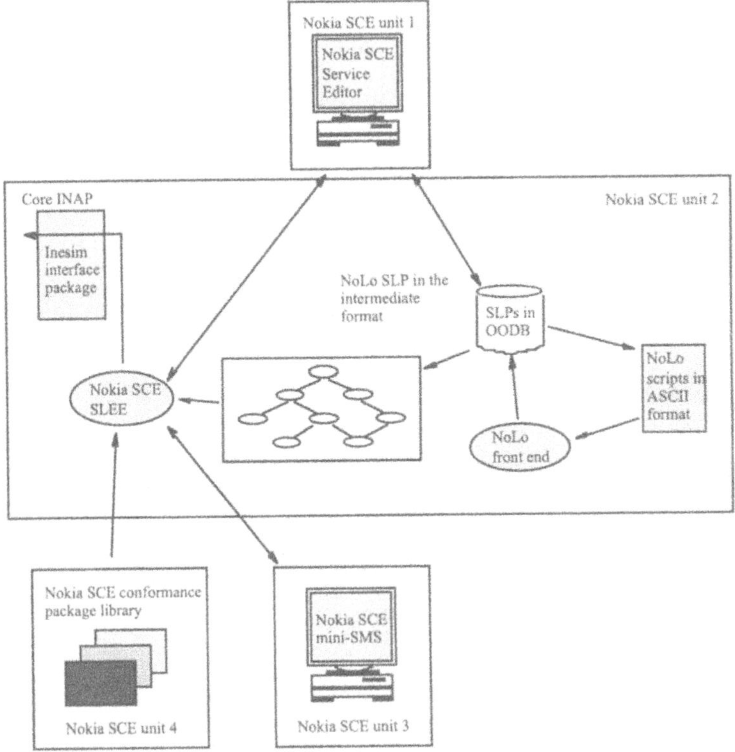

Figure 3 Overall NOKIA SCE architecture.

As can be seen, the editor has three interfaces to other parties of NOKIA SCE: one towards NOKIA SCE SLEE via socket based communication protocol, one towards NOKIA SCE Core database functionalities including NoLo interfaces and one is a computer-human interface towards the service designer. The sockets are needed in the communication between the NOKIA SCE Service Editor and NOKIA SCE SLEE since they will be executed in different processes. However, even though the editor and NoLo SLP database functionalities belong to different NOKIA SCE units they will be mapped to the same process in implementation.

3.3 NOKIA SCE Core

NOKIA SCE Core is the center of NOKIA SCE architecture. The core consists of four modules that are logically and functionally joined together. Together these modules provide the capabilities that are needed for service animation and testing. The modules are as follows:

- NOKIA SCE Core database (both runtime database and the one used for SLP storage).

- NoLo compiler front end and back end for ASCII NoLo scripts.
- NOKIA SCE SLEE with SIB implementations.
- Inesim interface package.

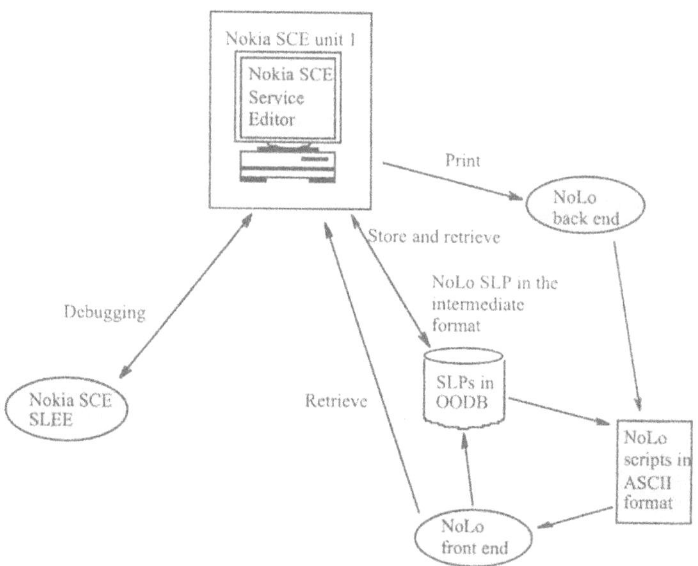

Figure 4 The interfaces related to NOKIA SCE Service Editor.

Both of the NOKIA SCE Core databases are modelled as object oriented databases. The reason for choosing OODB is twofold: on the one hand object modelling technique makes it possible to store tree structured objects, like service logic programs, into a database in an effective manner. On the other hand object database is a promising approach for implementing run-time datastore for service logic programs at least in the SCE itself. Of course, if runt-time database is object oriented in the SCE, and e.g. relational in the target SCP platform, a separate presentation transformation layer is needed in the database access components.

One feature of the NOKIA SCE architecture is the possibility to define services with both textual and graphical editor. In order to provide two kinds of service definition methods, two front ends of the NoLo-compiler are needed. Both front ends produce identical intermediate format of NoLo language. This intermediate language is then stored in the NOKIA SCE Core SLP storage database. There are tools for producing both textual NoLo scripts and graphical representations of intermediate format representations that are stored in the database. NoLo front end for graphical editor is considered to be conceptually a part of the NOKIA SCE Service Editor.

The execution of the service logic program in NOKIA SCE is the responsibility of NOKIA SCE SLEE interpreter. The commands of the NoLo language have been implemented in the SLEE so that there is an one-to-one mapping between NoLo commands and SIB implementations. The set of SIBs designed for NOKIA SCE architecture is flexible and extendible.

From software engineering point of view the structure of NOKIA SCE SLEE bases on parallelism achieved with usage of threads. This means that each SLP is executed in a thread of its own. This is beneficial in cases when NOKIA SCE SLEE runs on multi-processor computers.

NOKIA SCE SLEE uses as input the tree-like intermediate format of NoLo scripts. Thus there is no linear binary representation of NoLo scripts. The advantage of using a memory resident structure as an input format of an interpreter is the direct mapping between the graphical and the input representations. Thus it is possible to implement an easy debugging facility between the SLEE and the Service Editor. Furthermore, since the information is a graph structure that can be traversed, the interpreter is possible to be implemented in quite a straight-forward manner.

Inesim interface package provides a communication means towards the Inesim tool. The idea is that communication takes place via Core INAP protocol stack thus allowing the usage of different SSP simulators instead of Inesim tool. The interface between NOKIA SCE SLEE and Inesim tool can be implemented in a simplified manner; part of message parameters can be obtained from the default information base that is configuration specific.

3.4 NOKIA SCE Conformance package library

Even though IN standards for CS1 do not specify architecture of SCE they contain an idea of portable services. One of the goals of the NOKIA SCE concept is to provide a possibility to produce services that can be reused in several SCP platforms; for example the platforms may be delivered by Nokia, HP or some other supplier. The key portability tool in NOKIA SCE is the concept of conformance package (see figure 5).

The conformance package is divided into two separate logical parts. The first part is responsible for ensuring that the service is able to execute on the target platform. Usually this means that there exists a SLEE that can execute or interpret platform's native programming language. This implementation should also include the suitable set of SIBs implemented e.g. as macros or subroutines depending on the capabilities of the implementation language. The SIB library implementation is an additional component to the target platform's SLEE.

The second part of a conformance package is dedicated to service testing. The idea is that conformance package supplements NOKIA SCE SLEE with capabilities that are platform specific. This is due to the fact that from the functional point of view NOKIA SCE SLEE is the least common denominator of the functionality set needed for implementation of certain set of benchmark services. The terms "MIB" (Management Information Base) and "External channel" in the figure 5. refer to HP SCP specific capabilities that may not be present in other platforms.

It must be noted that there are service classes that cannot be reasonably ported to another platform. Most of such services perform operations that are not sensible in other environments.

For example certain SCP specific management related service logic programs might fall into this class. However, porting of the most common services can be considered worth the trouble. Furthermore, since service management issues are solved with very heterogeneous solutions in different platforms, they must be considered case by case anyway.

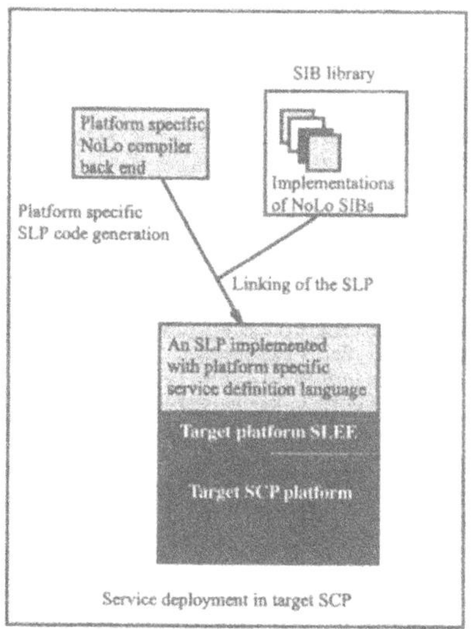

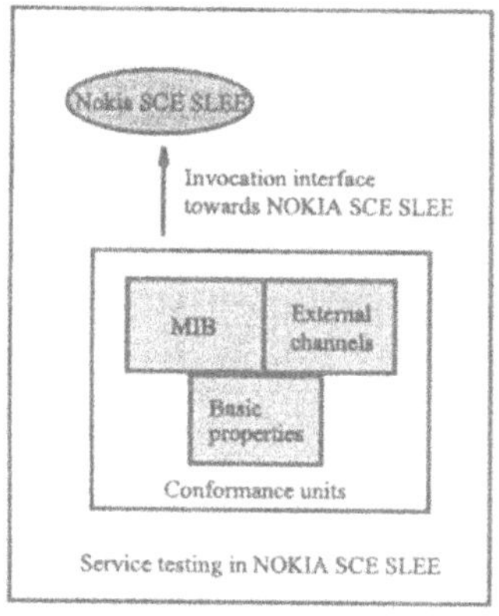

Figure 5 An example of a conformance package.

3.5 NOKIA SCE mini-SMS

According to current ideas about NOKIA SCE architecture, mini-SMS is not very interesting subject within the SCE architecture. The duties of mini-SMS are quite straight-forward: loading and stopping of SLPs, SLEE execution and network simulator management and data management.

4 THE NoLo-SERVICE DEFINITION LANGUAGE

NoLo-language is a simple scripting language that is not dependent of the application-area. This means that the structure of the language does not take any attitude to the question whether the language is intended for a certain application area or not. The functionality and the expression power of the language is based on the set of commands of which the language itself consist.

NoLo-commands are called SIB invocations. The set of SIBs is extendible. However, the extensions must be made in three locations: editor configuration file, NOKIA SCE SLEE and the corresponding conformance packages. The idea is that there exists a basic set of SIBs that can be used when creating feature entities by combining SIBs into macros. From these feature entities it is relatively straight-forward to create services. If the basic set of SIBs appears to be inadequate, new SIBs can be designed and implemented.

As said earlier, there are two possible methods to create NoLo-scripts: the graphical and textual. Both of these ways are equivalent in a sense that scripts created with one tool can be processed with the other tool. Thus the user has the benefits of both approaches without any trade-offs.

Due to the fact that NoLo-scripts are interpreted and stored in the intermediate format it is possible to produce Flexible Service Logic (FSL) scripts as defined in Bellcore generic requirements (GR-1280). In NOKIA SCE SLEE this is implemented plainly just adding the FSL script in the interpreted intermediate source format when the FSL invocation has been met. In the SCP platform level, the idea is quite similar: both NoLo scripts and FSL scripts are compiled to the native language presentation. From the structural point of view FSL scripts are not different from plain NoLo scripts, except for the routines needed for data definition.

5 DATABASE ASPECTS

When considering creation of portable IN services and database aspects, there are two things that must be taken care of: how run-time database operations can be designed to be portable, and what type of database should be used. These two issues set certain boundaries to the concept of portability.

In order to achieve the possibility to port service programs from a platform to another the basic set of database operations (Retrieve, Store, Delete, Insert) ought to be designed in such a manner that it would be relatively easy to change the database system that is used as a basis for database routine implementation. The figure below illustrates the idea of the design of routines.

According to the NOKIA SCE concept there can be unlimited number of basic SIBs and thus unlimited number of SIBs that access the database. Within each one of the database related SIBs there is an inner interface towards the common database access routines. In that common interface there is one thing that has to be decided: whether the database used is relational or object-oriented. This decision is imperative and must be done before compilation of the SIB implementations.

Below the common interface there is a presentation transformation layer that takes care of the actual database access. If database is object oriented, the access takes place via ODMG compatible interface. If it is relational, a data presentation transformation is performed, if needed. After that, database specific access routines are invoked.

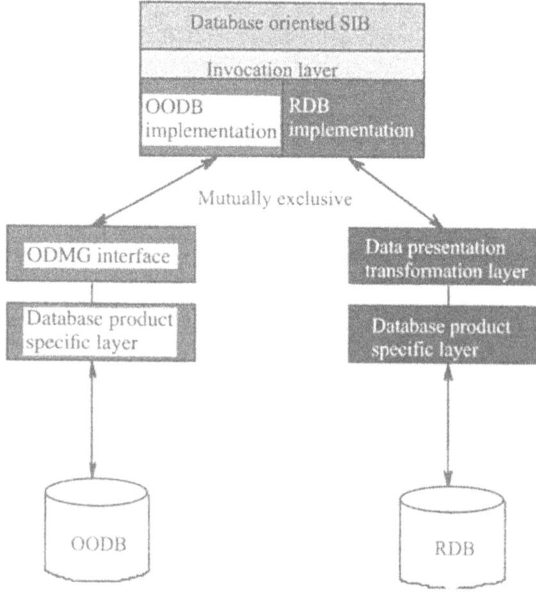

Figure 6 Database routine layers in a database oriented SIB.

6 CURRENT STATUS

A project that is heading towards implementing a lightweight SCE according to the NOKIA SCE architecture has started at Nokia Research Center. The project's aim is to achieve the specified research goals and both to evaluate the portability issues and to implement an exemplary conformance package. When the first phase of the project will finish, the second phase which concentrates on productization and portability will start. Precondition for the second phase initiation are the results received from the work done that far; if NOKIA SCE architecture will live to the expectations that have been set to it, the continuation of NOKIA SCE will take place.

7 SUMMARY

NOKIA SCE architecture is a concept of a lightweight SCE that is intended to supplement Nokia's IN product line as an easy-to-use service creation tool from the lower end of SCE product family. The main issue within the NOKIA SCE architecture is the possibility to define portable services in a relatively rapid manner either graphically or as a textual script. The NOKIA SCE approach is based on SIB philosophy as expressed in IN standards.

8 REFERENCES

Cattell, R.G. G.; ed.; "The Object Database Standard: ODMG -93"; Morgan Kaufmann Publishers, San Mateo, California, USA; 1994.

ETS 300 374-1; "Intelligent Network Capability Set 1 (CS1) Core Intelligent Network Application Protocol (INAP) Part 1: Protocol Specification"; ETSI; 1993.

Genette, M.; "The Service Creation Environment for Intelligent Networks"; Alcatel Telecommunications Review, 1st Quarter 1996; Alcatel Alsthom Publications; France; 1996.

GR-NWT-001280; "Advanced Intelligent Network (AIN) Service Control Point (SCP) Generic Requirements"; Issue 1; August 1993; NJ, USA; 1993.

Hino, K.; Tani, H.; Nishida, T.; "Object-oriented Service Creation Environment for Multimedia Communication Networks"; 3rd International Conference on Intelligence in Networks; ICIN; Bordeaux, France; 1994.

International Telecommunication Union, Standardization Sector; Q.1200 Series; ITU-T, Study Group XI; 1993.

Knight, C.; "Service creation from IN to mobile and broadband"; Intelligent Networks; Harju, J.; Karttunen, T.; Martikainen O.; ed.; published on behalf of IFIP; Chapmann & Hall, London; 1995.

Ricken, R.; Sevcik, M.; Eder, R.; "IN Service Creation - How to Make the Service Independent Building Block Concept Fly"; XV International Switching Symposium; Berlin, Germany; 1995.

Slutsman, L.; Lu, H.; Kaplan, M, P.; Faynberg, I.; "The Application-Oriented Parsing Language (AOPL) as a way to Achieve Platform-Independent Service Creation Environment"; 3rd International Conference on Intelligence in Networks; ICIN; Bordeaux, France; 1994.

9 LIST OF ACRONYMS

CS1	—	Capability Set 1
FSL	—	Flexible Service Logic
IN	—	Intelligent Network
INAP	—	Intelligent Network Application Part
Inesim	—	Intelligent Network Simulation Tool
MIB	—	Management Information Base
NoLo	—	Nokia Logics
SCE	—	Service Creation Environment
SCP	—	Service Control Point

SIB	—	Service Independent building Block
SLEL	—	Service Logic Execution Language
SLP	—	Service Logic Program
SMS	—	Service Management System
SSD	—	Service Support Data
SSP	—	Service Switching Point
ODMG	—	Object Database Management Group
OODB	—	Object Oriented DataBase
RDB	—	Relational DataBase

10 BIOGRAPHY

Mikko Kolehmainen was born in Raahe, Finland, on August 6, 1966. He received, in 1992, the M.Sc. degree in computer science from the University of Helsinki, Finland. After graduation he joined Nokia Research Center in 1993 and is currently working as a project manager in Communication Systems Laboratory. He has specialised in issues related to service creation in the area of Intelligent Networks. Currently he is also a graduate student in the University of Helsinki.

Part Three
Database Issues

11
Design Issues and Experimental Database Architecture for Telecommunications

Juha Taina and Kimmo Raatikainen
University of Helsinki, Department of Computer Science
P. O. Box 26 (Teollisuuskatu 23)
FIN-00014 University of Helsinki, Finland
Fax: + 358 0708 44441
E-mail: {juha.taina,kimmo.raatikainen}@cs.helsinki.fi

Abstract

Databases in telecommunications have become an important research area in recent years. The persistent and temporal information needed in operations and management of the telecommunication networks and services will be in databases. The current IN Recommendations of ITU-T (Q.1200 Series) imply that real-time transaction processing capabilities should be provided. We examine the design issues of real-time transaction processing in database architectures that can be integrated into any teleoperator's service provision architecture. We also introduce an experimental database architecture which is a real-time distributed object-oriented database architecture. The architecture can be used to implement the Service Control Function in the Intelligent Network and other database services needed in telecommunications.

1 INTRODUCTION

Databases will, already in the near future, have an important role in telecommunication. The databases will contain the persistent and temporal information needed in operations and management of the telecommunication networks and services. Such databases exhibit stringent reliability, availability, and performance requirements. Thousands of retrievals must be delivered in a second. Allowed downtime is only a few seconds in a year. The globalization of services implies that databases managed by different operators must cooperate. The interoperability requires that the database systems must be open, must use standardized protocols, must be protected against misuse and intruders, and must have a common logical view of information.

In this paper we examine some fundamental design issues that affect the feasible architectures of database systems which can be integrated into teleoperator's service provision architecture. We also introduce a database architecture for telecommunication use and its prototype. Our premises are primarily derived from the Intelligent Network Long-Term Architecture (IN LTA) framework as described in Recommendation Q.1201 [ITU 1993b]. The key premises are compatibility with the OSI Reference Model of Open Distributed Processing (RM-ODP) [ITU 1995] and with the Telecommunications Management Network (TMN) [ITU 1993a]. Since both RM-ODP and TMN are based on object-oriented modeling, an object-oriented database is our first presumption. Our second presumption is scalability which implies that the database system may be distributed.

The rest of the paper is organized as follows. In Section 2 we briefly discuss our premises — the ways how the IN architecture, together with the compatibility with RM-ODP and TMN, affects the database architecture. In addition, we summarize our assumptions about the future database needs in telecommunication and summarize requirements for the database architecture. In Section 3 we examine the design issues in real-time databases. In Section 4 we give an overview of the database architecture and of the prototype architecture developed in the Darfin[1]-project.

2 EXTERNAL INTERFACES AND DESIGN CONSTRAINTS

In addition to the current ITU-T recommendations in Q.1200 Series (*Intelligent Networks*), the Basic Reference Model of ODP and the concepts of TMN create the architectural framework into which the database system is to be embedded. The architectural framework sets the requirements for external interfaces that the database system may have to support.

2.1 IN Architectural Framework

The *distributed functional plane* in the IN architecture is described in terms of *functional entities*. In that model the database services are available in the *Service Data Function* (SDF). The database users (clients of SDF) are *Service Control Functions* (SDFs), *Service Management Functions* (SMFs), and other SDFs. Figure 1 outlines the part of the IN functional architecture involving database services. The internal structures of SCF and SDF in the figure are based on the figures 4-19 and 4-20 in ITU-T Recommendation Q.1214 [ITU 1994b], respectively. The outlined internal structure of the SMF and its impact on the SDF are based on the principles of OSI Management[ITU 1991-4]. Detailed description of database access in the IN, especially in the Capability Set 1 (CS-1), can be found in Raatikainen [1995].

2.2 Basic Reference Model of Open Distributed Processing

Open Distributed Processing is a joint standardization initiative by ISO and ITU-T. The Basic Reference Model of ODP [ITU1995] is a framework for standardization of ODP. The objective of RM-ODP is to create an architecture that supports distribution, interworking, interoperability, and portability. All these are important goals for future telecommunication.

[1] Research project *Database ARchitecture For Intelligent Networks* funded byTelecom Finland

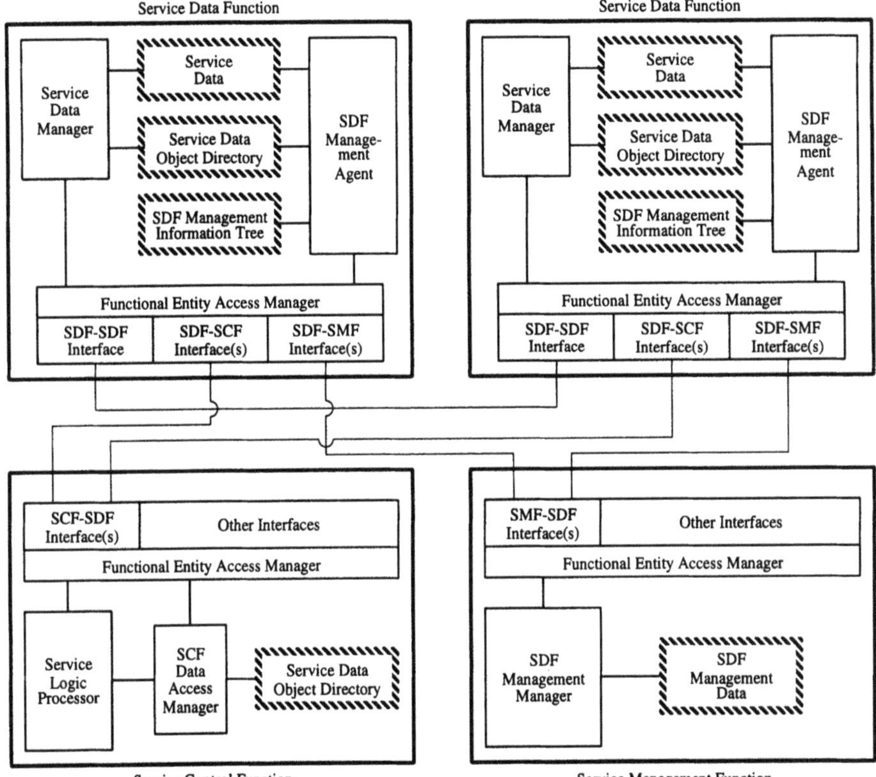

Figure 1 Elements of Database Servers and Clients in the IN Functional Architecture

The engineering viewpoint of RM-ODP defines a model for the infrastructure of a distributed system. The model, in turn, can be viewed as an abstract platform for ODP applications. The fundamental entities described in the engineering viewpoint are basic engineering objects, infrastructure objects, and channels. A channel provides the communication mechanism. In addition, the transparency functions are contained or controlled by a channel. The compatibility with RM-ODP requires that the database system must be accessible through an ODP-channel.

2.3 Telecommunication Management Network

Telecommunication Management Network is a generic architecture to be used for all kinds of management services. TMN is based on the principles of the OSI Management as specified in the Recommendations in the X.700 Series [ITU1991–4].

The fundamental idea in the OSI Management is that the knowledge representing the information used for management is separated from the functional modules performing the management actions. OSI Management is based on interactions between management applications that can take the roles of *manager* and *agent*. The interactions that take place are abstracted in terms of *management operations* and *notifications*. Management activities are effected through the manipulation of *managed objects* (MOs).

An agent manages the MOs within its local system environment. It performs management operations on MOs as a consequence of management operations issued by a manager. An agent may also forward notifications emitted by MOs to a manager. The agent maintains a part of the *Management Information Tree* (MIT), which is a dynamic database. The MIT contains instances of MOs organized as a hierarchical database tree.

Agents and managers exchange information using the services of *Common Management Information Service Element* (CMISE). CMISE uses the *Common Management Information Protocol* (CMIP) and utilizes the services of *Association Control Service Element* (ACSE) and *Remote Operations Service Element* (ROSE). In brief, the principles of OSI Management (and TMN) require that the database system contains the functionality of an OSI Management Agent.

2.4 Anticipated Database Needs

In Raatikainen [1994]we have identified five basic types of database operations needed in services and service features of IN Capability Set 1:

1 Retrieval of structured objects from persistent subscriber data objects. Some of the retrievals trigger a later update. For example, if Freephone service provides call routing statistics to the subscriber, an update is necessary.
2 User management actions that modify persistent subscriber data objects. These actions must be protected against unauthorized use. Some updates may also have to be verified against other criteria. For example, the subscriber must have the possibility to cancel or prohibit call forwarding to his or her number.
3 Verification of Personal Identification Number. For security reasons the PIN verification must be done in the database system when services are globalized. The actual PIN cannot be delivered into a foreign network.
4 Writing sequential log records.
5 Mass calling and Televoting. They need high-volume small updates that must be serialized. A special aspect of these updates is that they can be done in large blocks.

In addition to the IN services, future databases must support mobility and management. From a database point of view, the most challenging aspect of mobility is the maintenance of location information. This contradicts the traditional claim that most database operations needed in telecommunication services are reads. The management implies a twofold functionality. Firstly, the database and its elements are managed objects. Secondly, a management agent needs an efficient object-oriented database to maintain the Management Information Base.

2.5 Requirements for Database Architecture

In Taina [1995]we have identified the key requirements for an IN database architecture. The requirements that we identified are *data distribution, data replication, object orientation, object directories, multiple application interfaces, fault tolerance,* and *real-time transactions.*

Data distribution: A set of SDFs may cooperate to accomplish database tasks. An SDF may request information from another SDF in case needed information is not locally available. This implies data distribution among several SDFs. It is possible to implement the underlying database architecture without data distribution. In such an implementation all applications use a single database server. This differentiates logical data distribution between SDFs from physical data distribution among database nodes. The former is necessary, the latter is a database architecture decision. We have chosen a distributed database architecture. Our assumptions are that few SDF requests must consult other SDFs and that the request distribution between different SDFs is relatively uniform. Thus using a distributed architecture allows several transactions to be run in parallel on several database nodes and give better throughput. In a centralized architecture the database transaction processing can become a bottleneck. However, if the principle assumptions of few distributed operations and uniform request distribution between different SDFs turn out to be wrong then the distribution control itself may become a severe bottleneck. In such a case using a single database node for or IN services is a reliable alternative.

Data replication: In ITU [1994a] it is stated that the SDF contains customer and network data for real time access. This implies that the underlying database architecture must be able to answer requests in real time. We believe that currently most distributed operations are reads. Therefore, replication is an effective way to speed up distributed operations because it gives better throughput by allowing read operations to run parallel on several nodes compared to the extra overhead of replicated writes. For instance, service information is a good candidate for replication because usually all SDFs share the same image of different service profiles.

Object orientation: In ITU [1993b] it is stated that the best long term architecture for IN is probably object oriented.

Object directories: In ITU [1994b] the SCF holds a Service Data Object Directory. A similar concept is defined in SDF function model. This implies that the SDF architecture must support object directories.

Multiple application interfaces: Currently the exact interfaces to SDF are still evolving. Currently the only defined interface is CS-1 IN Application Protocol (INAP) that is defined in ITU [1994c]. According to Chatras and Gallant [1994] the next refinement of the SCF-SDF interface will be based on a subset of the Directory Access Protocol (DAP) in CCITT X.500 directory standard. Other possible interfaces to the SDF are X.700 management interface for management functions, and Open Distributed Processing (ODP) interface for distributed operations.

Fault tolerance: Real time access implies that data must be continuously available. The result of the implications is that the database architecture must be fault tolerant to a certain extent. Currently we believe that the maximum down time of an SDF must be at most seconds in a year. Yet we are not certain how much tolerance a customer has when a system failure occurs. The actual allowed down time may be minutes instead of seconds. Nevertheless a fault tolerant database architecture is a necessity.

Real-time transactions: A real time transaction has an explicit deadline that the transaction must meet. Although a real time access to data in SDF, as stated in ITU [1994a], does not directly imply that the underlying database architecture must support real time transactions, we believe that the most convenient way to support real time access to data is to use a database architecture that is specialized for real time transactions. A convenient database architecture allows time limits to transactions and to define actions if the deadlines are not met.

The list above shows the general outline of our Darfin DBMS architecture. It is distributed, object oriented, and supports real time transactions. Thus Darfin DBMS can be called a *Real Time Distributed Object Oriented Database Management System.*

Currently we are implementing a prototype database architecture that can answer some of the requirements listed above. With the prototype we hope to gain insight of the data access nature in IN environment. We are especially interested in statistics of the database usage, combining real time transactions and object orientation, and learning implementation methods for SDFs and other network elements using the database architecture. The prototype architecture is called Darfin Proto.

3 REAL-TIME DATABASES

Recommendation Q.1204 (*Intelligent Network Distributed Functional Plane Architecture*) [ITU 1994a] specifies that the *Service Data Function* (SDF) provides real-time access to customer and network data. This statement raises the question whether a real-time database is needed. Recommendation Q.1211 *(Introduction to Intelligent Network Capability Set 1)* [ITU1993c] augments the functional specification of SDF by requiring that SDF provides consistency checks on data. This raises the fundamental question whether it is possible to lower serialization requirements in order to gain better real-time transaction processing capabilities.

3.1 Principles of the Real-Time Database System

According to Haritsa et al. [1991] *Real-Time Database System* (RT-DBS) is a transaction processing system that tries to satisfy the explicit timing constraints associated with each incoming transaction. The timing constraint is usually given as a *deadline*. The ultimate goal of an RT-DBS is to maximize the fraction of transactions which meet their deadlines. This is in contrast to the goal of traditional database systems (DBMSs) which is to minimize the average response time.

At any time, all transactions in an RT-DBS can be divided into two classes: *feasible* transactions and *late* transactions. A feasible transaction still has a possibility to meet its deadline. A late transaction has either already missed its deadline or cannot any more meet its deadline. The RT-DBS executes feasible transactions until they complete in time or until the system detects that they can no more meet their deadlines. The policy of how late transactions are handled depends on the application needs. The two primary policies are: 1) to continue the execution with a low priority and 2) to abort the transaction without restart.

Real-time systems are usually classified into three categories according to the effects of missing a transaction deadline. *Hard* deadline transactions are those which may result in a catastrophe if the deadline is missed. A hard real-time system must guarantee that any transaction never misses its deadline. *Firm* deadline transactions are those the results of which are not any more of any value if the deadline is missed but have no serious consequences. *Soft* deadline transactions are those the results of which gradually decrease their value when the deadlines are missed.

The deadlines raise several difficult questions. How are deadlines assigned? Are deadlines firm or soft? It must be remembered that a transaction accessing the database is only a subtask in a telecommunication service. Unfortunately *subtask deadline assignment* is still almost an unexamined problem [see e.g. Kao and Garcia-Molina 1993]. The primary implication of transactions being subtasks is that the task can still meet (miss) its deadline even if the transaction misses (meets) its deadline.

Since the main goal of a firm or soft RT-DBS is to maximize the fraction of transactions that meet their deadlines, a key issue in transaction processing is *predictability* of transactions' execution and turnaround times. In a database system, data and hardware resource conflicts are an important source of unpredictability. Another significant source is the unpredictability of execution time caused by execution sequence depending on data values and by disk operations. Distributed databases have additional problems due to communication delays and site failures.

3.2 Designing an Real-Time Database System

One of the primary difficulties in designing an RT-DBS is the fact that the concurrency control of data access must be combined with timing constraints. The two primary approaches to arrange the concurrency control in database systems are based on *locking* and on the *optimistic approach*. Most commercial DBMSs are based on *two-phase locking* (2PL) [Eswaran et al. 1976]. This is due to the fact that under most operating circumstances locking algorithms give shorter response times than optimistic algorithms [see e.g. Agrawal et al. 1987].

Recent performance studies [Abbott and Garcia-Molina 1992, Haritsa et al. 1990, Huang et al. 1992, and Lin and Son 1990, among many others] indicate that the results obtained from traditional DBMSs do not hold in RT-DBSs. The relative performance of concurrency control algorithms in real-time database systems is heavily affected by several factors including policy with late transactions, a priori knowledge of transaction resource requirements, and the availability of resources. This leaves the floor open to several design issues that must be evaluated.

3.2.1 Design Issues in Concurrency Control

The major design issues in concurrency control are conflict detection and conflict resolution. In addition, the run policy is an important issue. Below we discuss these issues separately although they are not independent. We also name several mechanisms proposed in the literature.

Conflict Detection. Conflicts in data granule access that violates the correctness criteria of a transaction can be detected either before the granule access (*pessimistic approach*) or after the granule access (*optimistic approach*). In the optimistic approach the correctness is later verified at the certification time. The mechanisms proposed to be used in the detection include *locks, time stamps*, and *serialization graphs.*

When a lock-based scheme is used, a lock must be obtained before an access is allowed. In pessimistic approaches, such as various 2PL schemes, the lock is either shared (*read*) or exclusive (*write*). When an optimistic approach is based on locks, there are *weak* locks as well as normal (*strong*) locks. When a transaction has a weak lock, either read or write lock, on a granule, it only indicates that the granule is accessed. A weak lock conflicts with a strong write lock but is compatible with other weak locks (both read and write) and with strong read locks. Typically, the weak locks are converted to strong locks at the beginning of the certification time.

In some situations two different kinds of write locks may be profitable: an *update lock* (ordinary write lock) and a *replace lock* (blind write lock). A blind write lock is used when the locked item is updated only from an external source at predefined intervals. This kind of special situation occurs in mobile databases when subscriber location information is updated. Two *blind write* operations do not conflict with each other, since the old value is obsolete as soon as the new value arrives, and the write operations occur only on isolated objects. However, the operations do conflict with read operations.

Conflict Resolution. A conflict resolution mechanism is invoked once a conflict is detected. In the resolution at least one transaction is penalized through appropriate actions selected by the conflict resolution mechanism. The actions most commonly used are *blocking* (wait), *abort* (restart), *multiversioning*, and *dynamically readjusting the serialization order* (delayed commit or abort). It should be noted that the resolution mechanism is not independent from the detection mechanism. For example, if the conflict is detected after the granule access, the action must be abort.

Several conflict resolution mechanisms have been proposed in the literature. Variants of the 2PL include *no wait* [Tay et al. 1985], *wound-wait* [Rosenkrantz et al. 1978], *wait-die* [Rosenkrantz et al. 1978], *running priority* [Franaszek and Robinson 1985], *wait depth limited* [Franaszek et al. 1992], and *delayed abort* [Yu 1990]. In addition to the pure *optimistic concurrency control* (OCC) *broadcast OCC* [Kung and Robinson 1981] and a combination of pure and broadcast OCC [Yu et al. 1993] have been proposed. *Locking with deferred blocking* [Yu and Dias 1993] is a scheme that combines 2PL and OCC. Multiversioning opens a wide variety of conflict resolution mechanisms. A detailed comparison can be found in [Wu et al. 1993].

Run Policy. When abort is used as an action in the conflict resolution, a transaction may need to be restarted several times. The restarts create an effect known as *buffer retention* [Yu and Dias 1992]. In the first run it may be profitable to run a transaction that will be aborted to end so that all granules accessed will be in the buffer and the CC manager learns all the locks the transaction needs.

3.2.2 Integrating Concurrency Control and Real-Time Scheduling

There are four major problem areas when a real-time transaction scheduler is to be developed: *priority inversion, scheduling priority, overload management*, and *IO scheduling.*

Priority Inversion. A phenomenon called *priority inversion* arises when a low-priority transaction blocks a high-priority transaction due to the concurrency control. At least three different approaches have been proposed to solve the problem. In *priority inheritance* [Sha et al. 1990] the low-priority transaction inherits the priority of the high-priority transaction. In the *priority ceiling protocol* [Sha et al. 1991] blocking time of the high-priority transaction is bounded. The *dynamic priority ceiling protocol* [Chen and Lin 1990] is a modification of the previous based on earliest deadline first instead of fixed external priorities.

Scheduling Priority. *Earliest Deadline First* (EDF) is most widely used in the experimental real-time database systems as the scheduling priority. It can usually minimize the number of late transactions but not when the system is highly loaded [Liu and Layland 1973]. Another drawback in the EDF is that large transactions are discriminated. Alternatives to the EDF include *Least Slack First* (LSF), *fixed priorities, criticality of transaction, weighted priority scheduling*: see e.g. Abbott and Garcia-Molina [1992], Buchmann et al. [1989], Sha et al. [1991], Huang et al. [1989], Huang et al. [1991]. The fundamental design criteria are the heterogeneity of transactions and transaction characteristics available a priori.

Overload Management. If the EDF scheduling is used, then occasional periods of high load can significantly decrease the fraction of transactions that complete in time. In order to cope with the overload situations the scheduling needs an overload management policy. In Haritsa et al. [1991] an *adaptive earliest deadline* (AED) scheduling scheme was proposed. In AED, arriving transactions are placed in a HIT group or a MISS group depending on the system condition. In the HIT group the transactions are treated in the normal way but in the MISS group the transactions get low priority. In Pang et al. [1992] an *adaptive earliest virtual deadline* (AEVD) scheduling scheme was proposed. AEVD is an extension of AED that tries to take into account the fairness issue in an overloaded system.

IO Scheduling. When the primary copy of the database is on the disk, IO operations are extremely important for the RT-DBS. Traditional disk scheduling algorithms should be replaced by novel ones as in Sprunt et al. [1988] and in Abbott and Garcia-Molina [1990].

When a large fraction of the database can be held in the main memory buffers, the situation is completely different. We believe that in such cases hybrid databases (see the next section) should be used.

4 PROTOTYPE DATABASE ARCHITECTURE

Both Darfin DBMS and Darfin Proto consist of a set of database nodes that interact with each other. Every database node is an autonomous database and can be used alone.

A database node communicates with a set of applications, a mirror node, and the other database nodes in the distributed system. Applications, database nodes, and mirror nodes may be geographically distributed (see Figure 2).

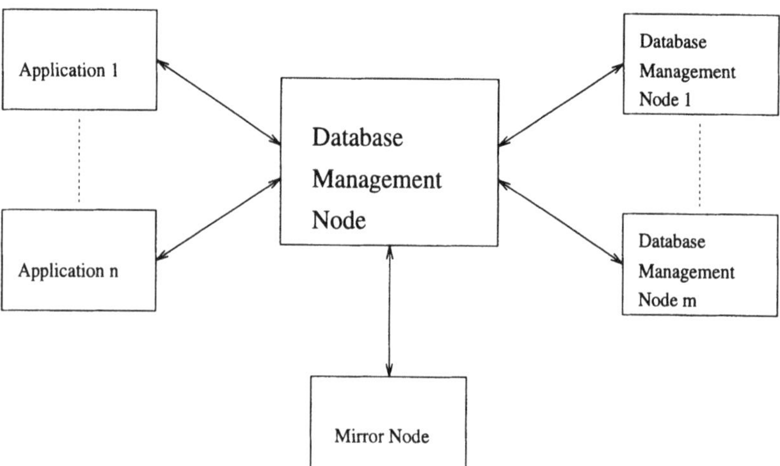

Figure 2 Overview of Darfin Distributed Database Architecture

A single application may communicate with several database nodes or only with a single node. If the application uses services of a single database node, then that node is the entry point to the whole distributed database. Then the actual distribution is internal: From the application's point of view the database architecture is not distributed. If the application uses services of several database nodes, then the application is responsible for taking care of distributed operations. For instance, an application may send a request to a single database node to add a new service to every customer, or it may send a request to every database node to update local information.

Every database node has a mirror node. All updates that are sent to the database node are also sent to the mirror node. In case of a system failure in the database node the mirror node changes its status into the database node. The architecture decision supports well fault tolerance but it may waste resources. All data is replicated at least to the mirror node. Usually this is not necessary. For instance, it is not necessary to store all IN database usage statistics into the mirror node.

4.1 Database Node Architecture

The architecture of a single database node consists of an Object-Oriented Database Management System (OODBMS), User Request Interpreter Subsystem (URIS), Distributed Database Subsystem (DDS), and Fault-Tolerance and Recovery Subsystem (FTRS) (see Figure 3). The URIS communicates with applications, the DDS with other database nodes, and the FTRS with the mirror node. The subsystems free the core database management system to be invisible to the network. It is possible to use alternate network communication methods by changing the subsystems, or it is possible to use a different core database architecture by changing the OODBMS.

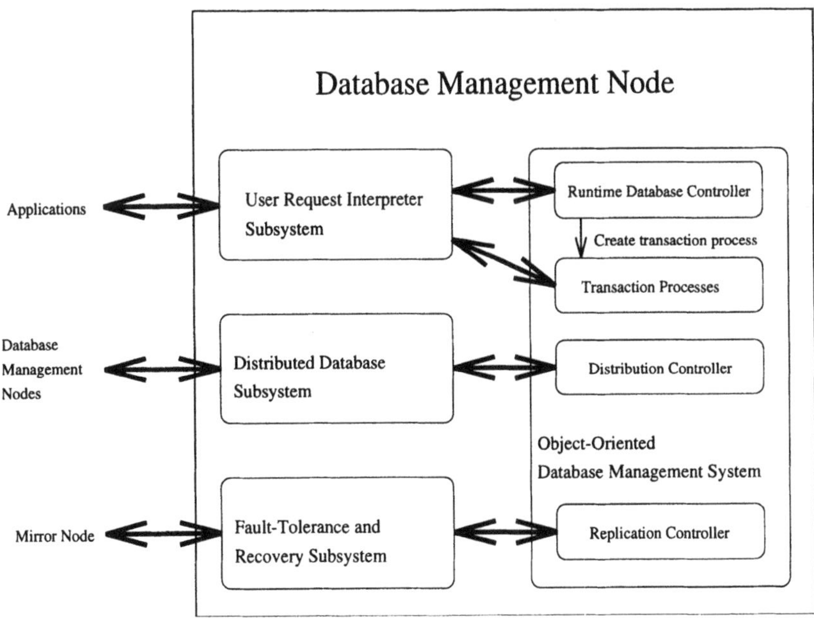

Figure 3 Database Node Architecture in Darfin

A database node system or subsystem may have its own processes. Thus the database node process architecture has two levels: at the operating system level the operating system scheduler schedules system and subsystem processes. At the database system level a system or subsystem schedules its private processes. Currently only the OODBMS has its private processes. If we used only the operating system level scheduler, we could not use different real time scheduling algorithms in the OODBMS unless we first implemented a suitable operating system.

Object-Oriented Database Management System (OODBMS). The OODBMS is the core element of a database node. It implements the actual database architecture and a logical object model.

From the application designer's point of view, the logical object model is the most important feature of the database architecture. It defines the entry points that the application can use to access data in the database and the appearance of objects in the database. The OODBMS implements the object model services and maintains persistent objects and object information.

We have chosen the ODMG-93 object database standard (Cattell [1994]) as the logical object model for our database architecture. The ODMG-93 standard consists of the object model, object definition language, object query language, C++ binding, and Smalltalk binding. Together they form a complete definition for object oriented database design, implementation, and use. We hope to benefit from future commercial database management systems that follow the same ODMG-93 standard.

The OODBMS architecture consists of a set of database processes that communicate with each other. The actual OODBMS implementation is a service for the database processes. The processes offer services to the subsystems of the database node, and the subsystems themselves offer services to applications, other database nodes (where again a subsystem takes care of processing the services), and the mirror node. The database processes are Runtime Database Controller (RDC), Distribution Controller (DC), Replication Controller (RC), and Transaction processes.

The RDC creates a new transaction process when the URIS informs it that a new transaction request has arrived. It is the main entry point for the URIS. After creation the new transaction process communicates with the URIS to send and receive data. The RDC continues to listen to the URIS for new transaction requests.

The DC is a transaction process that sends and receives distributed operations. It is responsible for hiding data distribution in the OODBMS. When a distributed operation arrives from an application, the DC forwards the operation to the DDS that is responsible for the actual network communication to other database nodes. The DC also serves DDS requests that are sent from other database nodes.

The RC is a transaction process that maintains data replication. It hides the replication method from the OODBMS. Every update operation to the database triggers the RC. It forwards the update to the FTRS that is responsible for mirror node maintenance.

A Transaction process receives an array of commands from the URIS and interprets the command line accordingly. It processes the commands and then uses the database services to gain results. The results are then forwarded to the URIS that forwards them to the requesting application.

User Request Interpreter Subsystem (URIS). All applications that use the services of the database node communicate with the URIS. It is responsible for maintaining network connections with the applications, translating their requests to the OODBMS, and forwarding the results from the OODBMS back to the applications. The URIS implements different language bindings to the OODBMS. When a new language binding is added, only the URIS need to be updated.

Distributed Database Subsystem (DDS). The DDS is responsible for processing and forwarding distributed operations between other DDS subsystems in other database nodes. The use of the DDS frees the OODBMS from low level distribution. In the OODBMS the DC is responsible for the distributed database operations, either operations from other nodes, or operations to the other nodes. The DDS is only a low level communication process.

Fault-Tolerance and Recovery Subsystem (FTRS). The FTRS is responsible for fault tolerance and recovery control. It communicates with the mirror node and with the RC in the OODBMS. When the OODBMS fails, the FTRS informs the mirror node to take control of the database node responsibilities. Then it informs the URIS to forward requests to the new node, and begins the recovery operation for the failed OODBMS. When the OODBMS is up and running again, the FTRS informs the mirror node that the actual database node is available. When the mirror node and database node are in the same state, that is when the mirror node has forwarded all updates that arrived while the database node OODBMS was down, the mirror node informs the FTRS that the system is stable again, and puts itself back to read-only state. The FTRS then informs the URIS that the requests should no longer be forwarded to the mirror node.

4.2 OODBMS Process Services

The most important element of the Darfin DBMS and Darfin Proto is the OODBMS and the services that it offers to the local database processes. The services that the processes may access are called the ODMG Interface Layer. The other deeper layers that implement the interface layer services are Physical Object Layer, Global Entity Layer, and Real-Time Core Layer (see Figure 4). A layer consists of a set of Manager Services that implement layer functionality.

ODMG Interface Layer. The ODMG Interface Layer is the service interface layer to the OODBMS services. It offers all services that the processes need. Currently the ODMG Interface Layer consists of three manager services: Schema Manager Service, Object Manager Service, and Directory Manager Service. Together the manager services offer a full ODMG compliant interface to the database.

The Schema Manager Service manages metainformation of objects and object classes. Every object belongs to a specific object class that defines the object structure.

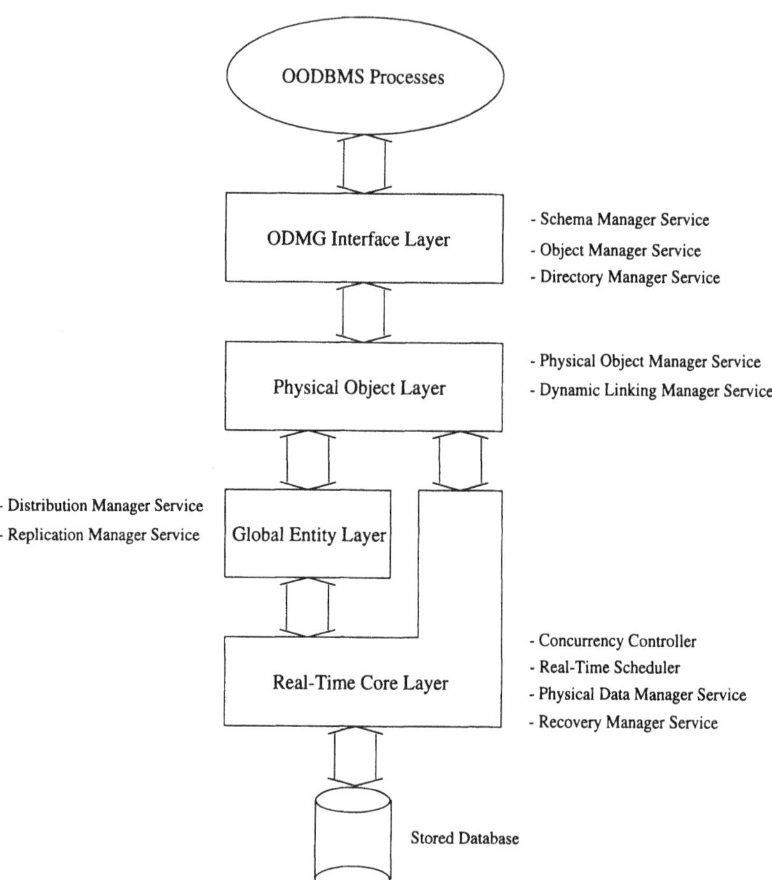

Figure 4 Layers of OODBMS Process Service in Darfin

The Object Manager Service manages objects. It offers the services that are defined to objects and types, such as accessing attributes or using operations.

The Directory Manager Service manages object directories. The principal object directory is based on object identity. There may be other directories based on names, attribute values, or conditions.

Physical Object Layer. The Physical Object Layer implements the services of the ODMG Interface Layer. On the Physical Object Layer level all objects are alike. The ODMG object model is no longer visible. In principle it is possible to use the layer and the lower layers in an object oriented database architecture that is not based on ODMG.

The Physical Object Layer has two manager services: Physical Object Manager Service and Dynamic Linking Manager Service.

The Physical Object Manager Service manages the physical objects. It offers services for object creation, deletion, access, object attribute access, and operation access. In the future the ODMG Interface Layer may need additional services. The Physical Object Manager Service is the core manager service for a logical object model. It uses services of the lower layers to implement the physical object managing.

The Dynamic Linking Manager Service is responsible for linking new executable code to the database when a new object operation is created.

Global Entity Layer. The Global Entity Layer offers services for distributed and replication operations. It is mostly used by the Distribution Controller and the Replication Controller. The layer has two manager services: Distribution Manager Service and Replication Manager Service.

The Distribution Manager Service manages distributed operations on objects. It is mostly used by the Distribution Controller. In addition, the Physical Object Manager Service uses the Distribution Manager Service when an access to an object in a different database node is needed.

The Replication Manager Service manages replication. It is mostly used by the Replication Controller but also all operations that change object's status are forwarded to the Replication Manager Service.

In the Darfin Proto both the Distribution Manager Service and Replication Manager Service only gather statistics of the service usages. The full Distribution and Replication s will be implemented in the Darfin TDMS.

Real-Time Core Layer. The Real-Time Core Layer implements the core operations of a real-time database management system. In the Real-Time Core Layer the processing model no longer deals with object orientation. This is dealt with in the Physical Object Layer. Instead, the Real-Time Core Layer maintains concurrency control, scheduling, physical data store maintenance, and recovery processing. The corresponding manager services are Concurrency Controller, Real Time Scheduler, Physical Data Manager Service, and Recovery Manager Service, respectively.

The Concurrency Controller is responsible for letting transactions to execute without breaking the isolation levels of transactions and objects. When the isolation levels are absolute, the transactions are serializable. The other extreme is to let transactions interfere with each other. Darfin DBMS supports various isolation levels depending on running transactions and referenced objects. We will use locking and the 2PL-HP algorithm, in which a higher priority process that claims a lock kills a lower priority process that holds the lock. The lower priority process is restarted if the deadline is not yet expired.

The Real-Time Scheduler schedules OODBMS processes. We will use a two level prioritized scheduling method. In the higher level, Runtime Database Controller, Distribution Controller, and Replication Controller have a higher priority than any of the transaction processes. In the lower level, the transaction processes have priorities according to their deadlines. We use the *Earliest Deadline First* (EDF) scheduling.

The Physical Data Manager Service manages the physical data storages: object data storage and metadata store. It is the only Manager Service that has access to the data stores. In Darfin Proto we use a pure main memory database architecture. All data resides in main memory, disks are used only for backup purposes. However, the physical data store architecture is not visible outside the Physical Data Manager Service. It is possible to use other data store architectures by changing the Physical Data Manager Service. The rest of the OODBMS architecture need not to be altered.

The Recovery Manager Service is needed when the OODBMS is recovering from a failure. It reloads the database to main memory and starts the necessary processes.

5 SUMMARY

When a database systems for telecommunication services is to be designed the most important issues include *real-time*. We believe that a Real-Time Database System will in the future have an important role in telecommunication. We anticipate that the both *firm* and *soft* deadlines will be needed.

Since the needs for database services have quite different characteristics, the database system should be quite flexible. We believe that the required functionality can most easily and economically be achieved by an *object-oriented* approach. The functional needs are not easy to fulfill but we believe that a distributed real-time object-oriented database architecture is able to do that.

The Darfin DBMS architecture consists of a set of database nodes that communicate with each other. Together the nodes implement a distributed database architecture. Each database node is a full real-time object-oriented database that follows the ODMG-93 standard. We believe that the architecture is flexible enough to answer the current and future IN requirements, as well as a lot of other telecommunications network requirements.

6 ACKNOWLEDGMENTS

The authors are thankful to the rest of the Darfin-team, particularly to Mika Rautila, Tapani Karttunen, and Harri Töhönen from Telecom Finland.

7 REFERENCES

[Abbott and Garcia-Molina 1990] Abbott, R. and Garcia-Molina, H. 1990. Scheduling I/O Requests with Deadlines: A Performance Evaluation. In *Proceedings of the 11th Real-Time Systems Symposium*. IEEE, Los Alamitos, Calif., 113–124.

[Abbott and Garcia-Molina 1992] Abbott, R. and Garcia-Molina, H. 1992. Scheduling Real-Time Transactions: A Performance Evaluation. ACM Transaction on Database Systems 17, 3 (Sept.), 513–560.

[Agrawal et al. 1987] Agrawal, R., Carey, M. J., and Livny, M. 1987. Concurrency Control Performance Modeling: Alternatives and Implications. ACM Transaction on Database Systems 12, 4 (Dec.), 609–654.

[Buchmann et al. 1989] Buchmann, A. P., McCarthy, D. R., Hsu, M., and Dayal, U. 1989. Time-Critical Database Scheduling: A Framework For Integrating Real-Time Scheduling and Concurrency Control. In Proceedings of the 5th International Conference on Data Engineering. IEEE, Los Alamitos, Calif., 470–480.

[Cattell 1994] Cattell, R. G. G., Editor. 1994. The Object Database Standard: ODMG-93, Release 1.1. Morgan Kauffmann, San Francisco, Calif.

[Chatras and Gallant 1994] Chatras, B. and Gallant, F. Protocols for Remote Data Management in Intelligent Networks CS1. In Workshop Record of IEEE Intelligent Network '94 Workshop, Volume 1. IEEE Communication Society, May 1994.

[Chen and Lin 1990] Chen, M.-I. and Lin, K.-J. 1990. Dynamic Priority Ceilings: A Concurrency Control Protocol for Real-Time Systems. The Journal of Real-Time Systems 2, 4 (Dec.), 325–346.

[Eswaran et al. 1976] Eswaran, K. P., Gray, J. N., Lorie, R. A., and Traiger, I. L. 1976. The Notions of Consistency and Predicate Locks in a Database System. Commun. ACM 19, 11 (Nov.), 624–633.

[Franaszek and Robinson 1985] Franaszek, P. A. and Robinson, J. T. 1985. Limitations of Concurrency in Transaction Processing. ACM Transactions on Database Systems 10, 1 (Mar.), 1–28.

[Franaszek et al. 1992] Franaszek, P. A., Robinson, J. T., and Thomasian, A. 1992. Concurrency Control for High Contention Environments. ACM Transactions on Database Systems 17, 2 (June), 304–345.

[Garcia-Molina and Salem 1992] Garcia-Molina, H. and Salem, K. 1992. Main Memory Database Systems: An Overview. IEEE Transactions on Knowledge and Data Engineering 4, 6 (Dec.), 509–516.

[Haritsa et al. 1990] Haritsa, J. R., Carey, M. J., and Livny, M. 1990. On Being Optimistic about Real-Time Constraints. In Proceedings of the 9th ACM Symposium on Principles of Database Systems. ACM, 331–343.

[Haritsa et al. 1991] Haritsa, J. R., Livny, M., and Carey, M. J. 1991. Earliest Deadline Scheduling for Real-Time Database Systems. In Proceedings of the 12th Real-Time Symposium. IEEE, Los Alamitos, Calif., 232–242.

[Huang et al. 1989] Huang, J., Stankovic, J. A., Towsley, D., and Ramamritham, K. 1989. Experimental Evaluation of Real-Time Transaction Processing. In Proceedings of the 10th Real-Time Systems Symposium. IEEE, Los Alamitos, Calif., 144–153.

[Huang et al. 1991] Huang, J., Stankovic, J. A., Ramamritham, K., and Towsley, D. 1991. Experimental Evaluation of Real-Time Optimistic Concurrency Control Schemes. In Proceedings of the 17th VLDB Conference. Morgan Kaufmann, San Mateo, Calif., 35–46.

[Huang et al. 1992] Huang, J., Stankovic, J. A., Ramamritham, K., Towsley, D., and Purimetla, B. 1992. Priority Inheritance in Soft Real-Time Databases. The Journal of Real-Time Systems 4, 2 (June), 243–268.

[ITU 1991–4] Recommendations in X.700-Series: OSI Management. International Telecommunications Union, Geneva, Switzerland.

[ITU 1993a] Recommendations in M.3000-Series: Telecommunication Network Management. International Telecommunications Union, Geneva, Switzerland.

[ITU 1993b] Principles of Intelligent Network. Recommendation Q.1201. International Telecommunications Union, Geneva, Switzerland.

[ITU 1993c] Introduction to Intelligent Network Capability Set 1. Recommendation Q.1211. International Telecommunications Union, Geneva, Switzerland.

[ITU 1994a] Intelligent Network Distributed Functional Plan Architecture. Recommendation Q.1204. International Telecommunications Union, Geneva, Switzerland.

[ITU 1994b] Distributed Functional Plane for Intelligent Network CS-1. Recommendation Q.1214. International Telecommunications Union, Geneva, Switzerland.

[ITU 1994c] Interface Recommendation for Intelligent Network CS-1. Recommendation Q.1218. International Telecommunications Union, Geneva, Switzerland.

[ITU 1995] Basic Reference Model of Open Distributed Processing — Parts 1–3: Introduction, Foundations, and Architecture. Draft Recommendations X.901, X.902, and X.903. International Telecommunications Union, Geneva, Switzerland.

[Kao and Garcia-Molina 1993] Kao, B. and Garcia-Molina, H. 1993. Deadline Assignment in a Distributed Soft Real-Time System. In Proceedings of the 13th Int. Conf. on Distributed Computing Systems. IEEE Computer Society Technical Committee on Distributed Processing, Los Alamitos, Calif., 428–437.

[Kung and Robinson 1981] Kung, H. T. and Robinson, J. T. 1981. On Optimistic Methods for Concurrency Control. ACM Transaction on Database Systems 6, 2 (June), 213–226.

[Lin and Son 1990] Lin, K.-J. and Son, S. H. 1990. Concurrency Control in Real-Time Databases by Dynamic Adjustment of Serialization Order. In Proceedings of the 11th Real-Time Systems Symposium. IEEE, Los Alamitos, Calif., 104–112.

[Liu and Layland 1973] Liu, C. L. and Layland, J. W. 1973. Scheduling Algorithms for Multiprogramming in a Hard Real-Time Environment. Journal of the ACM 20, 1 (Jan.), 46–61.

[Pang et al. 1992] Pang, H., Livny, M., and Carey, M. J. 1992. Transaction Scheduling in Mulriclass Real-Time Database Systems. In Proceedings of the 13th Real-Time Systems Symposium. IEEE, Los Alamitos, Calif., 23–34.

[Raatikainen 1994] Raatikainen, K. E. E. 1994. Information Aspects of Services and Service Features in Intelligent Network Capability Set 1. Report C-1994-45, University of Helsinki, Dept. of Computer Science, Helsinki, Finland.

[Raatikainen 1995] Raatikainen, K. E. E. 1995. Database Access in Intelligent Networks. In Intelligent Networks, J. Harju, T. Karttunen, and O. Martikainen (Eds). Chapman & Hall, London, UK, 173–193.

[Rosenkrantz et al. 1978] Rosenkrantz, D. J., Stearns, R. E., and Lewis II, P. M. 1978. System Level Concurrency Control for Distributed Database Systems. ACM Transactions on Database Systems 3, 2 (June), 178–198.

[Sha et al. 1990] Sha, L., Rajkumar, R., and Lehoczky, J. P. 1990. Priority Inheritance Protocols: An Approach to Real-Time Synchronization. IEEE Transactions on Computers C-39, 9 (Sept.), 1175–1185.

[Sha et al. 1991] Sha, L., Rajkumar, R., Son, S. H., and Chang, C.-H. 1991. A Real-Time Locking Protocol. IEEE Transactions on Computers 40, 7 (Jan.), 793–800.

[Sprunt et al. 1988] Sprunt, B., Kirj, D., and Sha, L. 1988. Priority-Driven, Preemptive I/O Controllers for Real-Time Systems. In Proceedings of Int. Symp. on Computer Architecture. ACM, 152–159.

[Taina 1995] Taina, J. 1995. Requirements Analysis for Database Services in Telecommunications. Report C-1995-17, University of Helsinki, Dept. of Computer Science, Helsinki, Finland.

[Tay et al. 1985] Tay, Y. C., Suri, R., and Goodman, N. 1985. A Mean Value Performance Model for Locking in Databases: The No-Waiting Case. Journal of the ACM 32, 3 (July), 618–651.

[Wu et al. 1993] Wu, K.-L., Yu, P. S., and Chen, M.-S. 1993. Dynamic Finite Versioning: An Effective Versioning Approach to Concurrent Transaction and Query Processing. In Proceedings of the nth Int. Conf. on Data Engineering. IEEE, Los Alamitos, Calif., 577–586.

[Yu 1990] Yu, P. S. 1990. Effective Concurrency Control With No False Killing. IBM Tech. Disclosure Bull. 33, 2 (July), 193–194.

[Yu and Dias 1992] Yu, P. S. and Dias, D. M. 1992. Analysis of Hybrid Concurrency Control for a High Data Contention Environment. IEEE Transactions on Software Engineering SE-18, 2 (Feb.), 118–129.

[Yu and Dias 1993] Yu, P. S. and Dias, D. M. 1993. Performance Analysis of Concurrency Control Using Locking With Deferred Blocking. IEEE Transactions on Software Engineering SE-19, 10 (Oct.), 982–996.

[Yu et al. 1993] Yu, P. S., Dias, D. M., and Lavenberg, S. S. 1993. On the Analytical Modeling of Database Concurrency Control. Journal of the ACM 40, 4 (Sept.), 831–872.

8 BIOGRAPHY

J. Taina has a M.Sc. (Comp. Sc.) from University of Helsinki. He is currently a graduate student in University of Helsinki working towards his Ph.D. dissertation on Real-Time Object-Oriented Database Management Systems.

K. Raatikainen has Ph.D. (Comp. Sc.) and M.Sc. (Comp. Sc.) from University of Helsinki. He joined University of Helsinki in 1986 and he is currently Assistant Professor at Department of Computer Science. He is also the leader of Darfin-project.

Part Four
Performance

12
Reusable Simulation Models for Performance Analysis of Intelligent Networks

Claes Wohlin, Christian Nyberg
Dept. of Communication Systems
Lund Institute of Technology, Lund University
Box 118, S-221 00 Lund, Sweden
Phone: +46-46-222 0000, Fax: +46-46-145823
E-mail: (claesw, cn)@tts.lth.se

Anders Larsson
Telia Research AB
Box 85, S-201 XX Malmö, Sweden
Phone: +46-40-105034, Fax: +46-40-105100
E-mail: aln@malmo.trab.se

Abstract

New services will be introduced at a much faster rate than today when Intelligent Networks are employed. It is important to control the process of introducing services and be able to forecast the consequences of a new service. This paper describes a method for performance simulation based on reusable simulation models which can be used for this purpose. The method builds on dividing the simulation model into three parts describing architecture, software of services and usage of the services respectively. Services and nodes can be described on different abstraction levels. If a new service is introduced all the old parts of the simulation program can be reused.

1 INTRODUCTION

The fast introduction of new services in the telecommunication networks leads to a number of problems. One of them is the problem of network performance as new services are introduced. The introduction of services must be made in a controlled manner, hence an analysis of the effect of the introduction of new services must be made prior to putting them into operation.

This paper presents a simulation method, which can evolve together with the network through reuse of simulation models. Thus allowing for performance analysis without having to formulate new models each time a new service is to be introduced.

A state of the art study of available simulation software for communication networks is presented by Law (1). There are some useful software packages available, but they do not support reusable models for performance simulation. Some tools include a software library, but they do not include support to reuse the models created by the user of the tools. This is also concluded by Saulnier (2), where they state that: "Simulation packages do not provide an effective infrastructure for reusing models ...".

2 PERFORMANCE ANALYSIS WHEN NEW SERVICES ARE INTRODUCED

The reusability of software, the centralization of service control and efficient design tools will make it possible to introduce new services in Intelligent Networks at a much faster rate than in traditional networks. This process must be properly controlled. A new service increases the load on the control systems of the Service Switching Points (SSP), the Service Control Points (SCP), the Signal Transfer Points (STP) and the signalling links. As a consequence of this, bottlenecks may be formed and the waiting times may get too long for both the new service and for the old services. Thus, before a new service is introduced, we ought to evaluate its consequences on the performance of the network. If we find that the load on certain nodes or links gets too high we can move services to other nodes to equalize the load, install new capacity in the network or perhaps even choose not to introduce a service.

A tool which evaluates the load on the control processors in the network and the delays experienced by the users can help us to answer questions like: where are the bottlenecks in the network and what is the highest allowed arrival rate of a new service. The input to such a tool would be the rate at which a service will be used, a description of the new service, a description of the present network and its services. The description of the new service could be on a high abstraction level to permit us to evaluate it even before it has been designed in full detail. But it should also be possible to use the detailed description. The services already present in the network may be described on a high abstraction level. It should also be possible to use different abstraction levels in the description of the nodes in the network just as well as for the services.

3 PERFORMANCE SIMULATION

3.1 Introduction

Simulation is a valuable tool for performance analysis, but it has some serious drawbacks, which must be overcome to make it really useful in an application which evolves rapidly, for example, with introduction of new services. Some of the major drawbacks are:

- the simulation model is based on one particular system or network, hence as the network or the available services changes the model must be changed as well. This can be hard as the model is not normally created to be changed. To overcome this problem, reuse of models must be considered to a much larger extent.
- the model used for performance simulation does not include the actual software description, hence relying on queueing models with little or no connection to the actual software. This becomes particularly crucial as new services are introduced and the delays in the network depend highly on the implementation of the service. The rapid introduction of new services does not allow for creation of good models of the services, instead the software design descriptions ought to be executed in the simulation model.

These two problems are normally not fully solved by available simulation methods and tools. Most available tools, for example, QNAP2 in Verán (3) and QNA in Whitt (4), use their own specific representation languages and they do not support inclusion of software descriptions as part of the simulation model. The objective of our approach was to include the software descriptions as part of the simulation model and that the simulation model should be formulated in the same description language as is used in software development. This objective was the major goal when developing a new approach to performance simulation, see Wohlin (5). Thus, this approach is adopted here. A summary of this work can be found in Wohlin (6). A brief summary of the concept is given subsequently.

3.2 A new approach to performance simulation

The new concept is based on the idea of dividing the simulation model into a number of models, which describe different constituents of the performance model. This objective is very similar to the one adopted in IN, i.e. a layering approach, hence it is believed that the proposed concept will be a valuable asset in performance simulation of IN.

Three models are defined to formalise the modelling and evaluation process. The mapping between reality and models is depicted in Figure-1. The models are denoted:

- Software Model
- Usage Model
- Architecture Model

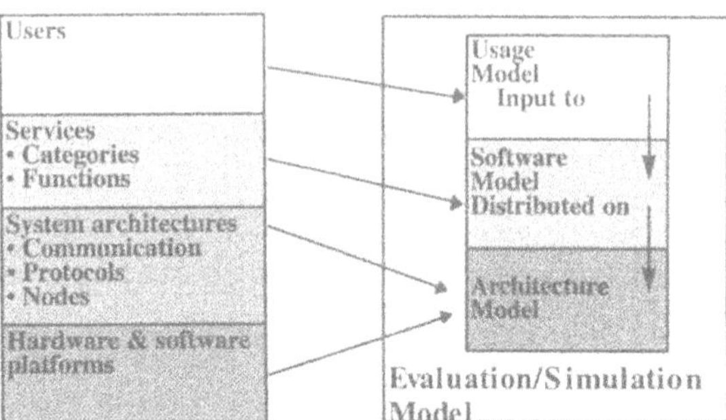

Figure 1 Mapping of the layers of users and system on the modelling concepts

The models, which will be explained below, are independent in the sense that the Software Model, the Usage Model and the Architecture Model are derived independently and they can be combined into a simulation model. The simulation is foremost intended to be used during the software design and hence supports re-design decisions. The Software Model and the Architecture Model are linked to each other through the distribution of the software units in the architecture. This means the Software Model is allocated to the Architecture Model in a way that describes the actual distribution of the software in the architecture. The Usage Model generates the input to the simulated system (Software Model allocated on the Architecture Model). Thus connecting the three models together into the Evaluation/Simulation Model.

3.3 Model descriptions

The models can be described as follows:

- Software Model
  The software descriptions (specification or design) are transformed to include the real time aspects of the software, which normally is not included in the software design. As part of the transformation the user is requested to add time consumption for executing different concepts of the software design. This addition of time consumption must be made based on prior knowledge or knowledge of the current system. The Software Model describes the application software, i.e. the features that the system provides to the user. The transformation can be made automatically based on a set of well-defined rules, see Wohlin (5). The Software Model includes aspects such as services and features (what is available to the external user), logical flow (the coupling between services) and relative service times (an execution time which does not refer to a particular platform).

- Usage Model
  The Usage Model describes the structural usage of the analysis object (which refers to the part of the system being evaluated) and it is complemented with usage intensities for different services. This type of model is similar to the ones used in certification of software reliability as discussed in, for example by Wohlin (7) and Whittaker (8). Therefore, it is believed that the concept originally proposed for performance simulation can be enlarged to incorporate reliability analysis as well. The Usage Model is concerned with identification of analysis object (which part of the system, i.e. hardware and software, is going to be analysed), user categories, usage states (the external observable states that a user may be in) and usage intensities. It must also be noted that the usage model describes the usage of the analysis object for all types of users, which includes humans as well as other systems or networks.
- Architecture Model
  The architecture is described in terms of a performance simulation model. The aspects to find are those governing the performance behaviour of the architecture. The objective here is to define a performance model of the architecture in the same description technique as has been used in the software design. This is also further discussed by Wohlin (5). The main rationale behind using the same language is the opportunity to easily map the software design descriptions onto a simulation model of the architecture. The Architecture Model includes aspects such as network architecture with interconnections, servers and resources, as well as operating system features, algorithms, for example scheduling, flow-paths and capacity.

3.4 Performance simulation using SDL

SDL is primarily intended as a specification and design language, but it has been and still is successfully used in performance simulation. SDL has been used for the latter purpose in a number of years, for example, at the department of Communication Systems, Lund University in a course on performance simulation and it is also used in the research at the department.

A major benefit of using SDL is that it is standardized and tool support is available. A particular benefit in our approach to performance simulation is the use of the same description technique for describing the services and for formulating performance models of the architecture and the user behaviour, which simplifies the work considerably. The use of one language for two purposes is not a prerequisite for the proposed simulation method, but it makes the combination of service descriptions with simulation models easier.

4 PERFORMANCE SIMULATION OF IN

4.1 Introduction

The previous section described an approach to performance simulation that uses a software, a usage and an architecture model. To be able to use this as a basis for a performance simulation method, that solves the problems when introducing new services, it is desirable that is should be possible to handle models on different abstraction levels. For example, old services can be described only by their service times on the different nodes in the network - perhaps found by measurements on the real network - and new services by a complete software description. Thus we run into several problems: what abstraction levels can be used in the software, usage and architecture models and which of these can be used at the same time. This also rises the question of interfaces. Suppose that two different abstraction levels for software are used at the same time. Is there an interface between the different parts of the complete simulation model that can handle this? Which abstraction levels are compatible with each other? And can an interface be designed that allows us to change the abstraction level of one part of the model without having to change the rest of the model?

4.2 Abstraction levels of models

The architecture model describes the nodes and signalling links of the network. The most abstract model of a node that seems to be meaningful is a queueing system and the least abstract model is a complete emulation of the hardware and the operating system.

The most abstract level of the software model is to describe a service by its service time on the nodes in the network. This level can be used for services that are already in use or for services that are not yet completely specified. It shall also be possible to use a complete software description of a service in the simulation program.

The usage model describes the environment of the network and should reflect both the behaviour of subscribers and the parts of the network that are not incorporated in the model. A usage description must receive all signals sent to it by its corresponding service description and react to the signals in a proper way. Thus, the abstraction levels of a service description and its usage description must be on the same abstraction level.

4.3 The interface between architecture model and the other models

As pointed out previously the interfaces between the different parts of the a model must be able to handle software, usage and architecture models on different abstraction levels. It must also be possible to change the abstraction level of a node or service description without having to change the rest of the model. This has one notable exception: when a service is changed its usage model may have to be changed.

To be able to change the abstraction level of a service without having to change the architecture model, the model is designed so that it does not need to know anything about a service except to which nodes it is allocated. Thus the signals used in the software description of a service must be packed into signals that can be interpreted by the architecture model. These signals do not exist in the original software. The discussion above indicates that the following three classes of signals are needed in the simulation model:

1. Signals that contain original signals. A node which receives such a signal uses the software distribution data to check if the software is allocated on itself in which case it sends the signal to the software or usage model. Otherwise it sends the signal to another node according to the routing table.
2. Signals exchanged between the software and architecture model that are used to mark a processor as free or busy. When the software model receives a signal of class 1 from the architecture model it computes the processing time of the jobs that is triggered by the signal. The software model executes according to the processing time and informs the processor when the execution is completed.
3. Signals internal to the architecture model, for example signals between processors in distributed systems.

5 EXAMPLE

5.1 The new service

To exemplify the methodology presented in the previous sections we have specified a new service which we intend to introduce in an Intelligent Network. The specification is a specification of the functional behaviour of the service, i.e. what is normally implemented in software. This new service is divided into two parts, SWA and SWB, corresponding to the Service Switching Function and the Service Control Function.

The service is invoked when a user sends a signal A to SWA. Upon receipt of this signal SWA sends a signal B to SWB. SWB performs some action and returns signal C to SWA. Finally a signal D is sent to the user, see Figure-2.

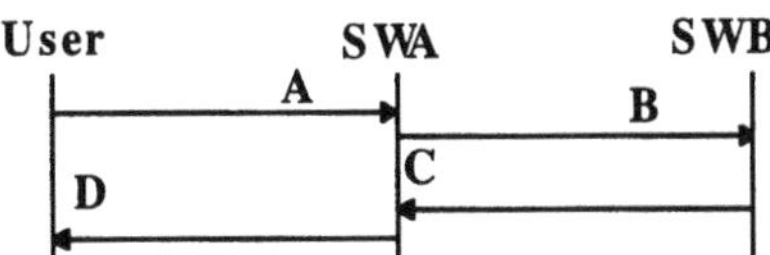

Figure 2 Message Sequence Chart for the new service. The signals A-D are "original signals" according to the terminology presented in the previous section.

Our new service is going to be installed in a simple Intelligent Network consisting of two SSPs and one SCP interconnected with #7-links. SWA should be installed at both SSP1 and SSP2 whereas SWB should be installed in the SCP.

5.2 The simulation model

The modelling concept states that the model should be divided into three parts: the Usage Model (UM), the Architecture Model (AM) and the Software Model (SM), see Figure-3.

- UM: Models of the users of the service.
- AM: Models of the SSPs, the SCP and the #7-links.
- SM: Model of the new service, i.e. transformed models of SWA and SWB, and models of existing services in the Intelligent Network.

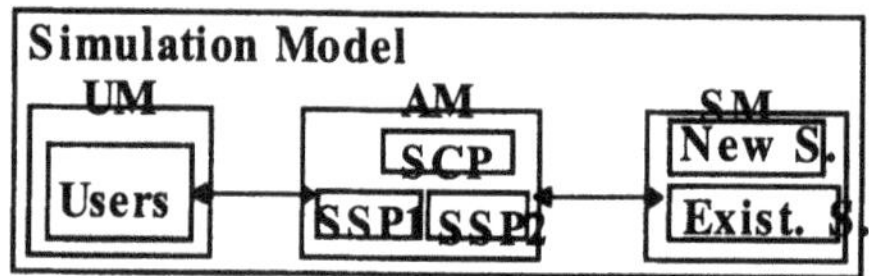

Figure 3 Overview of the simulation model. (Not all details are shown.)

This simulation model has been implemented in SDL, to allow for evaluation of the usefulness of the proposed modelling concepts. It was concluded that the approach is useful and that it is worth investigating further.

The transformed models of SWA and SWB are contained in the block New Service. SWA exists in two instances, one allocated to SSP1 the other to SSP2. SWB exists in one instance only and is allocated to the SCP. Original signals that are to be sent between software instances or between user instances and software instances are packed in a general signal SWSignal and are always sent via the Architecture Model. If the sender and the receiver of an SWSignal are allocated to different processors, for example if the sender is a software instance allocated to one of the SSPs and the receiver is a software instance allocated to the SCP, it has to be sent via a #7-link. In such cases the SWSignal is packed in a signal #7Packet and sent to the model of the #7-link which forwards the #7Packet to the receiving processor.

As an example we will look at what happens in the simulation model when SWA sends the signal B to SWB, see Figure-4. The signal B is packed in an SWSignal which is then sent to the processor on which SWA is executing, for example SSP1. SSP1 realizes that the receiver of this signal is allocated to the SCP so the signal must be sent via a #7-link to the SCP. The SWSignal is packed in a #7Packet and is sent to the model of the #7-link where it is delayed a time corresponding to the transmission time for this packet. Then the #7Packet is forwarded to the SCP where the SWSignal is unpacked and sent to the correct receiver, SWB. Before starting to execute SWB must have permission from the SCP to do so, this is accomplished with the signal Execute. Then SWB starts to execute. Apart from the logic already defined in the functional specification (e.g. the creation and sending of signal C) a delay is introduced corresponding to the time it will take to execute this transition on the SCP-processor. When this time expires SWB sends the signal ExecutionCompleted to the SCP.

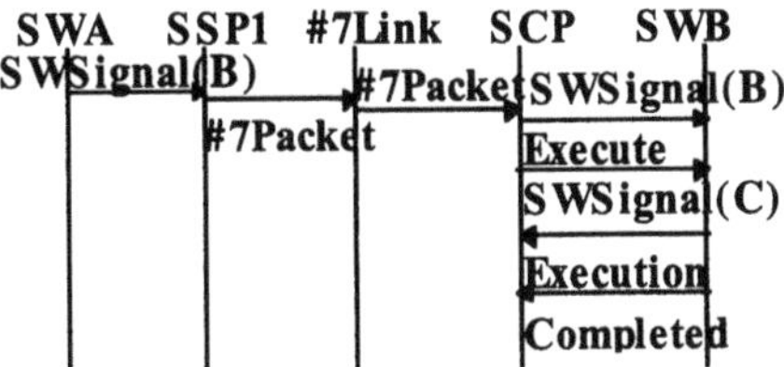

Figure 4 Message Sequence Chart, sending of signal B from SWA to SWB.

With this simulation model it is possible to study different performance aspects in the Intelligent Network such as the load on the SSPs, the SCP and the #7-links. The response time from a user's point of view can also be measured.

This simulation model can be reused if we for example want to introduce a new service. Such a service can be added in the Software Model without affecting the other services or the Architecture Model.

6 LONG-TERM RESEARCH OBJECTIVE

The long-term objective is to be able to work with performance models and service descriptions in SDL in at least three abstraction levels, see bullets below. The first level is:

- Software model and usage model without taking the architecture into account
  This type of analysis shall be used to identify software processes which either have long execution time or are called extremely frequently. The software processes having this characteristic ought to be further studied and perhaps even re-designed, since they probably will cause problems as they are introduced into the network.

In the long run the objective is to be able to handle different levels of abstraction of the models, i.e. several models on different abstraction levels of for example an SCP may exist. This means that it is possible to combine detailed models for some parts while others are described quite superficially. The major benefit of this is that it is possible to model, for example, critical nodes in the network in detail while other parts which are not critical for this particular simulation may be modelled on a much higher level.

Based on the above two other abstraction levels are identified, both of them include different levels of abstraction of the software and usage models. The abstraction levels of the software and usage models are: treating the software as an execution time based on measurements on existing services or taking the software specification in SDL directly. The two abstraction levels including different levels of abstraction of the architecture are:

- Simple architecture model
  The objective of the simple model is to identify the bottlenecks in the architecture, hence being able to identify parts which ought to be focused upon from a performance perspective.
- Focused architecture model
  The focused model means having a quite detailed model of some part of the architecture, primarily, of course, some part which has been identified as being problematic when analysing the simple architecture model.

The ability to handle different levels of abstraction for different model parts is essential as it allows for simulation studies focused on for example network bottlenecks or a specific problematic service.

7 CONCLUSIONS

A method based on reusable model components has been described. The primary objective is to formulate a method which can be used to analyse the intelligent network from a performance perspective as new services are introduced into the network. The method is based on three modelling concepts and their interaction. The models are: software model (service description transformed to incorporate execution time), architecture model and usage model.

The method provides a basis for:

- identifying software modules which will use too much execution time, due to long service times or frequent use. The early identification of these modules allows for re-design instead of identification of the problem as the service is put into operation.,
- studying the introduction of new services in an existing system (network),
- evaluating different distributions of software processes in an architecture,
- examining the ability of different architectures to execute a given software description,
- identifying system bottlenecks.

These possibilities are then further enlarged as the proposed method is supposed to cover analysis of reliability and feature interaction as well in the long run. Hence, the method supports:

- certification of reliability requirements
- identification of the most likely feature interactions

The latter two points are both based on the fact that users get more annoyed if frequently used services and features fail or interact in a way which is disturbing to the user, than if seldom used services do not work in accordance with expectation.

In summary, the method proposed will enable analysis and prediction of several critical quality attributes prior to introducing new services into a telecommunication network. Thus continuous control of the network and its services is supported.

8 REFERENCES

1 Law, A. M. and McComas, M. G., "Simulation Software for Communication Networks: State of the Art", IEEE Communications Magazine, March 1994, pp. 44- 50, 1994.

2 Saulnier, E. T. and Bortscheller, B. J., "Simulation Model Reusability", IEEE Communications Magazine, March 1994, pp. 64-69, 1994.

3 Verán, M. and Poiter, D., "QNAP2: A Portable Environment for Queueing Systems Modelling", Technical Report, INRIA, France, 1984.

4 Whitt, W., "The Queueing Network Analyzer", The Bell System Technical Journal, November, pp. 2779-2843, 1983.

5 Wohlin, C., "Software Reliability and Performance Modelling of Telecommunication Systems", Dept. of Communication Systems, Lund University, Report no. 106, Ph.D. thesis, 1991.

6 Wohlin, C., "Evaluation of Software Quality Attributes during Software Design", Informatica, Vol. 18, No. 1, pp. 55-70, 1994.

7 Wohlin, C. and Runeson, P., "Certification of Software Components", IEEE Transactions on Software Engineering, Vol. 20, No. 6, pp. 494-499, 1994.

8 Whittaker, J. A. and Poore, J. H., "Markov Analysis of Software Specifications", ACM Transactions on Software Engineering Methodology, Vol. 2, No. 1, pp. 93–106, 1994.

9 BIOGRAPHY

Wohlin, Claes - Dr. Wohlin is associate professor at the department of Communication Systems, Lund University, Sweden and acting professor at the Department of Computer Science at University of Linköping, Sweden. He has five years of industrial experience working within the field of software engineering. Claes Wohlin is currently responsible for the education and research in the area of software engineering at the department in Lund and he is heading a research group directed towards processes and methods for large scale software development in Linköping. His research interest includes methods and modeling techniques to achieve quality software, statistical usage testing and process improvement.

Nyberg, Christian - Dr. Nyberg received an MSc in Computer Engineering in 1987 and a Ph.D. in Electrical Engineering in 1992. Presently he works as an assistant professor at the Department of Communication Systems at Lund University. Most of his work has been on overload control of nodes in communication networks and on the performance impact of new services in Intelligent Networks. He has been a project leader for projects on load control sponsored by Ericsson Telecom and the National Board for Industrial and Technical Development, Stockholm, Sweden. He has published more than 25 papers.

Larsson, Anders - Anders Larsson received his M.Sc. degree in Electrical Engineering from Lund University in 1985. Since then he has been employed at different companies within the Telia (formerly Swedish Telecommunications Administration) group. His research interests include performance evaluation, simulation methodology and service management in intelligent networks.

13
Methods to synchronize the IN SCP's overload protection mechanism

Philip Ginzboorg
Nokia Research Center
P.L. 45, 00211 Helsinki, Finland
E-mail: philip.ginzboorg@research.nokia.com

Abstract

The Intelligent Network's (IN) architecture is based on Service Switching Points (SSP's) that switch telephone calls, and the Service Control Points (SCP's) that make part of the decisions about the routing and the charging of calls that are switched by the SSP's. To communicate, SSP and SCP use the IN Application Part (INAP) protocol. IN has a distributed mechanism for protecting the SCP from overload that uses a traffic filtering method called "call gapping". The synchronization of the load protection in IN is a problem because there is no reliable synchronization scheme in the INAP protocol. In the text I examine several solutions to the problem. Though formulated in terms of IN, the synchronization methods apply to any similar network, i.e. a network with a star topology, traffic throttles in the peripheral nodes, and a bottleneck in the central node.

1 SERVER'S OVERLOAD IN A CLIENT-SERVER SYSTEM

Let us start with the following example. Figure 1 shows two machines – a server and a client, that are connected by a signaling link. The server contains a database, and the client asks questions from the server by sending messages over the link. When the server receives a question it starts a transaction that results in an answer to that question; when the answer is ready, the server sends it to the client. If the server hasn't answered a question within 30 seconds the client either aborts the question, or tries to ask it again. Each answered question costs the client a penny.

If the server would have been an omnipotent machine, it would have answered every question immediately, and the plot of the answers rate against the questions rate would have looked like figure 2 a). Of course there is a limit to the amount of answers that the server can supply per a unit of time. Taking this into account, the response curve of the server should look like figure 2 b). However, that curve is still too good to be true. In reality, when the questions rate exceeds a certain threshold for an extended amount of time, the server becomes overloaded. When the server is overloaded more and more questions will result in less and less answers, as figure 2 c) shows. Such a degradation is caused by system resources being wasted. For example, the machine allocates more and more of the free memory for buffering the questions, so there is less and less working memory for computing the answers.

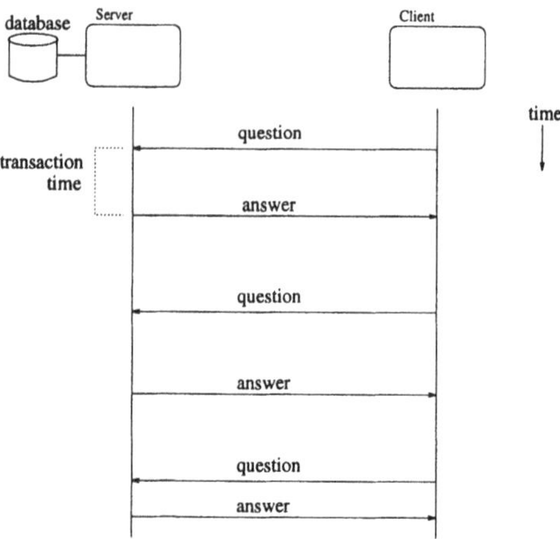

Figure 1 Two machines - a client and a server - are connected by a signaling link. The server has a database, and the client asks questions from the server. The communication protocol between the machines consists from the ``question'', ``answer'' message pair.

The threshold questions' rate is not fixed. Its value depends on how much of the machine's capacity can be dedicated to answering questions. For example, the threshold will be lower than usual during the updates of the server's database. The goal of any overload protection method is to change the response curve of the real machine so, that it will resemble the ideal curve in figure 2b). The closer to the ideal, the better the protection method. In the following paragraphs I describe a method in which the work of protecting the server from the overload is partly done in the client.

Suppose that to protect itself, an overloaded server sends a "filtering request" message to the client. The request contains two parameters: the upper bound on the questions rate, u [questions/s], and the duration of the filtering. When the client receives that message, it starts to filter the questions traffic so, that its rate will be at most u ; part of the questions are aborted with no answer, without ever reaching the server. The filtering lasts for the duration specified in the request message. While the client is filtering the questions traffic it can receive another filtering request. In this case the bound rate will be updated, and the filtering duration will be extended.

When using this protection scheme, the server has two difficulties. The first difficulty is how to choose the filtering parameters. Long duration and low u mean less questions and thus less money for the server. Too short duration and too high u may not reduce the traffic enough to protect the server from the overload. A natural first step in solving that difficulty is to divide the response characteristic into regions that are called "overload levels"; within each level the filtering parameters are constant (see figure 3) If at all times the server is able to determine the overload level of itself, then the operator who is responsible for the set-up of the machine can store the filtering parameters in a "level : duration, u" format. This is not a complete solution because the really difficult part of the problem is delegated to the operator. With a complete solution, the choice of filtering parameters should be fully automatic. In [1] Kumar describes an algorithm for periodically adjusting the parameters, based on the occupancy measures of the server.

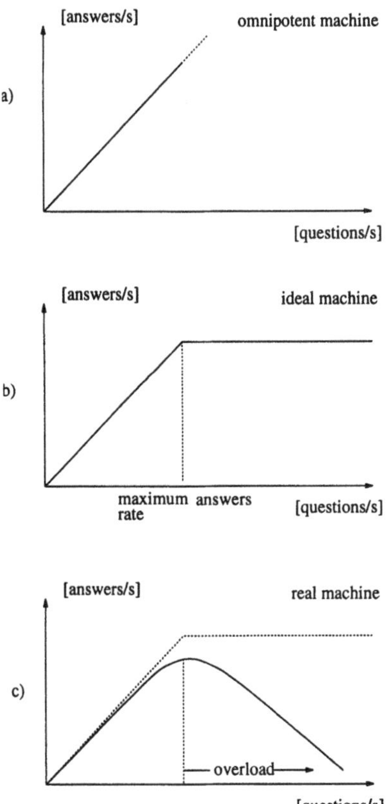

Figure 2 Response curves of the omnipotent, the ideal, and the real machines. a) The omnipotent machine will answer every question immediately, no matter how high is the questions per

second rate. b) The ideal machine has an upper bound on the answers rate. When the questions rate exceeds that bound, the answers rate stays constant; some of the questions remain unanswered. The ratio of the unanswered to the total amount of questions is called a ``grade of service''. c) Up to a certain threshold, the response curve of the real machine is close to the ideal. When the questions rate exceeds that threshold the machine becomes overloaded. In the overload state more questions result in less answers.

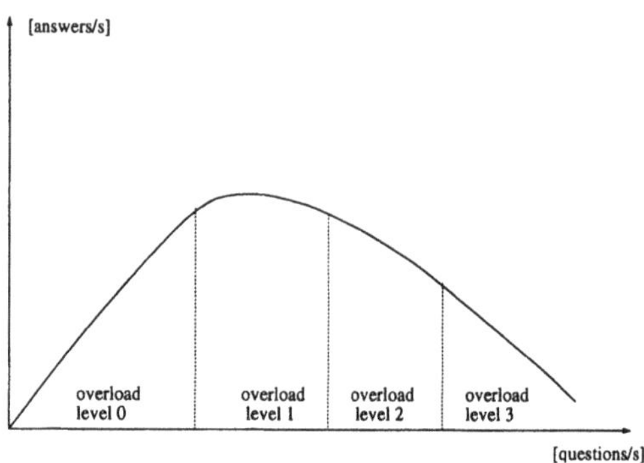

Figure 3 The response characteristic is divided into levels. Within each level the filtering parameters are constant.

Once the filtering parameters have been chosen, the second difficulty of the server is when to send, and when not to send filtering requests. The server should send the first request when it is close to becoming overloaded. After that the server should send a new request either when the duration expires, or when the filtering parameters change. The server shouldn't send a new request if the client filters questions correctly, with the right bound rate and with the right duration.

In our communication protocol, there is no feedback to the filtering request. After sending the request the server doesn't know if and how the client filters the questions traffic. If the client is the only source of questions, the server can solve the problem by monitoring the questions rate; when the monitored rate exceeds u, the server sends a new filtering request. If there are several machines that ask questions, monitoring the traffic from each needs a lot of book-keeping.

The second difficulty of the server is a synchronization problem. The server has to keep up to date, or synchronize a traffic valve that is in another, remote machine. Synchronization problems arise between a group of processes that communicate solely via message passing. A good treatment of synchronization in distributed computing is in Lamport's [2], and Jefferson's [3] papers.

The overload protection in the Intelligent Network (IN) works in a way that is very similar to the above example. In IN, the Service Control Point (SCP) is the server, and the Service

Switching Point (SSP) is the client. Synchronization of the overload protection in IN is difficult because there is no reliable synchronization scheme in the SCP-SSP communication protocol. In this text I describe several possible solutions to that problem.

Subsections 2.1 and 2.2 are a brief introduction to the IN and to the SCP's overload protection. Subsections 2.3 and 2.4 describe the difficulties in synchronizing the overload protection mechanism. The rest of the text is about how to solve the problem.

2 SCP'S OVERLOAD CONTROL IN THE INTELLIGENT NETWORK

2.1 IN architecture

The Intelligent Network's (IN) architecture is based on Service Switching Points (SSP's), and the Service Control Points (SCP's). SCP's and SSP's are connected via a Signaling System number 7 (SS7) network (figure 4) To communicate, SSP and SCP use the IN Application Part (INAP) protocol. In the SS7 protocol stack, the INAP layer is on top of the Transaction Capabilities Application Part (TCAP) layer (figure 5).

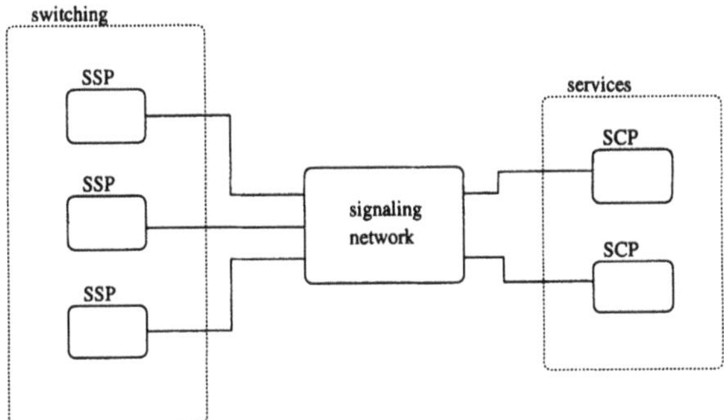

Figure 4 The Intelligent Network's architecture is based on Service Switching Points (SSP's) that switch the calls traffic, and the Service Control Points (SCP's) that make some of the decisions about the routing and the charging of calls that are switched by the SSP's. SSP's and SCP's are connected via a fast signaling network - SS7.

The SSP is usually a commercial telephone exchange with a modified call controlling software; the SCP contains the services' control logic and it has an access to the service database. In the IN, the calls' traffic flows through SSP's. The SCP makes part of the decisions that concern the routing and the charging of calls. During an IN call there may be one or more INAP dialogues between the SSP and the SCP. Each dialogue starts with an "initial detection point" (initial-dp) message.

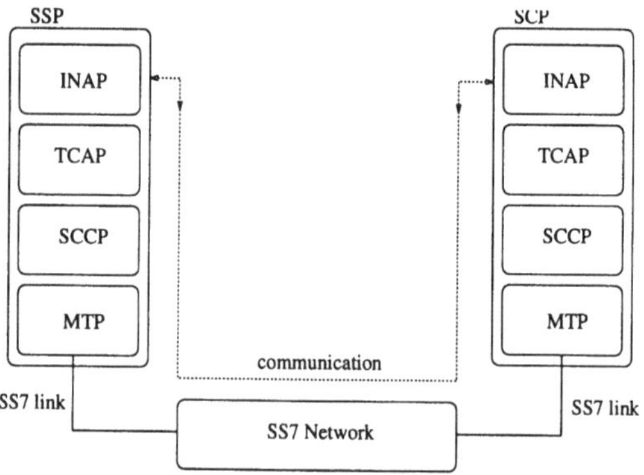

SS7: Signaling System number 7
MTP: Message Transfer Part
SCCP: Signaling Connection Control Point
TCAP: Transaction Capabilities Application Part
INAP: IN Application Part

Figure 5 SCP and SSP are connected via a SS7 signaling network. To communicate, SSP and SCP use the IN Application Part protocol.

Two examples of a service are the "free-phone", and the "premium-rate phone". A call to a free-phone number costs nothing to the caller; everything is paid by the called party that subscribes to the service. A call to a premium-rate number (for example a lawyer's office) costs a lot to the caller; the money, after a deduction of the subscription fee goes to the bank account of the called party that subscribes to the service. IN offers a centralized implementation of those services. During the establishment of the call, SSP sends the dialed number and other data to the SCP, and the SCP returns the destination and the charging method for that call. Besides the dialed number, the destination and the charging of the call may depend on other things; for example, they may depend on the location of the caller and on the calendar time. During the same call several services can be executed sequentially, one after the other. "Nesting" is another way to combine services; a nested service starts and ends during an execution of another service.

IN is a network with a star topology [1]. In a star network the nodes are of two kinds: central and peripheral. Peripheral nodes generate traffic that flows to the central node ; the central node is the computational bottleneck of the network. When IN contains more than one SCP, its architecture is a superposition of several star networks that share their peripheral nodes – SSP's (see figure 6).

[1] Besides IN, amultitude of other networks have a star topology. Three of the many examples are: a satellite that switches traffic generated by earth stations, a controller that processes data gathered by remote sensors, and a GSM Mobile Switching Center with its Base Stations.

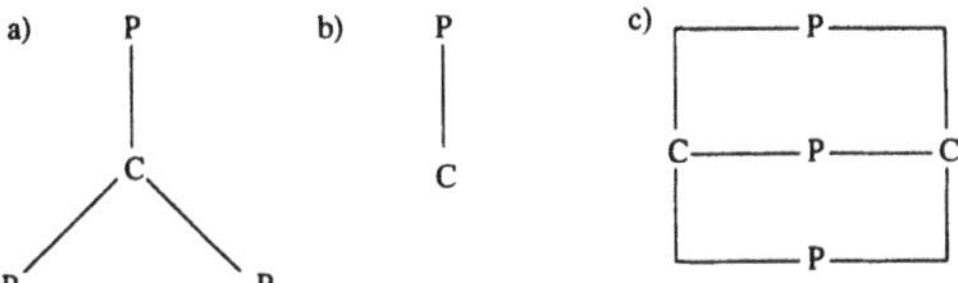

Figure 6 a) a star shaped network with three peripheral nodes; ``P'' denotes a peripheral node and ``C'' the central node. b) in the simplest case there is only one central node and one peripheral node. c) two star networks that share their peripheral nodes.

Conceptually, IN can be divided into three parts. The first part contains the SSP's and other nodes that switch the calls traffic. In the second part are the services that the network offers. In IN, both the knowledge on what the services do, and the knowledge on how to access the data that the services need are located in a small number of central nodes (SCP's). The third part is the SCP-SSP communication protocol that is the language in which the machines "talk". IN makes it easy to introduce and to modify services, but the use of any service has to be expressed as a sequence of INAP messages. In this sense the protocol defines the "intelligence" of the network.

2.2 Distributed overload control

When the traffic in the network is high, the SCP can get overloaded. IN has a distributed load control mechanism that uses "call gapping" method to limit the flow of messages towards the SCP. SCP detects the overload condition and the SSP's filter the excessive traffic. When the load control in the SSP is active, SSP rejects, or "gaps" a certain part of calls before they start a dialogue. SSP does not abort dialogues in the middle [2].

Suppose that we have an Intelligent Network with one SCP and two SSP's (see figure 7) The SCP contains a hierarchy of subsystems; each subsystem is located behind its controlling gate. All communication with the subsystem has to pass through the gate. The gate gathers statistics about the traffic, the condition of the subsystem, and the condition of the other parts of the SCP. From this data the gate computes the load level of the subsystem.

The normal load level of the subsystem is zero. When the load level changes from zero to one, the gate with a cooperation from SSP's, will try to limit the message traffic that is directed towards the subsystem. To limit the traffic, the gate sends a call gapping request (cg-request) to both SSP's. The message contains three groups of parameters (the following definitions are from section 7.3.6.1.1 of [4]). The first group is "gap criteria", the second group is "gap indicators" and the third group is "gap treatment".

Gap criteria identifies the portion of the traffic that is directed to the overloaded subsystem. Gap indicators define the maximum initial-dp's rate, u and the duration of the restriction [3] ;

[2] SSP can abort a dialogue in the middle, but not because of call gapping; for example, the reason can be that the SCP has failed to respond within a predefined period of time.

[3] cg-request contains the minimum interval between two consecutive initial-dp messages, I. The maximum initial-dp's rate, u is the inverse of this interval: $u = 1/I$.

between the arrival of the cg-request and the end of the duration, the rate should be at most this maximum (see figure 8) Gap treatment defines what to do with rejected calls. For example, the speech channel of a rejected call can be connected to a certain voice announcement, or to a busy tone. In addition, the cg-request message contains the "gap control" field. Gap control is either "SCP overload", or "manually initiated". When cg-request is due to the automatic load protection mechanism in the SCP, the value of the gap control field is "SCP overload". When cg-request results from a command of the SCP's operator, the value of the gap control is "manually initiated".

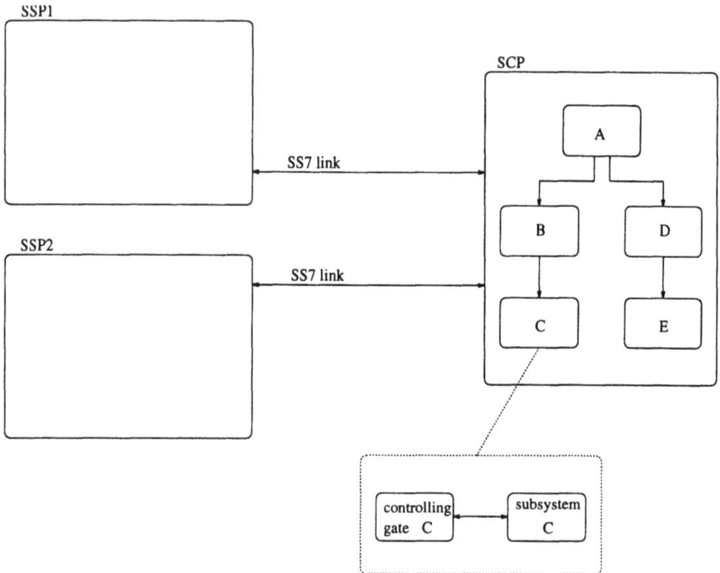

Figure 7 IN with one SCP and two SSP's. Each subsystem in the SCP is located behind its controlling gate.

Manually initiated cg-requests have higher priority than messages with the same gap criteria and "SCP overload" gap control field ([4] section 7.3.6.3.1). After the manually initiated message has arrived, and before the duration specified in it has expired, SSP ignores non manual cg-requests that have the same gap criteria. This allows SCP's operator to second-guess the machine.

When a cg-request arrives at SSP, SSP creates an "image" of the sending gate. The image works like a throttle-valve. With the aid of the gap criteria, the image identifies the portion of the traffic that is directed towards the overloaded subsystem, and limits the rate of this traffic stream. When the duration specified in the cg-request expires, SSP destroys the image (see figure 9).

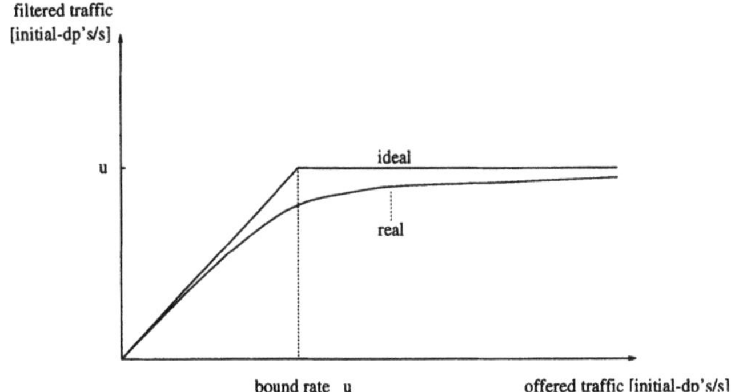

Figure 8 "call gapping" is a rate dependent method for filtering traffic. When the traffic rate offered by the network is less than the bound rate u, there is no filtering. When the traffic rate exceeds u, SSP rejects part of the calls, so that the traffic rate after filtering is u. The characteristic of the real gapper is a smoothed approximation of the piecewise-linear ideal.

The gate in the SCP is "static"; it exists all the time. The image of the gate in the SSP is "temporary"; SSP creates the image when it receives a cg-request and it destroys the image when the duration specified in the cg-request has expired. When the SSP receives a cg-request with gap criteria that is the same as the criteria of an already existing image, the gap indicators, the gap treatment, and the gap control of that image will be updated.

Another approach is to view the images in the SSP as static objects with two states: "active" and "passive". When the image receives a cg-request it turns active and starts to filter traffic. When the image is in the active state, it can receive several cg-requests from the gate in the SCP. Before the first cg-request has arrived, and after the duration specified in the last cg-request has expired the image is in the passive state.

When two of the subsystems in the SCP are overloaded at the same time, there are two images of their corresponding gates in the SSP's. As more and more subsystems become overloaded, the logical structure of the images in the SCP's starts to resemble the hierarchy of the SCP's gates (see figure 10).

2.3 The synchronization problem

When one of the SSP's starts a dialogue by sending an initial-dp message, the SCP has to decide if to send or not to send a new cg-request to that SSP. There shouldn't be a new cg-request if there is an image of the overloaded subsystem's gate in the SSP and if that image filters calls with the right set of parameters.

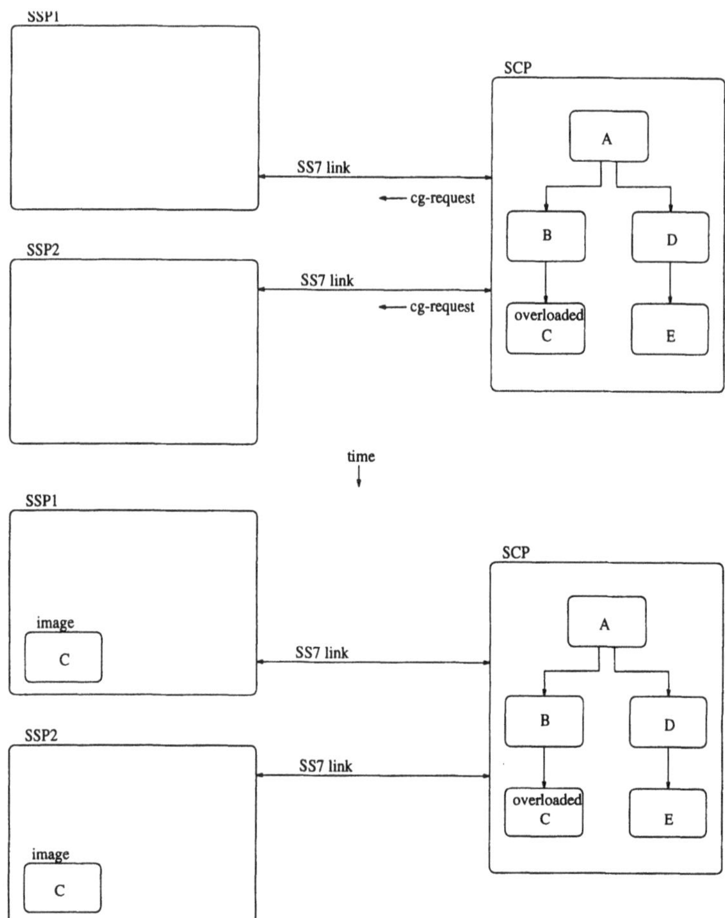

Figure 9 Subsystem C in the SCP is overloaded. By sending cg-requests, SCP has created images of C's gate in both SSP's. The images focus on the traffic that is directed towards the overloaded subsystem, and reduce that traffic by rejecting part of the calls. When the duration specified in the cg-request expires, SSP's will destroy the images.

Cg-request is a one–way message; there is no acknowledge message to it. The lack of feedback, together with the dynamic creation and destruction of call gapping entities in the SSP's makes the synchronization of the load protection mechanism difficult; SCP is "in the dark" as to what is the situation of the images in the SSP's.

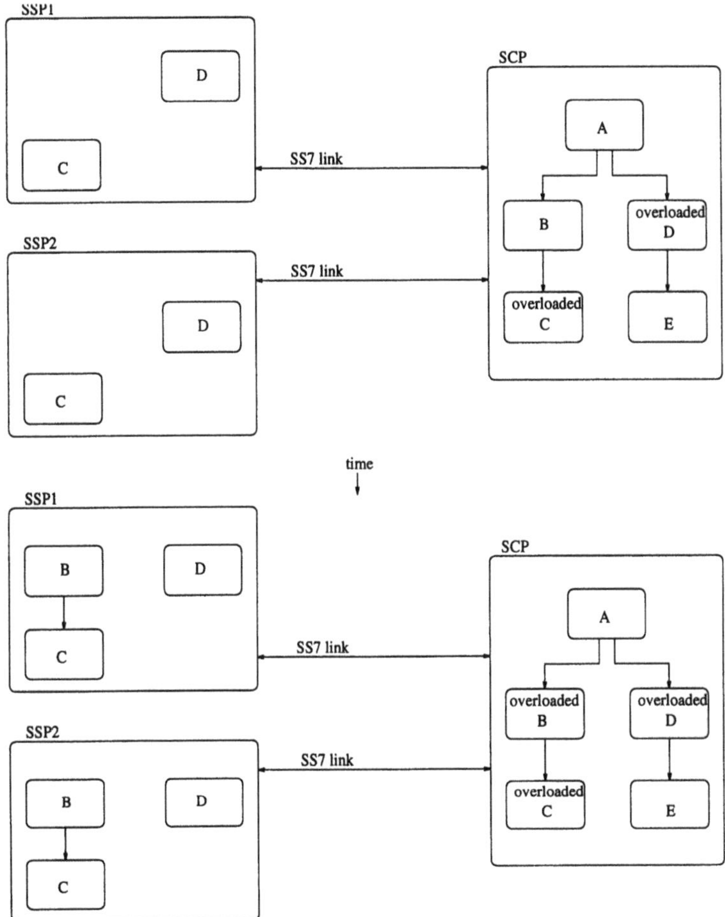

Figure 10 As more and more subsystems become overloaded, the logical structure of the images in the SCP's starts to resemble the hierarchy of the SCP's gates.

2.4 How the problem was solved in Tele's IN product

In 1989 Telecom Finland (Tele) has contracted NTC and ICL for developing SSP and SCP that where based on BELLCORE's AIN 1 set of standards ([5], [6]). In this product the problem was solved by repeating the same cg-request after every initial-dp message from an SSP. This solution causes frequent updates of the images in the SSP's. It also adds to the traffic load on the SS7 signaling link that connects SSP and SCP.

3 RANDOMIZED UPDATES WITH PROBABILITY p

An improvement over the previous solution is as follows. Instead of sending a filtering request after every initial-dp message, the gate that controls the traffic into an overloaded subsystem, will send a filtering request in response to a certain proportion, p of the initial-dp messages. A gate can have several images that are located in different SSP's, and to distribute the filtering requests between them evenly, we employ a random number generator. An initial-dp will trigger a filtering request, if the output from a random number generator is at most p. Let's denote by "rand" a function that returns a uniformly distributed random variable in the range [0, 1]. For every initial-dp message, the decision if to send or not to send a cg-request is done as follows:

$$\text{send cg-request} := \quad \text{rand} < p, \tag{1}$$

where rand, p ([0,1];

The reduced amount of update messages comes at a price of increased interupdate time. Let's denote the rate of initial-dp's from an SSP towards the gate by r [initial-dp's/s], and the average inter-update time, i.e. the time between consequtive cg-requests that the gate sends to the SSP, by d. Clearly

$$d = 1/(rp) \tag{2}$$

Our goal is to find out how to choose p, but before that we have to answer the following question: What are the constraints on the value of d?

To answer that question remember how a cg-request affects the SSP: if $r \leq u$, then the request will have little or no effect on it; if $r > u$, then r will be reduced to u for the duration specified in the request. It follows that the gate does not have to synchronize a SSP whose rate does not exceed u, and so we can restate the question in the following way: What are the acceptable values of d, if $r > u$?

This is an easy question to answer. Since r is bounded from below by u, d is bounded from above by $1/(up)$

$$\mathrm{d} < 1/(up) \tag{3}$$

To choose p, we take a sender with rate that equals the upper bound, u as a yardstick, and we decide on the average update time D for that sender. Then

$$p = 1/(uD) \tag{4}$$

If possible, the value of D should be an order of magnitude higher than $1/u$, and it should be lower than the gap duration:

$$\frac{10}{u} < D < \text{gap duration.} \tag{5}$$

The response time to the sender with rate r is $1/(rp)$, or $(u/r)D$ seconds. The higher the r/u ratio, the faster will the cg-request be sent.

For example, if $u = 10$ [initial-dp's/s], gap duration equals 24 seconds, and the chosen value of D is 2 seconds, then $p = 1/20$. To keep the SSP's up to date with a 2 seconds' accuracy, we need to respond with a cg-request to only 5 percent of the initial-dp's.

Note that if the network is faulty, and if the probability of loosing any message is l, then the actual rate of any traffic source shrinks by the factor of $(1 - l)$ compared to its nominal rate, and the interupdate time grows by the factor of $1/(1 - l)$.

In the rest of this section we will consider the case where p is a periodic function of time, and the case where p depends on the rate of the traffic that flows to the overloaded subsystem.

3.1 Periodic p

$p(t)$ is a periodic function of time with a constant period T, that is at most the gap duration, and with a mean value, $E[p(t)]$ that is at least $1/(uD)$.

$$p(t) = (t + T), \text{ where } T \leq \text{gap duration}; \quad (6)$$
$$E[p(t)] = 1/(uD);$$

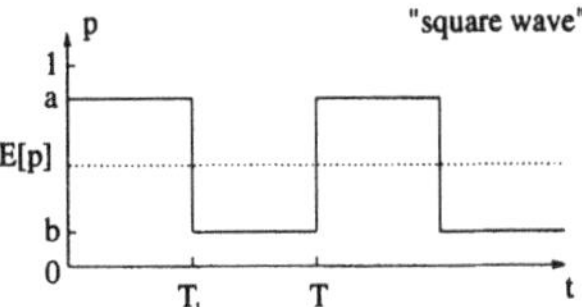

Figure 11 an example of the periodic p's shape.

For example, figure 11 shows a "square wave" p. Each period, T is divided into two parts. In the first part, $p = a$, and in the second part, $p = b$. The division into parts is defined by the value of time modulo T. In the first part, the time modulo T is between zero and T_1; in the second part, it is in the (T_1, T) interval.

$$p(t) = \begin{cases} a \text{ if } 0 \leq (t \bmod T) \leq T_1, \\ b \text{ otherwise}; \end{cases} \quad (7)$$
$$E[p(t)] = \frac{T_1(a-b)}{T} + b$$

In particular, when $a = 1$ and $b = 0$, the behavior of the gate is completely deterministic. The gate alternates between sending a cg-request after every initial-dp for T_1 seconds, and not sending any cg-requests until the next period starts. The constraint $E[p(t)] \geq 1/(uD)$ translates into the following condition on the ratio of T_1 and T

$$\frac{T_1}{T} \geq \frac{1}{uD} \tag{8}$$

A response to an excessive traffic rate from a source will come faster when p is near its peak. It is to advantage to choose the period of $p(t)$ so, that it will fit into the gap duration an integral number of times: gap duration = kT, $k = 1, 2...$, because then the end of the gap durations is likely to coincide with the time when $p(t)$ is at its peak.

We could also make p a function of the total traffic rate. For example, the gate will start to send cg-requests only when the total rate exceeds a certain threshold.

4 BROADCASTING

This solution is based on broadcasting the cg-request to every SSP and every SSP replying with an acknowledge message after it has either created or updated an image of the sending gate. The SCP keeps track of the acknowledgements, and after a time–out, retransmits the cg-requests to the SSP's that haven't sent an acknowledgement message. During the overload condition broadcasting is repeated periodically; the period is smaller or equal to the gap duration.

There are two problems with this solution. The first problem is that a reliable broadcasting is hard to implement. The second problem is that a fault can deactivate an image after an acknowledge has been sent.

5 RANDOMIZED BROADCASTING

The hard part of broadcasting is keeping track of the acknowledgement messages and resending the broadcasted message if necessary. If we omit that part, broadcasting is not as reliable, but much simpler to implement.

In a looseless network, that is, in a network with faultless nodes that always delivers every message to its destination, the gate has to broadcast only either when its parameters change, or when the gap duration expires. Suppose that the network is not looseless, and that the broadcasting is implemented without the acknowledgement part. To compensate for an occasional loss of some cg-requests the gate could broadcast more often, for example four times per gap duration.

Another possibility is that the gate will rebroadcast randomly, in proportion p of the incoming initial-dp messages. Each broadcast consists from n cg-requests, where n is the amount of SCP's in the network. To choose p we divide the right hand side of equation (4) in section 3 by n:

$$p = \frac{1}{nuD} \tag{9}$$

Here, D is the average interupdate time when the combined traffic rate from all SSP's is exactly nu.
For example, if $n = 5$, $u = 10$ [initial-dp's/s], gap duration equals 24 seconds, and the chosen value of D is 2 seconds, then $p = 1/100$.

6 THE STAMPING SYNCHRONIZATION METHOD

SCP maintains an integer that is called a "global stamp". Initially the value of the stamp is zero. When the parameters of any gapping gate in the SCP change, that gate increments the global stamp, reads its value, and stores it in its internal data. The stored stamp's value is called a "local stamp". The parameters of a gate change either due to an update, or because the subsystem behind the gate moves from one overload level to another. At any moment of time, the local stamp of a gate uniquely identifies that gate's current set of parameters.

```
function get stamp: integer                                          (10)
stamp := stamp + 1;
return stamp ;
endfunction;
```

Instead of being incremented with each update, the global stamp can be decremented, instead of holding an arbitrary integer the stamp can hold the current value of the SCP's computer clock. The important thing is that every time a gate needs a new stamp, it will get a unique value from a totally ordered set.

When sending a cg-request the gate will add the current value of its local stamp to it; when the SSP receives a cg-request, the message will contain the stamp of the sending gate. SSP copies that value from the cg-request to the data of the corresponding image. If a call has passed through the call gapping control of an image, the initial-dp message that is sent to the SCP will contain the stamp of that image.

The gate decides if to send or not to send a cg-request as follows. The initial-dp message can come with or without a stamp. If there is no stamp, there is no image of the gate in the SSP and the cg-request is sent. If there is a stamp, its value is either equal or different to the value of the local stamp. If the values are equal, the image of the gate is up to date; if the values are different, the gate sends a cg-request to the SSP.

```
send cg-request :=    ((no stamp in the initial-dp message) OR            (11)
                      initial-dp.stamp ≠ local stamp of the gate));
```

6.1 The SCP's subsystems hierarchy

ETSI standard defines four kinds of gap criteria ([4] section 7.3.6.1.1.) The criteria are: "called address", "service key", "called address and service key", and "calling address and service key". Called address corresponds to the dialed number. Service key identifies the IN service that is needed by the call. Calling address corresponds to the telephone directory number of the calling subscriber.

In subsection 2.2 I have described the internals of the SCP as a hierarchy of subsystems that are protected by gates. A "subsystem" is a part of the SCP that has its own unique gap criteria. A traffic stream that matches gap criteria *x*, can be a part of a larger stream that matches criteria *y* ; in this case *y* is a subset of *x*. The less precise the criteria, the more of the traffic it matches, and the higher in the hierarchy is the corresponding subsystem. For example, a sequence of calls to numbers that start with the digits "3146" is a part of a sequence of calls to numbers that start with the digit "3". It is that "part of" relation that imposes a partial order on the set of subsystems. A subsystem with criteria "called address = 3" is higher in the hierarchy than a subsystem with criteria "called address = 3146". This view of SCP's internals is what an external observer may deduce on the basis of SCP's filtering requests.

Our model of the SCP is a Directed Acyclic Graph (DAG), where the vertices are subsystems, and where there is a path from a subsystem "A" to a subsystem "B" if the traffic that flows to "A" includes the traffic that flows to "B". In DAG, the length of any path is bounded. In general, if there is a link from a subsystem "A" to subsystem "B", a transaction that executes at "A" either will end at "A", or it can continue to execute at "B".

6.2 Filtering calls that match the criteria of several images

When the SSP contains two or more active images of gates that lie on the same path we have to decide how to filter calls that are directed to the subsystems behind those gates. The question can be restated as follows. How to filter a call whose data matches the gap criteria of more than one image?

Example: SSP contains two images with gap criteria "called address". In the first image the criteria is "800", and the bound rate is 10 [initial-dp's/s]. In the second image the criteria is "80012", and the bound rate is 5 [initial-dp's/s] (see figure 12). When the subscriber dials "800123" we can filter the call either by the parameters of the first image, or by the parameters of the second image, or by the parameters of both images. Filtering by the parameters of the first image is wrong because then the subsystem behind the gate that has created the second image can get 10 [initial-dp's/s] instead of the specified 5 [initial-dp's/s]. Filtering by the parameters of the second image works in this case but it would be wrong if we would exchange the bound rates of the images. Filtering by the parameters of both images means that the call has to pass through the gap control of both, and if any of the images requires to gap the call, the call will be rejected. This is correct in all cases.

Note: according to the ETSI standard ([4] section 7.3.6.3.1), in the above example we should filter by the parameters of the second image.

Conclusion: When the data of a call matches the gap criteria of several active images, that call has to pass through the call gapping control of all those images, and if any image requires to gap the call, that call will be rejected.

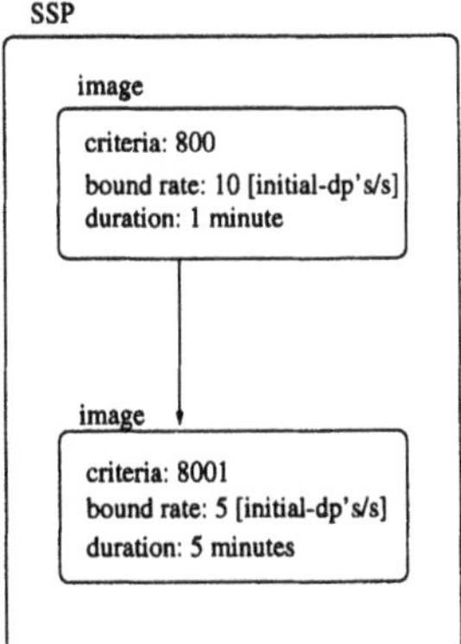

Figure 12 SSP contains two images. The first image filters calls with dialed number that starts with ``800..'', and the second image filters calls with dialed number that starts with ``8001..''. As long as those two images coexist, according to the parameters of which image should we filter calls with dialed number ``8001..''?

This rule has the following effect on the stamping synchronization mechanism. Instead of zero or one stamp, the initial-dp message has an array of zero or more stamps. The array contains the current stamp values of all images that the call has passed. Before a transaction starts to execute in a subsystem, the gate of that subsystem checks if the value of its local stamp is in the array; when the value is absent, the gate will send a cg-request to the SSP.

The length of the stamps array is bounded by the depth of the subsystems' hierarchy in the SCP. The processing capacity of the SSP also sets a limit on the length of the array. We can't afford to have a large amount of dependent images in the SSP; when a call has to pass through many images, the computational overhead is high. I assume that in practice the array will have no more then, say, thirty stamps.

6.3 Rollover of the stamp value

Because the physical size of the global stamp variable is limited, the value of the stamp can not be incremented forever. For example, if the size of the variable is one octet, its maximum value is 2^8 - 1, or 255; if the size of the variable is two octets, its maximum value is 2^{16} - 1 that is about sixty five thousands. When a gate will try to increment a stamp that is at its maximum, the value will not increase; it will jump back to zero, or "roll-over".

After rollover, old stamp values are reused, and it is possible that two or more gates will have the same local stamp. The frequency of this event depends on the relative speed with which the gates update their parameters. We can either ignore the rollover, or we can deal with the problem in the following way: a rollover will cause every gate to get a new stamp value, even though there hasn't been any change in the gates' parameters. The resulting cg-requests will deliver the new stamps to the images in the SSP's.

6.4 The effect of network transmission delay

During the time that the cg-request with a new stamp is on its way to the SSP, the SSP may send a number of initial-dp messages with the old stamp value. Each of those messages will cause another cg-request (see figure 13). The transmission delay depends on the amount of traffic. In Telecom Finland's IN implementation, the measured delay is between 0.1 and 0.2 seconds. The transmission time is composed from the time it takes to transmit the message between the machines, and the time it takes a message to propagate through the protocol stacks of the sender and the receiver.

For example, when the transmission delay is 0.2s, the time between sending a cg-request and receiving a first initial-dp with the same stamp is at least 0.4s. If the traffic intensity is 10 [initial-dp's/s], there will be four repeated cg-requests during that time.

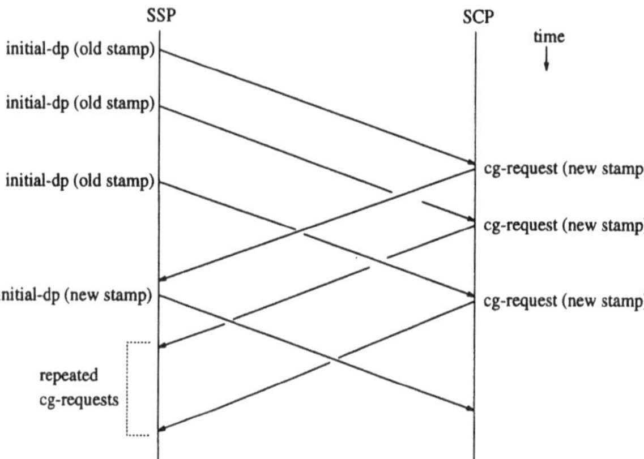

Figure 13 due to the network transmission delay, a cg-request may be repeated a few times

7 A COMBINATION OF THE STAMPING AND THE RANDOMIZED METHODS

We use the stamping method, but the gate will examine the stamps only in a certain proportion, p of the initial-dp messages. In section 3 I have shown that for a source with an average rate that exceeds u [initial-dp's/s], this guaranties a cg-request within $1/(pu)$ seconds on the average.

The advantage over a purely randomized update is that the gate doesn't send unnecessary cg-requests. If one of the stamps in the initial-dp message is equal to the local stamp of the gate, there is an image in the SSP that filters traffic with the right set of parameters. The advantage over pure stamping is that the stamp checking overhead is reduced by a factor of $1/(1 - p)$.

8 CONCLUDING REMARKS

I have described five ways to solve the problem of overload protection synchronization in the Intelligent Network. Though formulated in terms of IN, the methods apply to any similar network, that is, a network with a star topology, traffic throttles in the peripheral nodes, and a bottleneck in the central node.

When the SCP synchronizes by broadcasting, it initiates the filtering requests by itself. The other methods synchronize in response to the incoming traffic. Thus we can say that broadcasting is an "active" synchronization method, and that the other methods are "reactive". The intuition behind reactive synchronization is that the peripheral node (SSP) that loads the central node (SCP) the most, sends the highest proportion of messages to it, and thus it has the highest chance of receiving a filtering request.

The randomized reply, and the randomized broadcasting methods affect only the SCP, there is no need to modify the communication protocol between the SSP and the SCP. The other methods require small additions to the protocol.

For the broadcasting method, the addition is an acknowledgement to the cg-request. SCP should be able to broadcast a cg-request to all of the SSP's in the network, and to keep track of their acknowledgements. The problem with broadcasting is that a fault can deactivate the traffic throttle in the SSP after an acknowledgement message has been sent.

For the stamping method, there are two additions. The first addition is a stamp field in the cg-request message. The second addition is an array of zero or more stamps in the initial-dp message. The maximum length of the array is bounded by the depth of the subsystems' hierarchy in the SCP. This method is robust, it will work correctly in most circumstances.

For the randomized stamping method, the additions are the same as for the pure stamping. The advantage over pure randomized update is that there are less cg-requests. The advantage over pure stamping is that the stamp checking overhead is reduced.

Randomized reply is the easiest method to implement, and broadcasting is the hardest. Stamping and the randomized version of it are both reliable and not hard to implement. They are my methods of choice.

The discussion about the "answers vs. questions" throughput curve in section 1 is based on Kleinrock's paper [7]. The "gates and images" terminology is my own. The assumptions about the internals of the SCP, such as those made in subsections 2.2, and 6.1, are not restrictive; any SCP can be modeled in this way.

9 ACKNOWLEDGEMENTS

I'd like to thank Heikki Hämmäinen, Pekka Lahtinen and Richard Fehlman from Nokia Research Center for their comments, and also my collegues in Nokia Telecommunications who have reviewed this text. Special thanks to Brian Weitch from HUT's Ship Laboratory for his encouraging remarks.

10 REFERENCES

[1] Kumar A., Adaptive load control of the central processor in a distributed system with a star topology, IEEE Transactions on Computers, Vol. 38, No. 11, November 1989, pp. 1502 - 1512.

[2] Lamport L., Time, clocks, and the ordering of events in a distributed system, CACM 21, 7 (July 1978) pp. 558--565.

[3] Jefferson D., Virtual Time, ACM transactions on Programming Languages and Systems, Vol. 7, No. 3, July 1985, pp. 404--425.

[4] ETSI IN CS1 INAP Part 1: Protocol Specification, Draft prETS 300 374-1, November 1993.

[5] BELLCORE, AIN Release 1, Switching Systems Generic Requirements, SR-NWT-001123, May 1991.

[6] BELLCORE, AIN Release 1, Switching Systems Report, SR-NWT-001754, September 1990.

[7] Kleinrock L., On the modeling and analysis of computer networks, Proceedings of IEEE, Vol. 81, August 1993, pp 1179-1191.

11 BIOGRAPHY

Philip Ginzboorg received the B.S. degree in Electrical Engineering from Beer Sheva University, Israel, in 1989. From 1989 to 1995 he worked in Nokia Telecommunications on, among other things, line signaling, routing and overload control. Since 1995 he is a member of Communications Systems Laboratory in Nokia Research Center, Helsinki. His current work focuses on the control of services in broadband networks.

14
The Optimal Utilization of Multi-Service SCP

Tang Haitao and Olli Simula
Laboratory of Computer and Information Science
Helsinki University of Technology
Rakentajanaukio 2C
02150 ESPOO, FINLAND
Tel: +358 0 4513288, +358 0 4513271
Fax: +358 0 4513277, +358 0 4513277
E-mail: Haitao.Tang@hut.fi, Olli.Simula@hut.fi

Abstract

This paper introduces one method for the optimal utilization of the SCP which offers several kinds of services at the same time. There are three issues related to this method, i.e. the simulation of the traffic for the services, the approximation of the service-related system capacities, and the adaptation for the environment which is constructed mainly by the traffic for the different services. The adaptation (called SOM-D) is the main part of this paper. It consists of using Self-Organizing Feature Map to encode the time-varying environment and using one adaptive gap decoder to get the optimal gap sequences. Simulation is made to test the method. The results show that the system performance can be significantly improved by this adaptive method.

1 INTRODUCTION

Many system design problems are essentially resource management problems which can be done more efficiently if designer knows more about the on-going resource requirements. However, searching for the distributions of the on-going resource requirements is already very difficult sometimes, even if it may be possible. Then, approximation and adaptation are the logical results for dealing this kind of problems. It means that the suitable approximations of resource requirements and case-dependent system capacities as well as the needed adaptation method should be found. Based on those considerations, one method to increase the system efficiency of a Service Control Point (SCP) and its Quality of Service (QoS) is introduced in this paper.

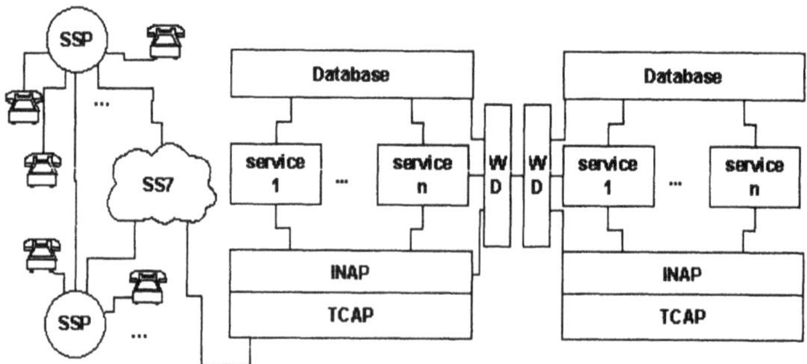

Figure 1 The Service Control Point

Figure 1 shows a SCP which supports several kinds of services at the same time. The WD (watchdog) in Figure 1 is the local manager of the SCP. It is introduced in [9]. The SCP is implemented as distributed processes inside a host computer. It is clear that the SCP is a time-sharing system.

There are many ways to model the SCP system. One may model it through the arrival of packets and the way of handling them (modeling at packet level). However, this way may not be quite suitable for the case of Figure 1. One reason is that it is very complex to model the sources (the caller) on the base of packet. Even though a call for certain service is closely related to certain sequences of packets, the caller (human being) will greatly affect system service rate by the delay of his manual operation. If one models the system through the arrival of calls and the way of handling them, the delay of human operation will be easier to trace and put into the account of modeling. Another reason is that complete tracing of packets seems an expensive way for the case of Figure 1 even if it is possible, which will use too much system resources so that the system capacity for real services is significantly reduced.

It is quite reasonable to use a straight forward way, modeling the system through the arrival of calls and the way of handling them (modeling at call level). This uses much less system resources than the packet level modeling, even through the call start, end, and service type of a call are needed to be checked from packet level. The parameters needed for modeling are also easier to measure in real situation. Then, the SCP system without any adaptive mechanism is modeled on the base of call as Figure 2.

Suppose that there are n different kinds of services in the system, named as service 1, service 2,..., and service n, in Figure 2. g_{h1}, g_{h2},..., and g_{hn} are fixed gaps in terms of maximal acceptable calls in unit time T for corresponding service. λ_1, λ_2,..., and λ_n are the total numbers of calls for corresponding service and arriving in unit time T; the numbers include dropped calls. The

system service rates for service 1, service 2,..., and service n are noted as μ_{h1}, μ_{h2},..., and μ_{hn}, respectively. μ_c is system capacity.
Since the system in Figure 2 uses fixed gaps, the optimal use of the system cannot be reached or the optimum could be reached for certain situation but not keeping the optimum when situation is changing. Therefore, using fixed gaps may not be a good way.

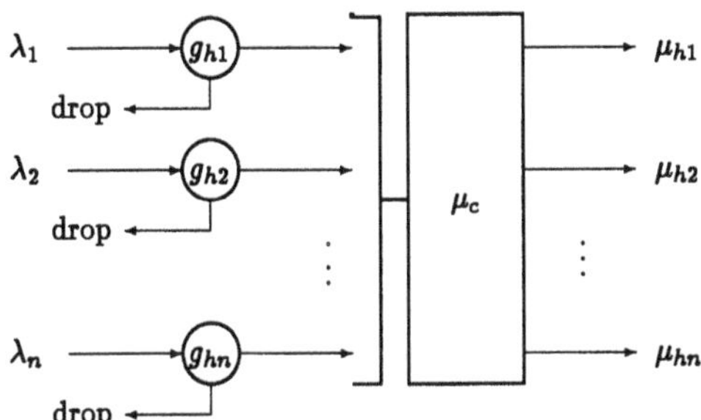

Figure 2 The model of the SCP without adaptation

If one tests the number of arriving calls for certain service in every unit time T, he can find that its distribution may not be stable. However, it is reasonable to believe [1, 2, 10], when the test is kept for days or even weeks or longer (T_t), certain relations can be found. The relations may be approximations of certain statistical relations. Figure 3 shows our simulation of calls for three different services in a day. Then, certain relations among the three different service calls are generated when the services are implemented in same host computer. In fact, those principles hold for any number of different services in the system.
By comparing the results from different services in Figure 3, one may be sure that, if he is able to encode the multi-dimensional relations into finite states, certain optimum can be reached through suitable decoding. The encoding and decoding processes are considered as an approximation of real situations. When situations change, the code book will change in order to adapt to the new situations and then, new optimum will be reached through the decoding. If the changes of situations are slow when compared to the updating rate of the code book, the newly updated code book can be used to approximate the situations in the coming period. In turn, the approximated optimum can be reached for the coming period. The optimum mentioned here is the global optimum in the period (T_t) between two successive code book updates.
The SCP system with adaptive gaps can be modeled as shown in Figure 4, which is also on the base of call like that in Figure 2; In Figure 4, g_1, g_2,..., and g_n are adaptive gaps which are the outputs of the gap-adapting block; μ_1, μ_2,..., and μ_n are service rates for service 1, service 2,..., and service n, which are affected by the adaptive gaps.

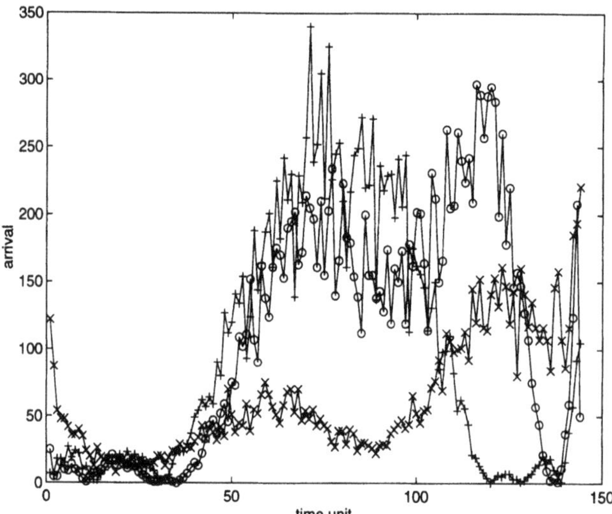

Figure 3 The simulated traffic for three different services in a day. X-axis is the unit time in a day and each unit time is ten minutes. Y-axis is the arrivals of calls for services. Line '-+-' is for service 1, line '-x-' for service 2, and line '-o-' for service 3.

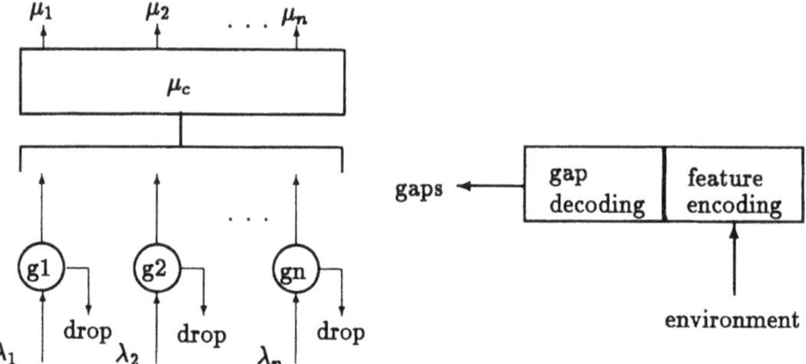

Figure 4 The model of the SCP with gap adaptation

2 THE ADAPTATION THROUGH SOM-D

In this section, there are three issues to be introduced: a kind of system capacity approximation, the adaptive encoding of the call environments through Self-Organizing Feature Map [3, 4], and the decoding for adaptive gaps of the SCP. The process of encoding and decoding together is called as SOM-D.

2.1 System Capacity

The system capacity (maximal number of acceptable calls in unit time T) for certain service is decided by the embodied capacity of the system and its service environments. In principle, it should be smaller than the embodied capacity of the system. Since we are interested in the system capacities for certain services, the latter will be simply called system capacity in the following context.

There could be many ways to measure the system capacity, e.g., from call level or packet level, etc. Even though each of the methods is approximative, some method could get the result which is closer to the real system capacity if certain conditions are considered.

The results from call level methods may not be good for a SCP system because the human behavior affects the call for service and it is changing with time, location, culture, etc. Therefore, for that case, the results may not be reliable and could be very expensive even if it is possible to measure through those methods. There are a lot of limitations which make those methods not workable practically.

Comparably, the methods from packet level seems more reliable for the results and more practicable. It is known that certain call for certain service is implemented by certain sequences of packets if looking from packet level. The sequences could have some change, but the results could not be possible to change too much if considering that, with very high probability, the call either is accepted and correctly processed or is refused back (drop out).

More reliable results can be obtained from the worst cases—there is no extra time delay existing, i.e. when caller SSP (Service Switching Point) receives some packet from the SCP, it can immediately return related packet to the SCP if needed. Certainly, the results are usually strict and have some distance from the system capacity. But, it is one of the most deterministic results which people could get.

One good side effect of using the strict capacity to approximate the system capacity is that some amount of system capacity will be not used. If the idle part is not too big (say, < 25% of the system capacity), it is good to be a safeguard for possible system congestion. The system delay can be reasonably low when accepting calls under the limit of those strict capacities. The experiences from real situations give people the confidence, it is better to put some safeguard for using the system capacity (i.e. let the system utilization factor $\rho < 1$).

One usable complement for the possible "too strict" could be the strict capacity updating; that is, when real call traffic exceeds the bound of the strict capacity, increase the strict capacity bound by a factor of γ, $1 < \gamma < 2$ and stop increasing before global service time delay starts to increase significantly. The newly found strict capacity could be set at 0.8 or 0.9 of the turning point. This method can decrease the distance between the strict capacity and the system capacity if needed.

Here is the way of measuring the strict capacities. It is to measure the strict capacities on the base that system serves one kind of service at a time. Through this way, the strict capacity for system to serve a certain kind of service alone can be found. The algorithm is the following:

(a) Find one single host computer for the SSP simulator which is not the SCP host and whose processing speed should be much faster than that for the SCP. If not much faster, the results should be adjusted on case-dependent base.
Let i=1.

(b) Choose the kind of service i in the SCP. The SSP simulator creates calls for the service and sends them to the SCP as well as keeps the traffic in a way, if one call is served by the SCP, a new call is sent to the SCP immediately. Simulator keeps this process by increasing m until the global service time of the system starts to increase significantly (the start of congestion). Then, hold this turning point (a little bit smaller than it) for the results measuring.

Continue the measuring for L unit time T. Let λ_{aij} be the accepted calls in j^{th} unit time T and λ^*_{ai} be the maximal acceptable calls in unit time T for service i. Then,

$$\lambda^*_{ai} \approx \frac{1}{L}\sum_{j=1}^{L}\lambda_{aij} \tag{1}$$

We know that the above test is on the base of ``one out and then one in''. Let μ^*_i be the system capacity for service i. If L is large enough, it is reasonable to assume,

$$\mu^*_i \approx \lambda^*_{ai} \tag{2}$$

Let i=i+1, if $i > n$, end (b); else iterate (b).

Now, the n strict capacities or maximal acceptable calls for the n services by the SCP are found. It is reasonable to say,

$$\mu^*_i \approx \mu_{C_i}, \textit{ (service i) (i=1, 2,...n)} \tag{3}$$

μ_{C_i} is the real system capacity for service i. Let $\mu^*_\alpha \approx min_i\{\mu^*_i\}$ and $\alpha_i = \frac{\mu^*_i}{\mu^*_\alpha}$. Then,

$$\mu_{C_i} \approx \alpha_i \mu^*_\alpha \textit{ (service i) (i=1, 2,... n)} \tag{4}$$

Thus, one may think that the SCP consists of the basic servers and each of them has μ^*_α service rate.

We know that the approximated system utilization has to be,

$$\sum_{i=1}^{n}\frac{g_i}{\mu_{C_i}} \leq 1 \tag{5}$$

Therefore, the upper bound for the maximal acceptable calls in unit time T can be,

$$\sum_{i=1}^{n}\frac{g_i}{\lambda_{ai}^*}\approx 1 \tag{6}$$

Because there are quite many possible solutions for (6), one should be able to find the solution which is the closest match of the environment at that time. The wanted solution can be found through (12) quickly. The details will be introduced in Section 2.5.

2.2 On Optimums of the System

Making Figure 4 as an example, in principle, one could find different system optimums from different cost functions which are built for different purposes, e.g., to get maximal global call throughput, to get maximal global call billing, or both of them, etc.

Suppose that one could find the following cost function;

$$\mathbf{E} = \mu\mathbf{a}^{\mathrm{T}} + \mathbf{Q}\mathbf{b}^{\mathrm{T}} \tag{7}$$

service rate: $\mu = (\mu_1, \mu_2,..., \mu_n)$
service cost: $\mathbf{a} = (a_1, a_2,..., a_n)$
staying cost: $\mathbf{b} = (b_1, b_2,..., b_n)$
average number: $\mathbf{Q} = (q_1, q_2,..., q_n)$

Then, try to find $min_\mu\, \mathbf{E}$ through (if it exists),

$$\frac{\partial \mathbf{E}}{\partial \mu} = \frac{\partial(\mu\mathbf{a}^{\mathrm{T}} + \mathbf{Q}\mathbf{b}^{\mathrm{T}})}{\partial \mu} = 0 \tag{8}$$

One may find that it is very hard (sometime even not possible) to find (7) or (8) from real systems as above. Its approximation may be very difficult to get, too. Thus, for the SCP case, directly trying this way is not suitable.

It seems that, if one can find certain approximation of the probability space of the system with its environments, the global optimums of the system could be approximated through certain adaptive method. The method should be directly based on the measured data from system. It makes the approximation and real time processing easier.

2.3 On SOM

The self-organizing feature map (SOM) is the Kohonen model [3]. It is a kind of unsupervised learning neural network and, in reality, it belongs to the class of vector coding algorithms [4]. It can learn the environment constructed by input vectors through itself.

The self-organizing feature mapping Φ has some usable properties as the following [4], (a) approximation of the input space, (b) topological ordering which relates to input patterns, (c) density matching which reflects the variations in the statistics of the input distribution. The property (a) and (c) are especially useful for the approximation of the probability space in the SCP case. For the details of SOM, please read the related materials in [3, 4].

As an example, a two-dimensional lattice of neurons of SOM is shown in Figure 5.

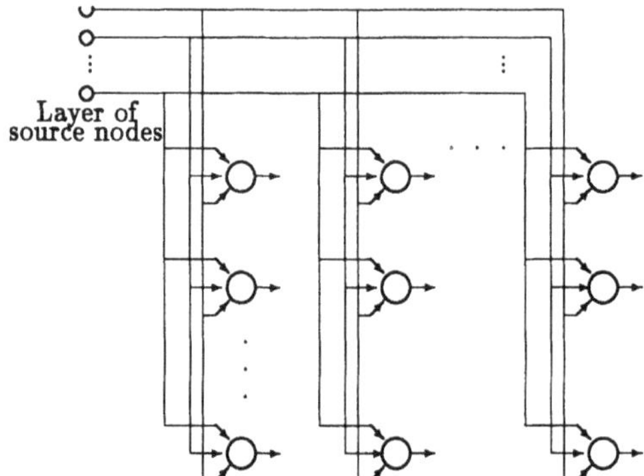

Figure 5 One example of two-dimensional SOM

The summary of SOM algorithm is listed as the following [4],

(A) Initialization. Choose random values for the initial weight vectors $\mathbf{m}_j(0)$. The only restriction here is that the $\mathbf{m}_j(0)$ be different for j=1, 2,..., N , where N is the number of neurons in the lattice. It may be desirable to keep the magnitude of the weights small.

(B) Sampling. Draw a sample $\mathbf{x}$ from the input distribution with a certain probability; the vector $\mathbf{x}$ is the sensory signal.

(C) Similarity Matching. Find the best-matching (winning) neuron $i(\mathbf{x})$ at time n, using the minimum-distance Euclidean criterion:

$$i(\mathbf{x}) = arg\ min_j \| \mathbf{x}(n)-\mathbf{m}_j \|, j= 1, 2,..., N$$

(D) Updating. Adjust the synatic weight vectors of all neurons, using the update formula,

$$\mathbf{m}_j(n+1) = \begin{cases} \mathbf{m}_j(n) + \eta(n)[\mathbf{x}(n) - \mathbf{m}_j(n)],\ j \in \Lambda_{i(x)}(n) \\ \mathbf{m}_j \text{ otherwise} \end{cases}$$

where $\eta(n)$ is the learning-rate parameter, and $\Lambda_{i(x)}(n)$ is the neighborhood function centered around the winning neuron $i(\mathbf{x})$; both of them are varied dynamically during learning for best results.

(E) Continuation. Continue with step B until no noticeable changes in the feature map are observed.

2.4 Encode Calls through SOM

Suppose that the results in Figure 3 are obtained through period T_t. Since customer behavior and situation are comparably "stable", it is reasonable to believe that the results tells the relations of global situations in the period T_t. If T_t ($1/T_t$ is updating rate of the code book) is comparably small and environment change is slower than the updating rate, it is also reasonable to believe that the newly updated code book can be used for the coming period, $(T_t + 1)$, in the way of "weekday for weekday and weekend for weekend, etc.". The environment changes in the coming period can be approximated as the quantitative changes of calls of all the unit time T in T_t. The call change can be quantized with a fuzzy coefficient β (9) and the training set S_t is constructed through (10). The t_h, t_d, and M in (10) are the time in a day, the time in a week, and the total number of sample vectors used. λ_{1j}, λ_{2j},..., and λ_{nj} are the measured call arrivals for service 1, service 2,..., and service n, respectively, in the *jth* unit time T.

$$\beta \text{ (finite set of positive values} \tag{9}$$

$$S_t = \{ ((\beta\lambda_{1j}, \beta\lambda_{2j}, ..., \beta\lambda_{nj}), t_h, t_d) \mid j = 1, 2, ..., M (\beta\} \tag{19}$$

The self-organizing feature map is constructed by two-dimensional lattice of neurons, e.g., in size of 64 x 64. It is properly initialized and then trained by the input sequence of the vectors which are randomly picked from the set S_t until the SOM reaches convergence phase. Then, the SOM can been seen as a code book for the environment. The environment in T_t is coded into the weights of the neurons. It is the optimal approximation of the environment in the sense of probability.

Now, the necessary conditions for constructing suitable decoding mechanism has been met, which produces the adaptive gaps. The gaps are optimally matching the traffic in global sense for the period T_t.

2.5 Decode for Adaptive Gaps

The idea of the SOM-D adaptation is shown in Figure 6. The algorithm of the decoding for the adaptive gaps is the following:

Let $\mathbf{m}_i$ be the weight vector of neuron i, $\mathbf{m}_i = (m_{1i}, m_{2i}, ..., m_{(n+2)i})$ and let $\mathbf{v}_j$ be one input vector from the environment in period (T_t+1), $\mathbf{v}_j = (\lambda_1, \lambda_2, ..., \lambda_n, t_h, t_d)$. Let $\mathbf{w}$ be the set which holds all the weights of the SOM.

(a) Input $\mathbf{v}_j$ into the SOM and then find the weight, $\mathbf{m}_j = (m_{1j}, m_{2j}, ..., m_{(n+2)j})$, which meets,

$$min_{\mathbf{w}} \| \mathbf{v}_j - \mathbf{m}_j \| \tag{11}$$

(b) decode them into the adaptive gaps for the unit time T^{jth} by,

$$g_k = round_{up}(\lambda^{*}_{ak}\,(\frac{m_{kj}}{\sum_{l=1}^{n} m_{lj}})), k = 1, 2, \ldots, n \tag{12}$$

(c) use these gaps for next unit time $T^{(j+1)th}$.
In fact, it is not needed to care the outlooking of the SOM for the adaptive coding and decoding. It is good for no need of extra-intervention.

The decoding in (12) has several characters. It constructs the strict upper bound by its gaps from decoding (see (6)) and optimally matches the environment in the sense of probability by its gaps. It gives optimal throughput and also prevents possible system congestion.

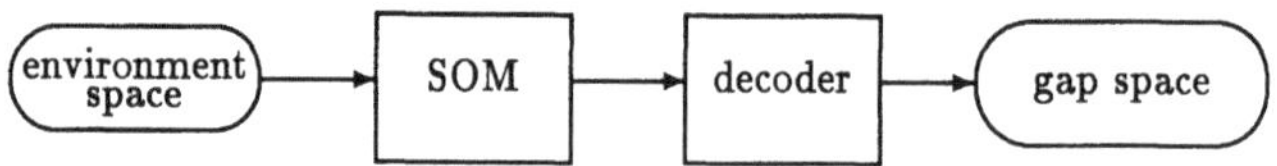

Figure 6 The principle of the adaptation method

2.6 SOM-D with Priority

The above SOM-D treats any service kind with equality in the sense of the fair share of system capacity. However, for some cases, e.g., emergency service, important televoting, etc., one has to decrease the load from other services by force if needed. It means that services should be tagged with different priorities for those cases. It also means that a compromise is required between the optimum and the wanted unfairness. This problem can be solved through the reconstruction of the gap space, etc. This part is not, however, introduced in this paper.

3 SIMULATION RESULTS

For simulation, three different kinds of simulated services are tested in our simulation SCP implemented in PC 486 with Linux OS. The traffic of ten working days is simulated for each service. Figure 3 shows one day's traffic. The first five days' traffic is used for training the SOM-D. Then, the next five days' traffic is used for the performance simulation of the SCP with SOM-D. One software packet is used for the SOM [6].
The SOM is updated every five days (omiting weekends at this moment). If the overall traffic distributions change slower than map update does, SOM-D works well. Otherwise, the fixed gaps can be used temporarily until SOM-D matches the call environment. The adaptive gaps are updated every ten minutes. Figure 7 shows the sequences of gaps in one day. One can see that the three sequences of gaps follow the potentials of the simulated traffic and compromise with each other in order to hold the upper bound of the maximal acceptable calls in unit time T for the call environment. The upper bound also prevents system from possible congestion.

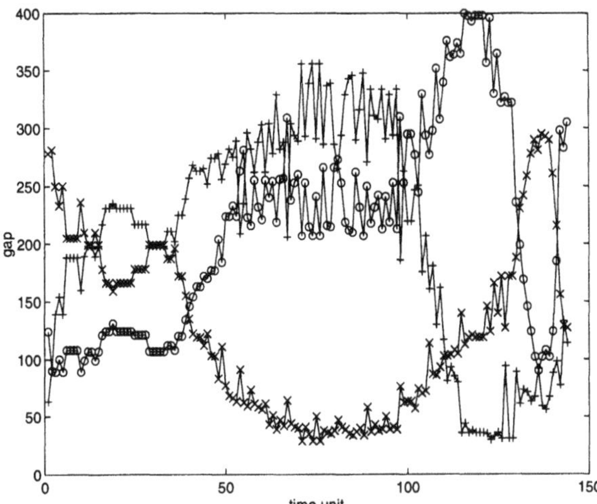

Figure 7 The sequences of the adaptive gaps for different services in a day. X-axis is the unit time in a day and each unit time is ten minutes. The gaps are updated every unit time. Y-axis is the maximal acceptable calls for the services at each unit time. Line '-+-' is for service 1, line '-x-' for service 2, and line '-o-' for service 3.

For performance simulation, different levels of traffic are simulated. Since no congestion can happen under the SOM-D, reasonably low system delay is guaranteed. Figure 8 shows the throughput improvement by using the adaptive gaps. For a better understanding of the improvement, the system delay, when the SCP has no any congestion control mechanism, is simulated with different traffic levels; it is also shown in Figure 8.

From Figure 3, one can find that the traffic in a day are unbalanced. It means that some amount of system resource is not used in certain times of the day, even through the system can be very busy in other times of the day. This kind of unbalance is caused by customers. Similar unbalance has been seen in many papers [1, 2, 5, 10], etc. It is very clear that one cannot count on it to improve the system performance. However, it is found that certain unbalances exist among the traffic levels from different service calls most of the time. The unbalances among the traffic levels offer us the opportunity to improve the system performance.

From Figure 8, one can find that, if using fixed gaps, the system throughput decreases far before the onset of the potential system congestion (relative traffic level 2.1 to 2.5, if happened). When using adaptive gaps, the system throughput is improved, the throughput band is expanded and it is closer to the onset of potential system congestion. When traffic is small, the improvement is small (the only possibility). The improvement increases with the increase of traffic until certain point. Then the improvement keeps a certain value (about 20% in Figure 8) and starts to decrease slowly. There may be two reasons that the improvement dose not increase linearly with the traffic. One is that there are still remaining differences between the approximation from SOM-D and the call environment. The other is that, when traffic increases after certain level, more and more calls have to be dropped in order to keep system from congestion. Thus, in fact, SOM-D functions as both optimal system utilization and flow control mechanisms.

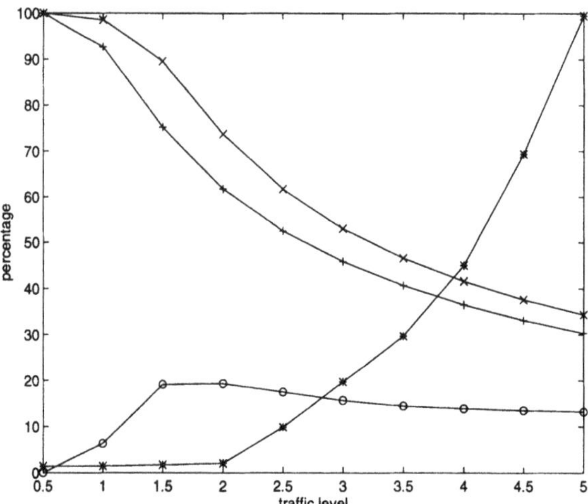

Figure 8 The performance improvement for the simulated traffic through SOM-D. X-axis is relative traffic level by comparing with certain amplitude of the simulated traffic. Y-axis at left side is percentage. For line '-x-' and line '-+-', the Y-axis means the percentage of real acceptable calls by comparing to the arrivals of calls for services (we call the percentage as effective throughput or, simply, throughput. It is a little bit different from the traditional throughput measurement). Line '-x-' is the throughput by using SOM-D. Line '-+-' is the throughput by using fixed gaps. Line '-o-' is the throughput improvement by comparing the throughput of SOM-D with that of fixed gaps. The y-axis at right side means system delay in second. Line '-*-' is the system delay which is measured when system has no any flow control mechanism used. We use it to help the analysis of the other results only.

In addition, there is still some difference between the improved throughput band and the onset of potential congestion (if no flow control mechanism used). The main factors for this difference are the unbalance of the traffic along the time and the distribution among the traffic levels. These two factors are decided by call environment (or the customers). It means that the call environment decides the upper bound for performance improvements. What we can do is to push the system performance closer to this upper bound.

4 CONCLUSION

Usually, when designing certain telecommunication systems, more attention is paid to using more powerful components than to the optimal operation of the system in case that the system performance is not good enough. We do know that the system capacity has to be increased in order to increase its performance under some situations. However, we have many experiences that, even if the average system utilization is very low, the performance of the system is sometimes starting to degrade very fast. Indeed, there may be many factors which cause this situation. System configuration, both static and dynamic, can be one of them. It means that only increasing system capacity may not be enough or so efficient and feasible.

In this study, one method for measuring service-dependent system capacity and one adaptive gap updating mechanism are introduced. The performance improvement is also proved through the simulation. In the developing of the methods, the capacity of SS7 is supposed to be sufficient for simplicity in the analysis.

From the simulation, it is demonstrated that system throughput can be improved and, at the same time, low system delay can be guaranteed as well as potential system congestion can be prevented; all these are done without necessarily using more powerful components. We hope this study can attract some attentions into the optimal management of some telecommunication systems, too.

We know that to upgrade the performance is also part of the efforts to increase Quality of Service (QoS) which is visible to customer. Then, the use of SOM-D for adaptive gaps helps to upgrade the serviceability performance of the SCP [7, 8].

Indeed, the SOM-D method adds extra load to the system, even if it is comparably small. Fortunately, in most cases, SCP is implemented with double systems, one active and the other standby at the time, in order to guarantee its service availability [9]. The processing load from SOM-D can be put into the standby system. The active system executes only gap updating. Other needed operations are the same as in the SCP without SOM-D. As a result, it means that hardly any extra load is added into the active part of the SCP, and thus, nearly no extra overhead is added.

5 ACKNOWLEDGMENT

We are grateful to Tapani Karttunen (Telecom Research Center/Telecom Finland Ltd) for his helpful comments on this paper and the valuable information from the cooperation between us and his group. We also like to thank Kimmo E. E. Raatikainen (Dept. of Computer Science/Helsinki University) for his valuable comments on the paper.

6 REFERENCES

[1] R. I. Wilkinson, "Theories for toll traffic engineering in the U. S. A.," Bell Syst. Tech. J. vol. 35, no. 2, pp.421-514, Mar. 1956.

[2] W. S. Hayward and J. P. Moreland, "Total network data system: Theoretical and engineering foundations, " Bell Syst. Tech. J., vol. 62, no. 7, pp. 2183 - 2207, Sept. 1983.
[3] T. Kohonen, "Self-Organizing Maps" Springer-Verlag, 1995.
[4] Simon Haykin, "Neural Networks," Macmillan College Publishing Company, Inc., pp. 397 - 434, 1994.
[5] Henry J. Fouler and Will E. Leland, "Local Area Network Traffic Characteristics, with Implications for Broadband Network Congestion Management," IEEE J. on Selected Areas in Communications, vol. 9, no. 7, pp. 1139 - 1149, Sept. 1991.
[6] Kohonen T., Hynninen J., Kangas J., and Laaksonen J., "SOM-PAK, The Self- Organizing Map Program Package, ver. 3.1," Mar. 1995.
[7] E.7IN1. ITU-SG2, "Draft Recommendation".
[8] CCITT Recommendations Q.1200 series, Intelligent Networks, Geneva, Switzerland (1992).
[9] Tang Haitao and Karkkainen Esa, "The Local Management for A Service Control Point," IFIP TC6 Workshop on Intelligent Networks, vol. I, pp. 45 - 55, Aug. 1994.
[10] EURESCOM P308, "Methods and Specifications for Tools to Dimension Intelligent Networks, Deliverable 2, IN Service Model and Traffic Performance Analysis", Volume 1 of 2: Main Part, March 1995.

7 BIOGRAPHY

Tang, Haitao was born in Kunming, China, Nov. 1961. He received the Master's Degree from Nanjing Institute of Telecommunications in 1986. Since then, he worked at Beijing Data Communication Co. as a researcher for designing network equipment until 1992. From 1993 to 1995, he worked as a researcher of Intelligent Networks at Telecom Finland Ltd. Currently, he is a researcher and a senior postgraduate student working towards his Ph.D. at the Laboratory of Computer and Information Science in Helsinki University of Technology. His current research interest is on the adaptive methods for telecommunication systems.

Olli Simula was born in Helsinki, Finland, on July 22, 1948. He has received the degree of Doctor of Technology from Helsinki University of Technology in computer science in 1979. Since 1974 Dr. Simula has been with the Laboratory of Computer and Information Science at the Helsinki University of Technology, being currently Associate Professor of Computer Science. During 1977-78 he worked one year as a Research Fellow at the Delft University of Technology, Delft, the Netherlands. His research interests include digital signal processing and computer architectures for signal processing as well as neural networks and their applications in process monitoring and analysis.

Part Five
Standards and IN

15
The Telecommunication Standardisation Process: Can it be 'Reformed' to Support 'De-Regulation'?

Richard Hawkins
Science Policy Research Unit
Mantell Building, University of Sussex, Falmer, Brighton
UK BN 1 9RF
Tel: +44 1273 678165, Fax: +44 1273 686758
E-Mail: r.w.hawkins@sussex.ac.uk

1 INTRODUCTION

The current conditions under which the telecommunication industry is 'governed' - in the dual sense of being both 'controlled' and 'enabled' - are described in a plethora of often ill-defined and frequently contradictory terms. One of the most ill-defined terms is 'de-regulation', implying that the role of regulation is diminishing, and that the quantity of regulations is lessening - implications that are contrary to fact in most instances. The widespread acceptance of 'de-regulation' as an operational concept in the telecommunication industry, however, has some especially significant ramifications as the technical configuration of both public and private networks becomes more decentralised - i.e. as the 'intelligence' controlling individual network functionalities becomes distributed throughout the network. One assumption supporting the 'de-regulation' concept is that technical co-ordination can now best be achieved through an industry-led standardisation process, rather than through the formal controls of a public administration or regulator. Timely, non-proprietary standards, it is argued, will keep the network environment 'open' to potential new market entrants. The distribution of network 'intelligence' throughout the public network, however, can only be achieved through the widespread deployment of specialised computer applications. This also distributes much of the control over the development and evolution of standards, as, increasingly, public network operators and equipment suppliers must share this control with computer and software vendors.

All of this raises three immediate problems. *First,* although the technology bases of the telecommunication and computer sectors have converged, it is uncertain that the commercial cultures of the two sectors have converged to the extent that co-operation between them in defining standards will be possible or productive. *Second,* the computer sector does not have a particularly strong record of producing and implementing non-proprietary networking standards. Indeed, many of the interconnection and interoperation problems that for years have bedevilled the private computer networking environment may well migrate to the public network arena. *Third*, as standards for public and private network environments converge, it will become more and more difficult to distinguish the attributes of what we now recognise as the 'public network' from amongst the myriad of private networks aimed at closed user groups.

This paper uses these three problems as the context for examining critically the relationships between technical standards, regulation, and the evolution of new network forms. The objective is to assess the industrial and political motivations for reforming the standardisation process, and to outline some of the problems that ensue.

2 TECHNICAL STANDARDS AND REGULATORY REFORM

Prior to examining the dynamics of standards and standards-making in the 'intelligent' network environment, it will be helpful if we first clarify some basic features of the regulatory environment for telecommunication, and then examine briefly some of the primary relationships between standards, regulation, and public policy.

2.1 The 'phantom' regulatory taxonomy

If we look critically at the terms 'regulation', 'de-regulation', and 're-regulation' in a European context, it soon becomes clear that the first term is historically ambiguous and that the latter two are redundant to a considerable extent. It was noted above that 'de-regulation' is already something of a misnomer as the amount of regulation is not decreasing, even if much of its focus is shifting from administrative and operational matters to commercial practices and market structures.

The term 're-regulation', however, is even more ambiguous, in that it carries the tacit assumption that 'regulation', by some broadly accepted definition, is already in place. Arguably, however, there is a difference between the rationales of regulating a private monopoly in order to protect the public interest from potential abuses of a privileged market position (the US rationale), and the appropriation of an entire industry by government administrations, on the premise that the only appropriate forum for the exercise of monopoly power is the public sector (the 'PTT' rationale).

Ostensibly, the European PTT system was directed at providing a 'public service', and it can be questioned whether 'regulation', in the common sense of 'market regulation', was ever actually applied. Indeed, it is probably more useful in the present circumstances to eliminate the 'de-regulation' and 're-regulation' terminology altogether. Perhaps the least ambiguous description of the European situation is that a set of regulatory institutions is now being constructed in order to 'regulate' for the first time a set of new or evolving commercial and industrial relationships.

In this process, the role of standards and the effectiveness of standards-making mechanisms is becoming pivotal to the regulatory agenda, and subject to intense scrutiny. Indeed, the use of standards in a 'de-regulation' context is a contradiction in terms, in that, by nature, standards are a type of 'regulatory' tool. At issue is the *nature* of the act of regulation through standardisation, and its relationship to public policy goals.

2.2 Standards, regulation and policy

The imposition by a government authority of mandatory reference to industry technical standards - either as developed within the industry concerned or by some outside body - is an explicit act of 'external' industry regulation in that it restricts and/or directs the future actions of the implementors. However, even the use of entirely voluntary standards also constitutes an 'intervention' in a process that might otherwise have a different outcome. Voluntary standardisation is therefore also an act of regulation. Moreover, in institutional terms, both forms of regulation by standards can be closely linked. An industry can elect, for example, to develop and apply a voluntary standard in preference to having one imposed upon it by some external entity. Or, a public authority can agree to accredit industry-developed standards on the understanding that the industry actors are the only source of reliable information and expertise concerning a particular technology.

The problems begin to occur when the process of developing voluntary standards by industry becomes linked with government regulatory frameworks and public policy objectives. Voluntary standards can be used to achieve public policy goals, but there is no natural affinity between these goals and voluntary standards (Breyer 1982; Reddy 1990). Indeed, to pursue public policy goals with voluntary instruments is often to invite additional measures of uncertainty as to the possible outcomes.

Voluntary standardisation is not a random process, but it is frequently subject to a range of internal and external pressures that are beyond the direct control of governments and publicly accountable regulatory agencies (Hawkins 1995). In terms of policy expectations, a 'bad' outcome from an industry standardisation process is often compounded in that, by delegating this degree of regulatory power to the private sector in the first place, the publicly accountable body may lose a considerable measure of ability to correct outcomes that are suboptimal in terms of achieving policy objectives (Baggott 1986).

Voluntary standards occupy an especially troublesome position as tools of industrial policy. The aim in this instance is to achieve much more than technical co-ordination and the attendant economies of scale and scope that are commonly assumed to flow from technical co-ordination through standards (Lecraw 1984; Hawkins 1995a). As industrial policy tools, voluntary standards are expected also to define and consolidate discrete technology initiatives, and to commit producers and consumers alike to co-ordinated trajectories of technological development.

European Union (EU) standardisation policy in the information and communication technology sector has always had a strong industrial policy focus (Barry 1990). In the telecommunication sector, this has yielded a limited number of positive outcomes of which the Global System for Mobile communications (GSM) is probably the most prominent. It has also resulted in two very negative general consequences. First, the EU has imposed virtual production quotas on the European standards development mechanism, which, while congruent with policy aims for the sector, are not always compatible with the capabilities and strategies of European firms. This has resulted in a proliferation of standards that no one wants to use. Second, confusion over the industrial or political origin of the standards programme, and over the eventual implementation status of the standards produced (i.e. voluntary or mandatory), has led to doubts about their technical content and quality (Hawkins 1993).

2.3 Standards and the changing focus of telecommunication regulation

In examining the changing relationship between standards and regulation in today's rapidly evolving network environment, it is important to identify two basic regulatory phases. In the *first phase*, regulation (in a variety of forms and by a variety of definitions) was based on the assumption that monopolistic and oligopolistic structures were the only logical paradigms for the provision of telecommunication services.

Thus, prior to the pressures for technological and structural change that began in earnest in the 1980s, most telecommunication 'standards' were, in effect, little more than internal procurement specifications as worked out between individual monopoly public network operators and preferred, often vertically integrated, equipment supply firms. The exceptions occurred at the international level, where technical 'recommendations' were agreed in the international consultative committees (the CCITT and CCIR) of the International Telecommunication Union (ITU), and in various regional associations of public network operators like the European Conference of Posts and Telecommunications Administrations (CEPT).

This process was highly 'internalised' within the telecommunication sector, and direct participation was exclusive to recognised national public network operators. The result was the preservation of national idiosyncrasies in the technical configuration and service base of the public network.

The *second phase* of regulation flowed from such major events as the divestiture of AT&T, the privatisation of British Telecom, liberalisation measures in Japan, and the official embrace by the EU of the principle of competitive, pan-European telecommunication markets. In this case, the policy focus shifted from preserving monopoly structures to officially discouraging them. This new rationale was bolstered by a sustained period of rapid and radical technical change in the industry, and the resulting commercial potential of a much expanded service base.

Acceptance of the principle of liberalisation, however, has not been matched by practice. With few exceptions, most European countries preserve some variant of the PTT system. With respect to the technical co-ordination of networks, however, public policy-makers have for the first time actively begun to promote the elimination of administrative and technical idiosyncrasies in national public networks. Pressure is now applied for the development of more standards that refer to general network attributes on both sides of an international interface, and that are orientated to international application from the outset.

Thus, the standards development mechanism is now 'externalised' to a considerable extent. During the 1980s, 'regional' standards organisations began to appear that allowed for direct, independent participation by telecommunication equipment suppliers as well as network operators, and by many non-traditional actors in the sector - computer firms, software developers, consultancies and so forth (Hawkins 1992). The current process is now mostly harmonised with the principles of voluntary consensus standards development as practised in other major industries.

Officially, this new regional regime is directed at making the ITU process more efficient and responsive. The three major regional bodies - T1 in the US, the Telecommunications Technology Committee (TTC) in Japan and the European Telecommunications Standards Institute (ETSI) - are committed in principle to feeding co-ordinated regional positions into the ITU processes. The new ITU Standardisation Bureau, however, now allows for direct participation by the same kinds of constituencies as represented in the regional bodies. Promises to maintain the ITU as the highest international telecommunication standardisation authority - a rather vague distinction in today's technological and service environment - still beg the question of how to differentiate the respective 'missions' of the ITU and the regional bodies to an extent sufficient to prevent costly duplication of their respective activities.

There have been diverse reactions to this new environment. National public network operators face an acute dilemma. On the one hand they have a considerable interest in promoting standards that continue to protect their established sources of revenue, or that give them advantages over new entrants in expanding the service base. On the other hand, increasing opportunities to become involved in international market ventures provide incentives to opt for more 'open' network structures.

For very different reasons, the reaction of non-traditional actors in the telecommunication sector to the new regime is also mixed. There is support in principle for the opening up of the process, but dissatisfaction with its responsiveness, and with the residual domination of the process by incumbent actors (OECD 1995). The result is frequently that standardisation programmes of relevance to public network environments, but in which actors in the computer sector are the technology leaders, are siphoned off into private computer industry consortia that can be less accessible and responsive to participants from the public network arena.

The incentive to support the regime centered in T1, TTC and ETSI is perhaps strongest among established telecommunication equipment suppliers. High and costly R&D intensities, related both to existing and new product lines, are increasing the pressures to open up new international markets. These new markets are usually much less vertically integrated than long-established markets, thus increasing the incentive for equipment suppliers to agree on international standards.

The real question is whether 'standards', as developed in the present environment, are in fact a new phenomenon in telecommunication. In many respects, the 'standards' as produced by ETSI or T1 are instruments of a different order to the old CCITT or CEPT 'recommendations', and they reflect a fundamental re-ordering of many institutional relationships in the industry.

3 STANDARDISATION AND NETWORK EVOLUTION

As networks become more 'intelligent', in the sense that the service provision and network management functions become resident in decentralised computer applications, the role of standards expands from merely enabling interconnection and interoperation, to determining the conditions under which this distribution of 'intelligence' will be configured and to what effect. In this context, the source of the standards - public network operators, equipment suppliers, computer vendors and so forth - becomes a new and critical factor. This point can be illustrated by looking at virtually any new development in network technology, but space permits only a look at a couple of key examples.

The Asynchronous Transfer Mode (ATM) represents the first step in an approach to networking that might eventually eliminate altogether the requirement for large, centralised switching facilities. In the process, ATM and its successors could also eventually circumvent the 'intelligent network' architecture as being presently implemented by many national public network operators. Although originally conceived in the public network arena and formally defined in the CCITT, ATM has appeared first in private data networking applications. The lead standardisation body is now the ATM Forum, a US-based private consortia in which the principal actors are (mainly US) computer and software vendors. Arguably, primary control over perhaps the most significant philosophical change in the configuration of large networks, public and private, has already been wrested away from the 'traditional' telecommunication sector.

The second example is Computer-Telephony Integration. Although CTI is made possible in the first instance by CCITT Signalling System 7, control over the protocols for PC-based CTI applications has been concentrated in the computer sector, and current standardisation initiatives are being spearheaded by a handful of specialist companies. In 1993, the US firm, Dialogic, succeeded in organising an international advisory council around its nominally proprietary Signal Computing System Architecture (SCSA), and now claims that over 350 firms world-wide support this standard. SCSA is independent of hardware and software vendors, and claims now to support a full range of telephony services in distributed computer network environments. Links between CTI, integrated broadband networks, and ATM-style network configurations portend major changes in the service and control environments of the public network, but public network actors appear to be influencing this process in only minimal ways.

The prospect of a network with fully distributed switching and management functionalities creates a new kind of standardisation problem if some semblance of a 'public network' environment is to survive. The historical situation has been for private networks to begin as 'add-ons' or 'overlays' to the public network infrastructure. We may eventually be faced with an environment that is an accumulation of basically private networks, within which a level of non-discriminatory 'public' access can only be assured by regulatory means, tied closely to assuring the provision of non-proprietary standards for key network interfaces.

4 CHANGING PERCEPTIONS OF THE STANDARDISATION PROCESS

Traditionally, standards developers and implementors alike have viewed the process in terms of '*milestones*'. Standards were perceived generally as technical documents that appeared at controlled intervals, and that defined agreed common approaches to technology implementation at distinct developmental stages. In telecommunication, the 'milestone' approach was perhaps appropriate to the old analogue technical environment that tended to evolve in slowly paced and relatively discrete increments, and in which there was close co-operation between operators and suppliers over long periods of time with respect to network planning and design.

The evidence that 'milestones' were actually observed in this sector, however, is tenuous. The fact, for example, that the CCITT recommendations were worked out in a small, concentrated expert community, meant that they could be 'phased in' to ongoing network planning processes while still in draft form, and that flexibility could be built in to the recommendations in order to accommodate existing 'national conditions', at least in the larger telecommunication markets.

In the very dynamic digital environment facing the industry today, the 'standard' has become a very much more 'fluid' instrument of technical and commercial co-ordination. In essence, standardisation should now be perceived much more as a 'process' rather than as a 'state'. Increasingly, industry actors become involved in standardisation not to define discrete conditions as fixed in time, but, as illustrated in Figure 1, to determine on a dynamic basis the benchmark below which the parallel development of technology is perceived to be inefficient, and/or technology-based competition is perceived to be redundant (Tassey 1991; Hawkins 1995b).

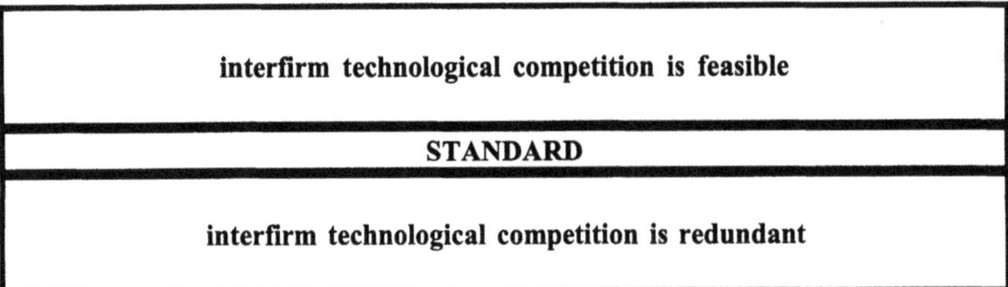

Figure 1

In the current commercial and technological environment, it makes more sense for firms to support a '*mapping*' approach to standards. In an environment of technical change, this approach is orientated to the maintenance of levels of inter-firm agreement on what constitutes the generic aspects of the technology. Thus, although 'milestones' still serve a function, it is now more accurate to perceive standards as 'living documents' and standardisation as an ongoing and dynamic process of information exchange between competing firms, rather than as a process orientated completely to achieving specific and documentable agreements (OECD 1995).

4.1 Pressures on the standardisation process

Given that many telecommunication standards have always been prone to flexible implementation conditions, the most significant new factor is that sponsorship of these standards now emanates from a much greater and more diverse range of sources. This results in pressures from many directions for 'reform' of the process. The problem in many instances is that there is a lack of perspective concerning (1) the limitations on what the standardisation process is actually capable of achieving, (2) the internal conditions necessary for the process to yield successful outcomes of any kind, and (3) the nature of the institutional relationships between standardisation organisations, industry and government. As a result, the outcome of 'reform' to date has been mostly confusion.

Pressures related to 'de-regulatory' assumptions and expectations have encouraged greater participation in the standards process, but, correspondingly, have also made consensus agreements more difficult to achieve (Besen 1995). Moreover, public sector expectations of the process - especially on the part of the EU - have been overwhelmingly 'milestone' orientated (European Commission 1990).

The 'standards problem' has been perceived, by industry and government bodies alike, in terms of the 'need for speed' in the face of rapid change in technological and market conditions. Whereas few would argue with the premise that standards-making can be improved somewhat by organisational adjustments, concentrating these adjustments on project management objectives, and the production of increased quantities of documents runs contrary to the nature and purpose of standardisation in dynamic technological and commercial environments as described above.

Part of the rationale for the proliferation of private standards consortia in the computer sector is that by concentrating participation among key industry actors, and by restricting the scope of a consortia to a limited range of technologies, positive outcomes from the standardisation process should appear more quickly, and rapid implementation of the standards should be assured. Problematically, however, evidence is accumulating that consortia are in many cases no more or less efficient than committees in the already established standards development organisations, and that uncertainty about the balance of sectoral inputs into the consortia leads to 'legitimisation' problems as the standard enters the market.

Most importantly, the mere proliferation of standards organisations at national, regional and international levels, has created a major co-ordination problem, and there is evidence that this is leading to a reluctance on the part of individual firms to maintain or increase their support of voluntary standards initiatives (OECD 1995). The point is rapidly being reached when the costs of co-ordinating non-proprietary standards are perceived to be approaching the costs of dealing with the very proprietary standards that they are supposed to supplant or prevent.

5 CONCLUSION

As is so frequently the case, the seemingly 'obvious' question in the title of this paper, turns out to be the 'wrong' question. The seeds of its demise lie first in the ambiguity surrounding the term 'de-regulation', and then in the questionable tacit assumption that 'self-regulation' through technical standards is necessarily a sufficient (much less complicated and more responsive) alternative to formal regulation where the basic configuration of the public network is at issue. The matter of 'reform' is similarly ambiguous in the light of the many contrasting perspectives that exist concerning the quite radical change that has already occurred in the standards-making mechanism for telecommunication.

As the structure of the telecommunication industry began to be widely questioned in the early 1980s, and as the goal of a 'liberalised' regime began to be promoted by a number of governments, the basic *problématique* was articulated in terms of how to maintain the technical integrity of the public network in the face of the rapid entry of new actors and technologies into the formerly tightly controlled environment of the public network. The public sector concern at that time was that 'the market' might not yield the appropriate standards in a timely enough way to support the new regulatory objectives focused on encouraging liberalised conditions for entry into telecommunication markets. As it turned out, the problem is not that industry does not produce the standards, but that they produce them in numbers that are so great, and in organisations that are so numerous, that implementation of the standards becomes uncoordinated.

In order to maintain any semblance of a public network environment in the future, the challenge is to re-link the production of standards with the planning and co-ordination of networks in the new regulatory environment. For this to be accomplished, the fiction that voluntary standardisation can somehow act as a surrogate for more direct forms of regulatory action must be discarded, and the internal and external regulatory functions of standards must be recognised and brought into balance. In this endeavour, a new role for public sector intervention must be defined, one that is focused on the task of encouraging mediation within the diverse and widening spectrum of actors and interests in the contemporary telecommunication industry.

6 REFERENCES

Baggott, R. (1986) 'By Voluntary Agreement: The Politics of Instrument Selection', *Public Administration*, Vol. 64, Spring, pp. 51-67.

Barry, A. (1990) 'Technical Harmonisation as a Political Project', in G. Locksley, (ed.), *The Single European Market and the Information and Communication Technologies*, London: Bellhaven.

Besen, S. M. (1995) 'The Standards Processes in Telecommunications and Information Technology', in R. Hawkins, R. Mansell and J. Skea (eds.), *Standards, Innovation and Competitiveness: the Politics and Economics of Standards in Natural and Technical Environments*, Cheltenham: Edward Elgar, 1995, pp. 136-146.

Breyer, S. (1992) *Regulation and its Reform*, Cambridge Mass.: Harvard Univ. Press.

European Commission (1990) Commission Green Paper on the Development of European Standardization: Action for Faster Technological Integration in Europe, COM(90) 456 final, Brussels, 8 October.

Hawkins, R. (1995) Standards for Communication Technologies: Negotiating Institutional Biases in Network Design', in R. Mansell and R. Silverstone (eds.), *Communication by Design: The Politics of Information and Communication Technologies*, Oxford: Oxford University Press (forthcoming).

___ (1995a) 'The Public Sector Role in the Development of Information Technology Standardization Strategies', *OECD STI Review,* Summer, (forthcoming).

___ (1995b) 'Standards-Making as Technological Diplomacy: Assessing Objectives and Methodologies in Standards Institutions', in R. Hawkins, R. Mansell and J. Skea (eds.), *Standards, Innovation and Competitiveness: the Politics and Economics of Standards in Natural and Technical Environments*, Cheltenham: Edward Elgar, pp. 147-159.

___ (1993) 'Changing Expectations: Voluntary Standards and the Regulation of European Telecommunication', *Communications & Strategies*, No. 11, 3rd Quarter, September, pp. 53-85.

___ (1992) 'The Doctrine of Regionalism: A New Dimension for International Standardization in Telecommunication', *Telecommunications Policy,* Vol. 16, No. 4, May/June, pp. 339-353.

Lecraw, D. J. (1984) 'Some Economic Effects of Standards', *Applied Economics*, Vol. 16, pp. 507-522.

Organisation for Economic Co-operation and Development (1995) *ICT Standardisation in the New Global Context*, Paris: OECD Directorate for Science, Technology and Industry, Committee for Computer and Communications Policy, DSTI/ICCP(95)2, 29 March.

Reddy, M. N. (1990) 'Product Self-Regulation: A Paradox of Technology Policy', *Technological Forecasting and Social Change*, Vol. 38, pp. 49-63.

Tassey, G. (1991) 'The Functions of Technology Infrastructure in a Competitive Economy', *Research Policy*, Vol. 20, pp. 345-361.

7 BIOGRAPHY

Dr. Richard Hawkins is Fellow in the Centre for Information and Communication Technologies, Science Policy Research Unit, University of Sussex. His academic research focuses on standardization processes and network interconnection issues. He has undertaken research and consultancy for the Government of Canada, the British Standards Institution, the European Commission, the UK Department of Trade and Industry, and the London Metropolitan Police Service. He has been consultant to the OECD on computer and telecommunication standards, and on EDI, and is currently advisor to the OECD High Level Group of Industrial Experts on Electronic Commerce.

16
Formal Description of Services*

Yuzhang Liu, Fangchun Yang, Junliang Chen
Beijing University of Posts & Telecommunications
BOX 206#, Beijing, 100088, P.R. of China
Tel: +86 10 2282007, Fax: +86 10 2022770
E-mail: yzliu@bupt.edu.cn

Abstract

On the basis of reviewing the basic concept of IN and life cycle of the service, this paper gives the concept of SLB, and formally describes the service in the syntax, behavior, and semantics way, which provides an approach to resolve the service interaction in the service specification phase.

Keywords

IN, SCE, Service Interaction.

1 INTRODUCTION

Service is a concept of Intelligent Network that the telecommunication network provides to the users. A service provides stand-alone functionality for some commercial purposes.

The aim of IN is to introduce services quickly and deploy the services onto the network rapidly. To realize this, ITU-T defines four layered IN Concept Model (INCM). The first plane of INCM is the service plane. The service plane represents an exclusively service-oriented view. The IN service is also called the supplement service which is based on the POTS (Plain Old Telephone Service) and adds value to customers. The second plane is the global functional plane (GFP), where the service independent building block (SIB) is defined and used to construct a service.

In the IN recommendation CS1 (Capability Set 1), ITU-T has defined 25 services, some of them are widely used, such as Freephone service, Account Card Calling service, Call Forwarding service, and Originating Call Screening service etc.

The service has its own life cycle (the life cycle of the service is shown in Figure 1.)

* Sponsored by Doctoral Research Fund of the State Education Commission of China.

- Requirement: Specify the requirement and behavior of a service by formal language, such as SDL, MSC, TTCN.
- Creation: Create the service according to the requirement of the service by SIB-based language in Service Creation Environment (SCE), including service testing and service validation.
- Deployment: Deploy the service to telecommunication network, then the software of this service is downloaded to the physical nodes which control and manage this service.
- Subscribed: The service is subscribed by the customers, and the customers customized the service.
- Invocation: The service is activated/deactivated by the users.
- Termination: The service is terminated and withdrawn by network operators.

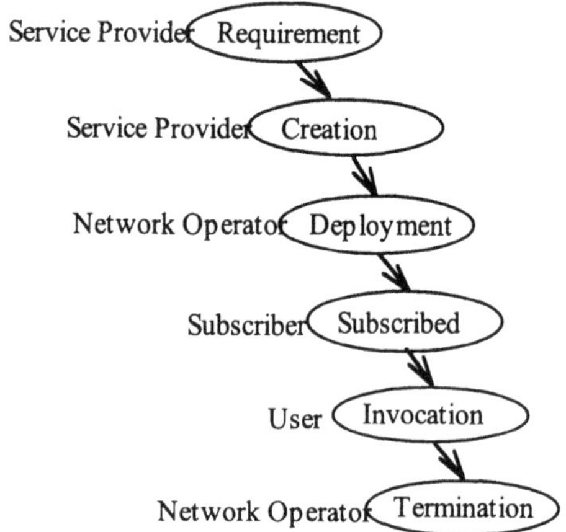

Figure 1 The life cycle of the service

The appearance of the IN technology, with its rapid and abundant service development, has stressed the problem of interactions between services. The service interaction has become a main obstacle of the service creation and IN technology. The difficulty should be resolved in such way:

- Formal definition of the services, including the syntax, behavior and semantics of the services in the requirement and creation phases.
- Dynamic negotiation in the deployment, the subscribed and the invocation phases.

This paper proposes a method to formally describe the service from the syntax, the behavior and the semantics views and provides an approach to handle the service interaction at the specification level. The method evolved from an IN project at BUPT IN research supported by the 863 project of China. We have implemented an SCE tool to create services by SIB-based language and successfully download these services into SCP to execute. Now we are engaged in the research on the service validation, detecting and resolving the service interaction.

2 THE CONCEPT OF SERVICE LOGIC BLOCK AND THE EXECUTION OF SERVICES ON IN

The IN service can be created by the SIB-based language. One service is constructed by several executable chains of SIBs. Among those SIBs, there is a special SIB called BCP (Basic Call Process) which provides the basic call processing, implements the function of connecting and disconnecting the call at proper time and holding the call instance data for call processing. The BCP is described with a finite state machine, the basic call state model (BCSM). The BCSM consists of states that represent the processing done by exchanges, transitions and detection point. Detection Points (DP) are points in call processing where the SSF can determine if a request to the SCF should be reported according to the service control logic. The detection points have two types, trigger detection point (TDP) that is the static point and armed when the service is deployed and event detection point (EDP) that is the dynamic point and armed when the service is executed.

The service logic program (SLP) described by SIB-based language is shown below:

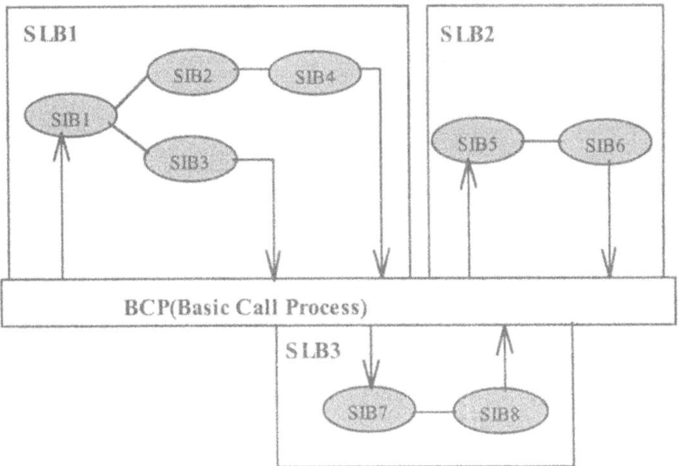

Figure 2: The Service Logic Program of one Service

The SLP of a service consists of several SLBs (Service Logic Block). The SLB is a chain of SIBs that starts from BCP and finally returns to BCP (i.e. in Figure 2, the service has three SLBs). Each SLB of a service is invoked by one detection point, either TDP or EDP. The relationship between O_BCSM (originating BCSM) and SLB is shown in Figure 3.

During the execution of a service, control is switched between SSF/CCF and SCF. Firstly the call is processed in SSF/CCF, when a detection point is encountered, the control is switched to SCF and one SLB of this service is executed, after the SLB is finished, the control returns to SSF/CCF.

The introduction of SLB has following advantages.

- Subdivide the function of a service, such as the Freephone service (shown in Figure 4) has two SLBs. One SLB implements the function of One Number (one feature of the Freephone service), and the other SLB implements the function of Reverse Charge (the other feature of the Freephone service).
- SLB can be used to handle the service interaction. The interactions between services always come from the interactions between the SLBs of different services, such as the Freephone service and Account Card Calling, the two services have the SLBs which are invoked from the same EDP (O_Disconnect).

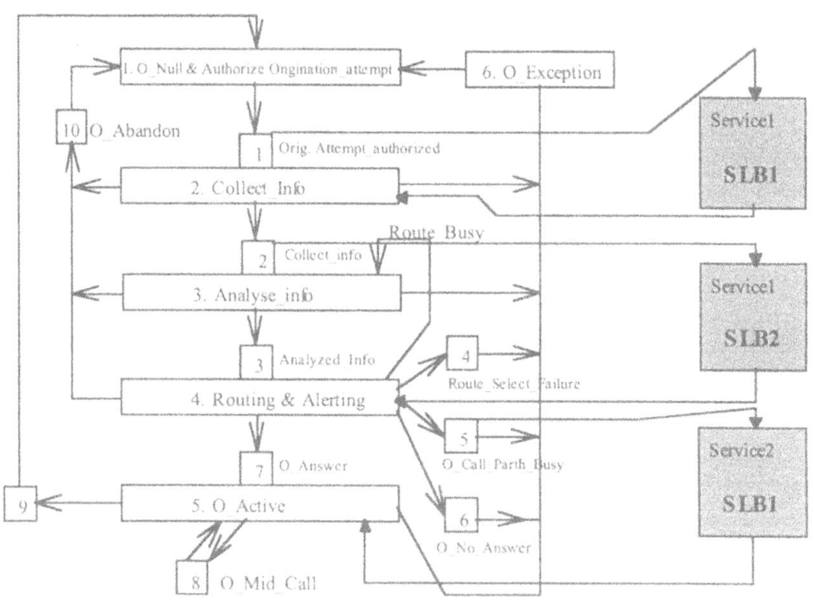

Figure 3 The relationship between O_BCSM, Services and SLBs

The SLB of a service is different from the feature of the service. The feature describes part of the functionality of the service, it can hardly be described by SIBs, while the SLB describes the part of execution of the service (always also represents the part function of the service) and it can be formally represented by SIBs (a chain of SIBs which starts from BCP and returns to BCP), so we can define:

$$\text{Service} = \Sigma \text{ SLBs}$$

In our IN project, we realize ten SIBs defined by ITU-T in CS1 (*Algorithm*, *Charge*, *Compare*, *Log*, *Screen*, *SDM*, *Translate*, *UI*, *Verify*, *BCP*), and realize three SIBs defined by Australia (*Connect*, *EDPRequest*, *ReleaseCall*), and we implement two SIBs defined by ourselves (*Start* SIB used to choose the number from the whole number that user dials; *Charge2* SIB used to charge dynamically).

A typical example of IN services is the Freephone service which is widely used all over the world. The SLP of the Freephone service in our IN product is shown in Figure 4. The *Start* SIB chooses the formal number from the whole number that is reported from BCP; and *EDPRequest* SIB arms the EDPs (O_Abandon, O_Disconnect). *Screen* SIB examines if the formal exists in SDF; if NoMatch, *UI* SIB notifies the user that the number does not exist and *ReleaseCall* SIB disconnects the connection; if Match, *Translate* SIB translates the formal number into a destination number. *Queue* SIB creates and maintains a queue for every destination number invoked, if the destination number is free, then *Log* SIB records the call data, and *Connect* SIB connects the call; if the queue is full or the time expires, *UI* SIB notifies the user and *ReleaseCall* SIB disconnects the connection.

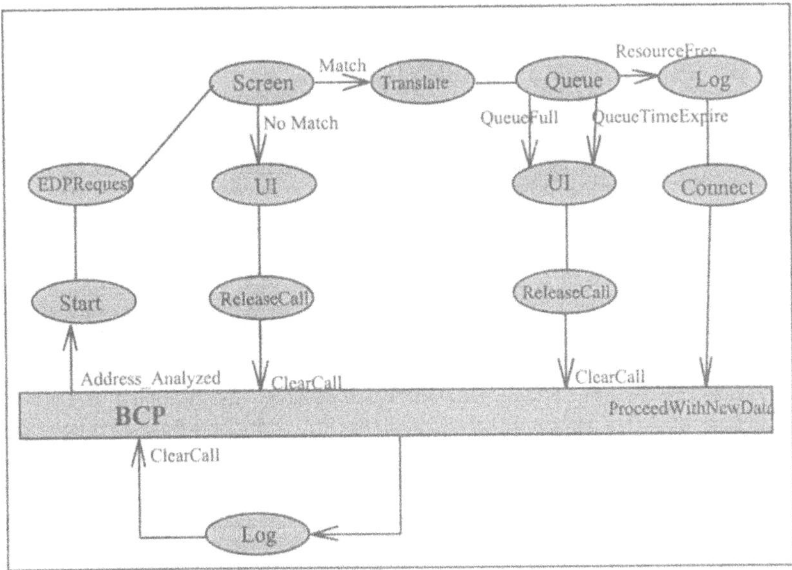

Figure 4 The SLP of the Freephone service

During the execution of the Freephone, firstly the user dials the prefix number 800 and formal number (for instance 12345), the number is analyzed by CCF and it is sent to SCF as a TDP. In SCF, an SLP is invoked. This SLP arms other DPs, retrieves the formal number from

SDF. If the formal number does not exist, an announcement is sent to the user; otherwise the SDF translates the formal number into a destination address according to the location of caller and the time, the SLP is hung and the control returns to CCF. The call continues, while the user is connected to the destination address and starts conversation. At the end of the conversation, when the user hooks on, an event is captured by SSF/CCF and sent to SCF as a detection point. In the SCF, the SLP is resumed to handle the charging (reverse charge). During the execution of the Freephone service, the control is switched between the SSF/CCF and SCF. The control switched between SSF/CCF and SCF is shown in Figure 5.

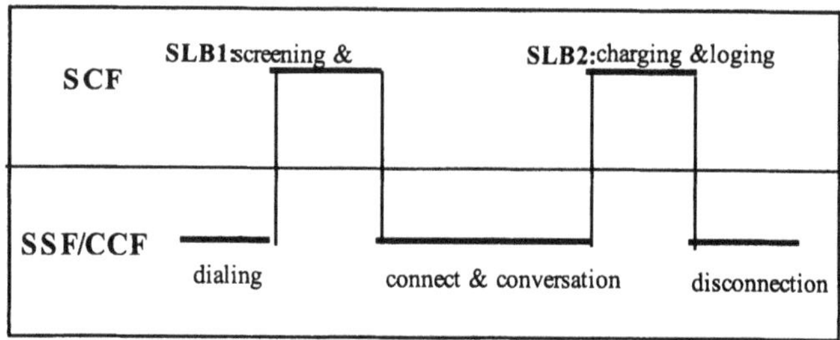

Figure 5 control switched between SSF/CCF and SCF

3 THE SYNTAX OF THE SERVICE

The service is considered as SLP (service logic program, the software processing in SCF) and the database in the SDF. To describe the syntax of a service, it is necessary to describe the syntax of SLP and service-related databases. The SLP is constructed by several chains of SIBs. Each SIB has its own data, including SSD (Service Support Data) and CID (Call Instance Data).

During the execution of a service, the service needs data to record the user information. The data is stored in the service-related database in the SDF. So, the syntax of a service consists of two parts:

Syntax_Service = *SLP* £« *DB*

SLP refers to the service logic program which describes the execution of a service, and *SLP* consists of several executable chains of SIBs (SLB). *SLP* is defined as 5-tuple £º $<Q, E, O, T, q_0>$.

Q refers to the set of SIBs defined in the GFP.

Q = {*BCP*, *Queue*, *Charge*, *Algorithm*, *Verify*, *SDM*, *ReleaseCall*, *Translate*, *UI*, ...}

E refers to the environment of each SIB, which describes the service data of this SIB. The element of E is different for different SIBs and also different for the same SIB under different situations.

O is the outlets of the specified SIB, such as the *Screen* SIB, having two outlets: *Match* and *NoMatch*.

T refers to a function: Q¡ÁE¡Á$O \rightarrow Q$ ¡ÁE

$q_0 \in Q$, refers to the initiating and terminating states of all service logic program.

DB refers to the service-related database stored in the SDF. In the syntax definition of a service, only the logic profile of the database is defined. The BNF definition of *DB* is as follows:

DB ::= { database_definition }*
database_definition ::= {database_fields}$^+$
database_fields = fields_name + fields_type
fields_name = string
fields_type = Integer | String | ...

The definition of SLP and DB of the Freephone service is as below:
The SLP definition of the Freephone service is: $SLP_{fph} = < Q_{fph}, E_{fph}, O_{fph}, T_{fph}, q_0 >$.

Q_{fph} = { *BCP, Start, EDPRequest, Screen, UI, ReleaseCall, Translate, Queue, UI, ReleaseCall, Log, Connect, Log*}
E_{fph} = { *<Start, SSD, CID>*, *<Screen, SSD, CID>*, ...}
O_{fph} = { *Screen_Match, Screen_Nomatch, Queue_Queuefull, QueueResourcefree*, ...}
T_{fph} is omitted.
q_0 = *BCP*.

The description of databases of the Freephone service is omitted.

4 THE BEHAVIOR OF THE SERVICE

From the user viewpoint, the service is an entity which implements the function of how to subscribe this service, how to connect the calling party to the called party, how to interact with users. In the requirement phase of life-cycle of the service, the behavior of this service can be specified by the service provider. The service provider can describe a service by formal language, such as SDL or MSC. The Freephone service is described by MSC below.

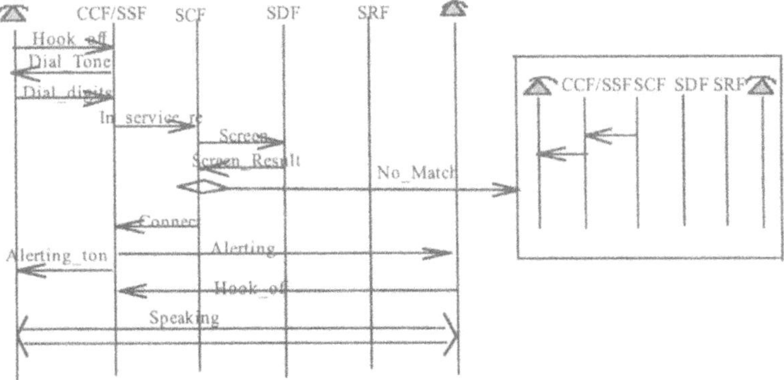

Figure 6 The behavior of the Freephone service described by MSC

For the IN viewpoint, the behavior of a service can be considered as a sequence of information flows between different function entities (FE) and the function entity actions (FEA) which are executed in FEs.

The behavior of a service can be described below £°
Behavior-Service =<*TDP*, *EDP*, *Behavior-Specification*>
TDP refers the trigger detection point of a service.
EDP refers the set of event detection point of a service.
Behavior-Specification is a description of a service based on the SDL language.

The mapping of SIBs in GFP (Globe Function Plane) to DFP (Distribution Function Plane) is the information flows and function entity actions. Figure 7 shows the mapping of SIB in GFP to FEA/IF in DFP.

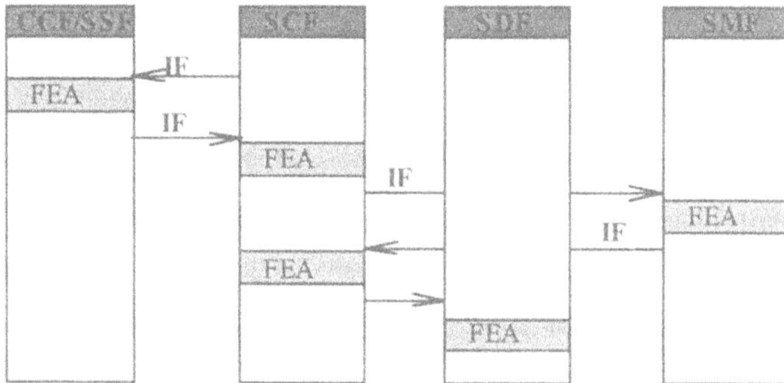

Figure 7 Mapping of SIB in GFP to FEA/IF inDFP

So, we can say: SIB = IF + FEA. IF refers to the information flows between FEs (including SCF, SDF, SSF, SRF, SMF).

We can deduce the *Behavior-Specification* of a service from the syntax of the service automatically by combining the behavior of each SIB, and the *TDP* and *EDP* can also be deduced from the outlets of *BCP* SIB in the *SLP*, so we can deduce the service behavior from the syntax of the service.

The behavior of a service shows the function of the service, and it is useful for the service simulation, validation and detection of service interaction.

- In the service simulation, we can simulate each FE by a process and simulate the FEAs of each SIB by software, according to the IF between different FEs, and the execution of a service can be simulated. The service provider can tell if the execution of a service is identical with his expectation.

- In the service validation, we can compare the behavior of a service written by the service provider in the specification phase to that deduced from the syntax, and tell if they are identical.
- In detecting the service interaction, the detection points and SDL specification are also useful to detect the interaction between services. For example, the Call Waiting service and Call Forwarding on Busy service have interactions because those two services have the same trigger detection point.

5 THE SEMANTICS OF THE SERVICE

The semantics of a service describes the purpose of the service, or the purpose of subscriber to subscribe this service. For example, the subscriber of the OCS (Originating Call Screen) service wants to screen automatically all numbers on his screen list from his handset. The subscriber of the Call Forwarding service wants to redirect his incoming call to the specified number.

The semantics of a service may separate or partly separate from the syntax or behavior of the service. It is impossible to be deduced from the syntax of a service automatically. It must be written by the service provider when the service is created. The semantics of a service can be described by the assertion of logic.

For example the semantics of the service Call Forwarding and OCS (Originating Call Screen) are described below. Call Forwarding is the service that the subscriber can redirect the incoming call to another number when he is busy. OCS service is the service that the subscriber can screen the outgoing call by the screen list.

The semantics of Call Forwarding is:

$$\{ \exists x \, \forall y \, \forall z \, (\mathrm{Call}(y, x) \wedge \mathrm{Busy}(x) \wedge \mathrm{Reroute}(x, z) \rightarrow \mathrm{Call}(y, z)) \}.$$

The semantics of OCS is:

$$\{ \exists x \, \forall y \, (\mathrm{Screen}(x, y) \wedge \mathrm{Call}(x, y) \rightarrow \mathrm{Block}(y)) \}.$$

There should be some basic assertion which describes the POTS service, such as the proposition of Call(y, x) means the user y makes a call to x; Busy(x) means user x is busy; Reroute(x, z) means that call to x is redirected to z. The new assertion used to describe the characteristic of new services must be defined by some basic concepts, such as Screen(x, y) means number y is screened on x's handset; Block(y) means the call to y is blocked.

Also, they are some pre_conditions and post_conditions of each SLB. The pre_conditions describe the assumption of the SLB, while the post_conditions describe the execution result of the SLB, such as the modification of call instance data. So the semantics of a service can be described below:

Semantics = *ASSERTION* + *CONDITION*
CONDITION = {SLB_Condition}*

SLB_Condition = Pre_condition + Post_condition

The semantics can be used to detect the service interaction. Now we give an example of how to detect an interaction from the OCS service and CF (Call Forwarding) on busy service. The OCS screens all outgoing calls in the screen list of the handset, while CF on busy redirects the incoming call to the specified number when the handset is busy. If user x (his telephone is also x) uses OCS to screen number y in his handset, but another user z can use CF on busy to redirect incoming call to y, and then he dials himself (dials number x) to get the effect to call to y.

To describe the two services, we define a proposition *call*(a, b, c, d), where a refers to the calling number, b refers to the dialed number of caller, c refers to the called number and d refers to the destination number.

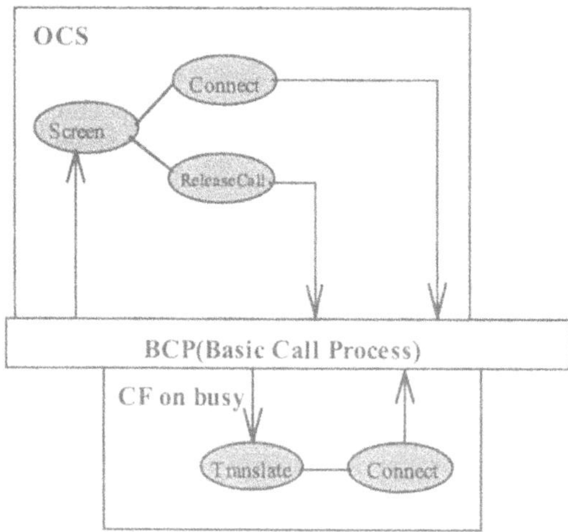

Figure 8 The SLP of the OCS Service and the CF on busy Service

The SLP of the two services is shown in Figure 8. The Pre_condition of OCS is: for the proposition of *call*(a, b, c, d), $c = d$, the called number must be equal to the destination number. During the execution of the call, user z dials the number x, the proposition is *call*(x, x, x, x), after invocation of CF on busy, the proposition is changed to *call*(x, x, x, y), and when the OCS is invoked, the *call*(x, x, x, y) violates the assumption of OCS. Thus, the interaction is detected.

6 CONCLUSION

IN gives approach to rapidly and economically create services. As everyone, every company can create services. It is difficulty to deploy the services from different venders into the whole telecommunication network and make them co-operatable, and it also adds the difficulties for the service management and resolving the service interaction. The formal description of services, in the syntax, behavior, semantics way can standardize the service in the specification and creation phases. This may help for the services cooperating from different venders and different network components, and provides a way to resolving the service interaction, also it can support user for the service simulation and service validation in the service creation phase in SCE.

7 REFERENCES

A ITU-T, CS-1, Recommendation Q 12XXseries.
B Feature Interaction In Telecommunication System, L.G. Bouman and H.Velthuijsen, IOS Press.
C A Practical Approach to Service Interaction, Eric Kuisch, IEEE Communication Magazine, August 1993.
D Intelligent Network Service Creation Environment, Yuzhang Liu, ISIB'95, Beijing, China.
E The Research and Implementation of Intelligent Network, Wang Bai, Doctoral Thesis of BUPT, P.R. of China, 1995.

8 BIOGRAPHY

Yuzhang Liu received a Bachelor degree in computer science in 1992 from the university of Tianjin, China, and graduated from Beijing University of Posts and Telecommunications (BUPT) and received the Ph.D. degree in electronics and communications science in 1996. From 1994 to 1996, he takes part in project of Intelligent Network in BUPT, and have developed the tool of Service Editor, Service Analyzer and IN Emulator which can emulate the whole IN system and service execution. His research interests include Intelligent Network, Feature Interaction, Software Engineering.

Fangchun Yang received the MS in computer science, Ph.D. in communication and electronic system from BUPT in 1986 and in 1990 respectively. Currently, he is a professor of National Lab of Switching Technology and Telecommunication Networks at Beijing University of Posts and Telecommunications (BUPT). His Research interests include intelligent network, software engineering. He has been responsible for more than ten national projects and collaboration abroad since 1990. At the moment, he is in charge of a national key project to develop IN project in China.

Junliang Chen graduated from the Department of Telecommunications, JiaoTong University, Shanghai, China, in 1955. From 1957 to 1961, he studied as a graduate student at the Moscow Institute of Electrical Telecommunications, Russia. From 1979 to 1980, he was a visiting scholar at the Departments of Electrical Engineering and Computer Sciences of the University of California, Berkeley. From 1980 to 1981, he was a visiting scholar at the Department of Computer Sciences of the University of California, Los Angeles. In 1989, he also visited the University of Bristol. Currently, he is a professor at the Beijing University of Posts and Telecommunications (BUPT), and a member of the Chinese Academy of Sciences and Chinese Academy of Engineering. His research interests include switching systems, telecommunication networks, software, and fault-tolerant computing.

17

Intelligent Network Service Creation: an ITU-T CS-1 view

Sehyeong Cho, Choongjae Im, JeongHun Choi
Electronics and Telecommunications Research Institute
161 GaJung, Yusong, Taejon, 305-606 Korea
Tel: +82-42-860-5630, Fax: +82-42-861-2932
E-mail: (shcho, cjim, jhchoi)@dooly.etri.re.kr

Abstract

ITU-T, the international standardization body, recommends a 4-plane model of intelligent network. This paper describes how we used the model for rapid prototyping, and discusses the merits and drawbacks of the intelligent network as described in the ITU-T Q.1200 series recommendations, in particular, CS-1, in terms of automated service creation. We focus on the middle two planes, the Global Functional Plane (GFP) and the Distributed Functional Plane (DFP). We also discuss formal constraints on SIBs and the relationship to automated service creation.

1 INTRODUCTION

ITU-T is currently standardizing the advanced Intelligent Network (IN) and also studying service creation. ITU-T developed the Intelligent Network Conceptual Model (INCM) to provide a framework for the design and description of each IN Capability Set (CS) and target IN architecture. With the INCM, IN is modeled by using four planes, each an abstract view of IN-structured network: Service Plane (SVP), Global Functional Plane (GFP), Distributed Functional Plane (DFP), and Physical Plane (PHP)[1]. We shall not be concerned too much with the service plane in this paper, and will focus mainly on the middle two planes. We shall be discussing the merits and drawbacks of the INCM, in particular, IN CS-1, in terms of modeling and implementation of an IN-structured network as well as service creation in the framework of INCM.

Section 2 describes our prototype implementations. Some SW engineering issues will be touched upon. Section 3 will discuss pros and cons with CS-1 in terms of service creation.

Section 4 summarizes the findings and discusses the future of function-based service creation and control.

2 PROTOTYPING THE ITU-T IN

The prototypes described in this section represent part of a series of rapid prototyping activities, that are intended to give a comprehensive evaluation on the ITU-T recommendations, to give a hands-on experience with service creation and service control, and to create the actual implementation of IN as the last step in the prototyping. The prototyping started from the service plane, then moved to the global functional plane, then back to SVP, then to DFP, and so on. This is a very intuitive way of mapping the prototyping cycle to the INCM [3]. Experience and information gathered at lower planes are fed back to higher planes, in order to reflect to the local standard and to implementation.

Since recommendations for the service plane had very little to say about how to specify services, we used a conventional method of using a service requirement specification template (with the service name, the prose description, the usage, network capability requirements, and so on).

The GFP describes standard reusable units of service functionality, referred to as service-independent building blocks (SIBs), independent of how the functionality is distributed in the network. SIBs can be combined with global service logic on the GFP to realize services and service features. The concept of SIBs is of primary interest to service designers and serves as an abstract service modeling tool. These SIBs may be used in service creation processes, though IN CS-1 did not address this aspect and provides no standardized capabilities to support such processes [2]. In the global functional plane, all network functions are represented as if they are provided by a point entity. The way how these functions are actually performed in the network is not disclosed at this plane. Information hiding of this nature is extremely important because it implies those who work at this level don't have to have expert knowledge. One such class of people are service logic designers. Relieving them of the burden to be network experts enables quicker service design, as would be required in order to cope with the increasing number of services and tight time-to-market.

For the purpose of initial modeling of the intelligent network and putting the standard SIBs into test, the "global functional" concept helped us a great deal by reducing the complexity down to a manageable size. The reason is all the same: we didn't have to know the detail of communicating multiple entities. For instance, a user interaction SIB can be realized by a process in a computer (not a networked entities) by using audio device for announcement and keyboard for entering digit strings. In our case, we didn't even have to write a complex program, but instead we took advantage of the power of the inference engine of a production system to do the reasoning [10]. With a production system, the domain knowledge is encoded as assertions that represent contingent facts and rules that represent the "law of nature" in the domain of interest. The "programmer" does not specify how a task is performed. For instance, SIB firing rule can be represented as follows (with a little syntactic sugar):

If SIB A finishes action at logical output X
 and an arrow connects from X to the logical start of SIB B,

then SIB B starts execution.

This rule fully encodes the semantics of SIB chaining, in other words, dictates the meaning of a SIB connected to another SIB by an arrow, and that, in a machine-executable form. The following is an excerpt from a Global Service Logic specification that uses Distribution SIB.

```
(make   SIB
        ^TYPE Distribute
        ^SIB-ID sib1
        ^state    inactive
        ^SSD-ID              ssd-1)
(setf   (distribution-type ssd-1) 'C)  ;;; time of day
(setf   (numberof_branches        ssd-1)   2)
(setf   (timeofday ssd-1)
        '(((0.0)(17.30))(17.30)(24.0)))) ;;; Midnight to 5:30PM -> branch 1
                                         ;;; 5:30 to Midnight -> branch 2
...
(make   GSL-CHAIN
        ^from-sib sib-1    ^to-sib sib-2
        ^logical-out 2)
(make   GSL-CHAIN
        ^from-sib sib-2
        ^to-sib sib-3
        ^logical-out Error
        ^logical-in  Clear-call-POR)
```

We built a graphic editor for creating GSLs and entering service data (see Figure 1) , and a translator that translates the graphical representation of GSLs into the form of rules and facts. These are fed to the rule-based GSL simulator. User-noticeable events such as ring, tone, announcement, etc. are given to the call/service animation engine to simulate the audio/visual effect. User interaction is input by simulated keypad and mouse, then fed to the translator, to be represented as a fact. Put together, they constitute a service logic design and validation tool.

In the distributed functional plane, the network is viewed as discrete logical groupings of functionality called functional entities (FEs). FEs realize service functionality by executing functional entity actions (FEA's). The distributed functional plane hides physical implementation details such as supporting protocols, or the physical location of functional entities, and the like, thus making it easier to analyze and to model. As services are represented by combinations of SIBs at GFP, each SIB is realized as one or more Functional Entity Action (FEAs) performed by Functional Entities (FEs). Some FEAs represent local actions, while others represent communication, or "information flows" among FEs.

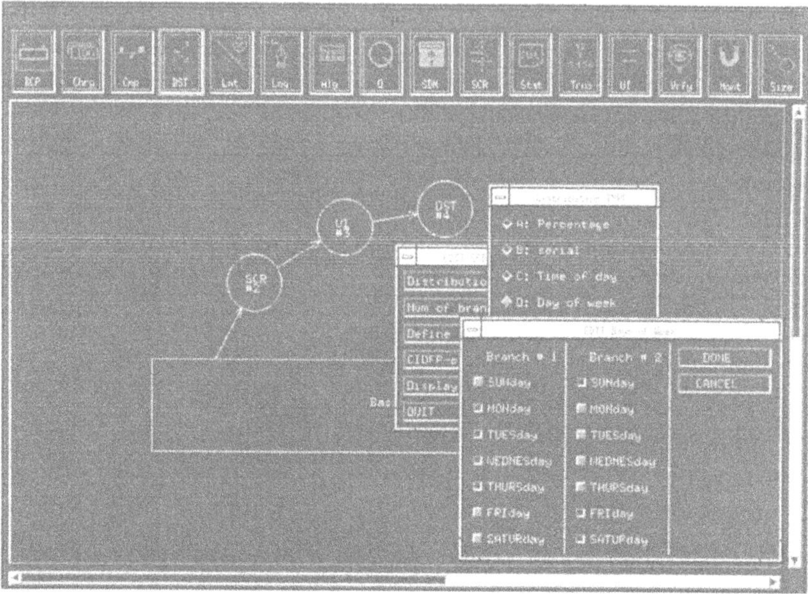

Figure 1 Editing a GSL.

A service logic at the DFP prototype, which we call a distributed service logic (DSL), is composed of instructions in a proprietary applications programming interface (API) that modeled the functional entity actions without the actual underlying transport, such as TCAP. The DSLs in turn are created by translation from GSLs, which are created by a service logic editor implemented in the first stage of the prototyping.

Most part of the translation went straightforward, but we did have trouble, which we shall allude to in the next section. Also, in order for service simulation, we needed some information that seem to pertain to the physical plane, such as database schema or SSP trigger information such as DP criteria.

Each DSL is statically and dynamically checked for verification. In static verification, we check for static errors of functional entity actions and the information element of information flow. After static verification, the distributed service logic is executed in the simulation environment with SCF simulator (a sort of virtual machine that executes the instructions in the DSL), a SSF/CCF simulator, SDF simulator and an SRF simulator. Figure 1 shows a snapshot from the simulation process of the Universal Personal Telecommunication (UPT) service.

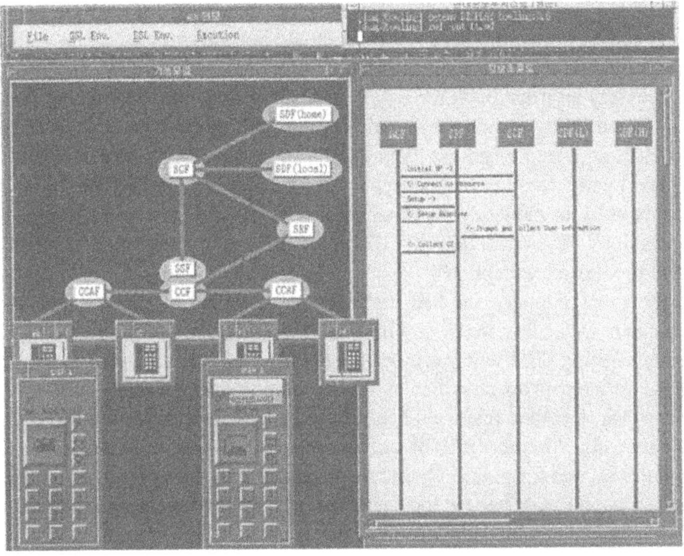

Figure 2. Service Simulation process of DSL.

3 ITU-T SIBS FOR SERVICE CREATION

As we alluded to in the previous section, the viewpoint distinction provides a easier way of service creation. A service designer draws a global service logic, without the need of knowing all the network detail - no information flows, no protocols, no database. All the dirty work is done by a machine - hopefully, and service logics and databases etc. are automatically produced. However, there are things to be resolved, before getting the dream come true.

One of the conceptual problems is that the basic call process is too much DFP-oriented. Even though basic call process is practically very different from other SIBs, the global functional plane should have hidden such detail. Currently, with CS-1, the global functions are conceptualized as "basic call process," which represents the functions provided by CCF/SSF, and "other SIBs," which represent the functions provided by SCF and other functional entities. That is, users of the recommendation Q.1213 must somehow understand the interaction between SSF and SCF, which is an obstacle for non-experts to using SIBs for service creation.

One way of getting out of this trouble is to reorganize the set of POI's and POR's into a set of SIBs. For instance, "Continue with new data" POR can be replaced by "Connect to a new party" SIB, depending on the context. This way, the viewpoint based on distributed functions can be replaced by the global viewpoint involving user(s) and network, where the network is viewed as a point entity that provides functions necessary for services.

POIs and PORs bear the name because it is presupposed that the BCP initiates (triggers) the service logic and the control will eventually "return" to BCP. However more than one POIs (or PORs) can exist, which entails there should be points at which the (suspended) service logic resumes execution. They differ from POIs semantically.

Then there is the problem of parameter mapping between GFP and DFP (to PHP, of course). Even though some information need not (and should not) be revealed at the global functional plane, there should be at least a source of information that can be used to make some inference to derive necessary information at the lower planes. For example, End-of-call POI in BCP indicates that "a call party has disconnected." In the GFP there is no way of distinguishing a service which is activated when the calling party hangs up from a service which is activated when the called party does (T_Disconnect vs. O_Disconnect). This might not be a problem if GSLs are only used for modeling purpose, but is a problem if they are to be used for formally specifying services. Physical details that should be filled in at deployment time or by a service management process are exceptions.

With current methodology, the SIB instances are specified by service support data (SSD) and Call Instance Data. The distinction between those two are unnecessarily rigid, restricting the flexibility of using SIBs. For instance, in a *Screen* SIB, the screen list indicator (or the list name in CS-1 refinement) is classified as service support data, therefore cannot be changed at runtime. However, there are many applications (e.g., UPT) which require that the database ID change dynamically. *Compare* SIB is another example, where comparison is allowed only between a constant and a variable. Rather than syntactically fixing the type of data (CIDFP or fixed value), it would be better for the user to be able to declare the type when instantiating a SIB.

Data transparency used to be a problem (GSL needed to use screen SIB twice, one for finding out the DB and the other for screening) but is hoped to be solved by using directory services based on X.500.

Error management is elsewhere pointed out to be not useful [5]. Some SCE implementors [6] make use of explicit handling of error cases, but we believe a few default error handling will suffice for all practical purposes.

In order to use GSL for service creation, a GSL should be a formal specification, even if it intentionally hides many network details. However, current SIB specifications lack the formality in data. This is especially true for SIBs that relate to data handling such as *Screen*, *SDM*, and *Translate*.

The DFP is supposed to be independent of physical plane, but sometimes it is not the case. In ITU-T recommendation [8], information flows for user interaction is shown to be varying according to the physical location of IP.

User interaction poses another problem, which is optimization problem. When *UI* SIB is used more than once, straight forward translation into DFP tends to show that connection to SRF can be either unnecessarily maintained or unnecessarily disconnected when another use follows. This is even more complicated when used in combination of if-then-else's such as screen SIB. Peephole optimization [9] should take care of simpler cases, but optimizing complex cases are for further study.

Perhaps the most significant potential obstacle to service creation is the non-monolithic nature of SIBs. Even though the SIBs are supposed to be monolithic by definition, some are only conceptually so. Figure 3 depicts two cases of combining SIBs into service logics, the first monolithic case, and the second non-monolithic. Mapping to distributed functional entity actions are described in simplified pseudo-SDL (state symbols omitted).

In the latter case, it is not possible to compose a service logic program from the specifications of the SIBs in a straightforward manner. The following definitions will be used for characterizing certain properties of SIBs. We shall be concerned only with the service logic for SCF. We will assume, for the sake of simplicity, the SCF action for a SIB is defined as a linear sequence of FEA's (This will be true for each service instance).

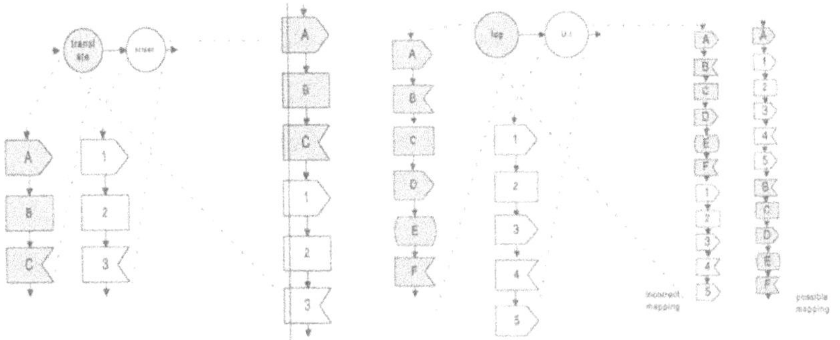

Figure 3-a combining two monolithic SIBs.

Figure 3-b combining two SIBs, one of them non-monolithic.

Definition 1 [Monolithic SIBs: type 0]:

Let a sequence of actions $\sigma_1, \sigma_2, \ldots \sigma_n$ define the (functional entity) actions necessary for SCF to provide the function defined for SIB *i*. SIB *i* is defined to be *monolithic* if in any combination of SIBs (that is, in a GSL) including an instance of SIB *i*, the sequence $\sigma_1, \sigma_2, \ldots \sigma_n$ will be executed contiguously (i.e., without any action δ_k defined for SIB *j* intervening). We shall call such SIBs as type 0.

If all SIBs are monolithic, a GSL specification itself is the execution sequence in terms of distributed functions, and therefore synthesis of service logic programs from a GSL specification is rather straightforward (see Figure 3-a). However, monolithicity is too stringent a constraint in order for the set of SIBs to define the network capability of interest. Actually some SIBs in ITU-T CS-1 are found non-monolithic. Among CS-1 SIBs, *Charge* and *LogCallInformation* are such examples. Figure 3-b illustrates the point. With definitions 2 and 3, we shall try to relax the constraint.

Definition 2 [Monolithicity of a part of a SIB]:

Let a sequence of actions $\sigma_1, \sigma_2, \ldots \sigma_n$ define the (functional entity) actions necessary for SCF to provide the function defined for SIB *i*. A contiguous sub-sequence $\theta = \sigma_k, \sigma_{k+1}, \ldots \sigma_l \ (1 \le k \le l)$ is defined to be *monolithic* if in any combination of SIBs (that is, in a GSL) including an instance of SIB *i*, the sequence $\theta = \sigma_k, \sigma_{k+1}, \ldots \sigma_l$ will be executed contiguously (i.e., without any action δ_m defined for SIB *j* intervening).

Definition 3 [Quasi-monolithic SIBs: type 1 and type 2]:

1. SIB i is defined to be <u>*quasi-monolithic*</u> if condition 1 or 2 holds: In any combination of SIBs (that is, in a GSL) including an instance of SIB i, all actions which are defined for SIB i are divided into contiguous partitions $\theta_1,...,\theta_k$, (i.e., $\theta_1 = \sigma_1,\dots,\sigma_{i_1}$, $\theta_2 = \sigma_{i_1+1},\dots\sigma_{i_2}$, $\cdots$, $\theta_k = \sigma_{i_{k-1}+1},\dots\sigma_n$) such that: 1) each θ_m $(m = 1,\cdots,k)$ is monolithic, 2) θ_i precedes θ_{i+1} for all $i = 1,...,k-1$ and 3) there exist syntactically identifiable preconditions for each θ_i , $i = 2,...,k$. We will call such SIBs as type 1.
2. In any combination of SIBs (that is, in a GSL) including an instance of SIB i, all actions which are defined for SIB i are divided into contiguous partitions $\theta_1,\cdots,\theta_k$, (i.e., $\theta_1 = \sigma_1,\dots,\sigma_{i_1}$, $\theta_2 = \sigma_{i_1+1},\dots\sigma_{i_2}$, $\cdots$, $\theta_k = \sigma_{i_{k-1}+1},\dots\sigma_n$) such that: 1) each $\theta_m (m = 1,\cdots,k)$ is monolithic, 2) θ_1 precedes $\theta_2,\cdots,\theta_k$, and 3) $\theta_2,\cdots,\theta_k$ are mutually independent and can be performed concurrently, each driven by an asynchronous event. We will call such SIBs as type 2.

Translating a GSL into distributed functional entity actions are relatively straightforward, provided that SIBs are monolithic or quasi-monolithic. A type-1 SIB can be broken into k pieces and reordered according to the "syntactically" identifiable preconditions. Type-2 SIBs can be implemented by using $N+1$ service logic program processes, one for the leading monolithic sub-sequence, $N-1$ for the concurrently executing sequences (possibly after the SIB meets the logical end), and one for coordinating the N SLP's. Figure 4 depicts these two situations.

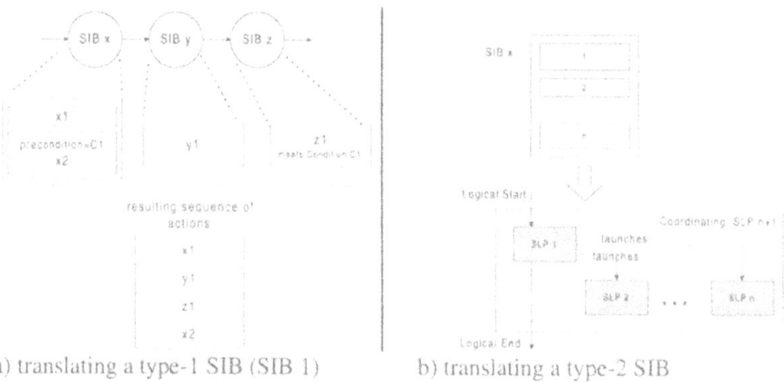

Figure 4 Translating SIB sub-sequences into SLPs.

Currently all CS-1 SIBs seem to be at least quasi-monolithic. However in the future when we extend the SIB sets or the capability sets, it would be advisable to be careful either to check such constraints, or extend the constraint beyond, within manageable complexity. Currently the most obvious extension to the SIB constraint would be to allow $\theta_2,..,\theta_k$ in definition 3 to be quasi-monolithic in turn. The runtime snapshot would look like Figure 5.

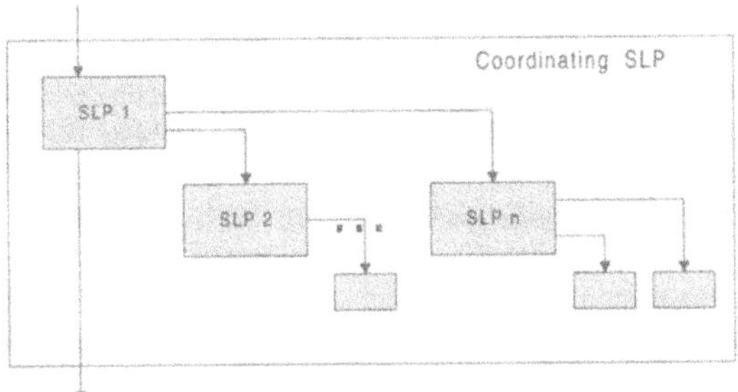

Figure 5. The most obvious extension of the SIB constraint

4 CONCLUDING REMARKS

The relation between INCM and rapid prototyping was described in this paper in terms of our experience in a series of rapid prototyping. We also touched on some issues in CS-1 (and beyond) in relation to service creation. Despite some drawbacks discussed in this paper and others, the 4-plane model provides a good basis for service creation in that it enables proper information hiding for people involved in, and a relatively sound basis for automatic translation from GSL to lower-plane representations, such as SLPs.

It must however be stressed that extension of SIBs and/or capability sets must take precaution as to the characteristics of those SIBs in order to use them for automated service creation. The foregoing classification of SIBs is by no means complete or exhaustive, but it merely suggests the need of such a formal analysis of the relation between SIB characteristics and the complexity of automated service creation.

Recently, SIB concept is being challenged by object-oriented school, since IN SIBs are not object-oriented, but basically functional. It might not look fashionable, considering that OO is such a buzz-word these days. However, unless OO is a panacea, there are domains that is suitable and domains that are not. In general OO paradigm applies well to domains where the world consists of "things," such as a management domain. In case of IN, the main concern is functions, not objects. Thus chances are, you might end up creating objects which are a bunch of functions in disguise.

Functional paradigm is not dead yet, but we should provide some formal guidelines as to how to safely extend the capability in order to provide a service-creatable platform for CS-2, CS-3, and beyond.

5 REFERENCES

[1] ITU-T draft Recommendations Q.1203, 1993.

[2] James J. Garrahan, Peter A. Russo, Kenichi Kitami, and Roberto Kung, " Intelligent Network Overview," IEEE Communications magazine, Vol. 31, No. 3, pp. 30-36, March 1993.

[3] Sehyeong Cho, "A Rapid Prototype of IN Using a Production System," Intelligent Network '94 Workshop, Ramada Renaissance Hotel, Heidelberg, Germany, Section 212.2, May 24-26, 1994.

[4] Masaya Akihara, Keiichi Shimizu and Shuji Ito, "An Implementation and Evaluation of Service Creation," Intelligent Network Workshop '95, Congress Center, Ottawa, Canada, May 9-11, 1995.

[5] Ty Chang, "Discussing the Weaknesses of the Standard Service Independent Building Blocks," 3rd International Conference on Intelligence in Networks, Bordeaux, France, pp. 67-72, Oct. 11-13, 1994.

[6] Martin H. Petruk, "Experiences in applying the ITU CS-1 Intelligent Network Conceptual Model (INCM) to service Creation," 3rd International Conference on Intelligence in Networks, Bordeaux, France, pp. 134-139, Oct. 11-13, 1994.

[7] ITU-T Recommendation Q.1213, May 1995

[8] ITU-T Recommendation Q.1214, May 1995

[9] Aho A. et al, Compilers - principles, techniques, and tools, Addison Wesley, 1986

[10] Lee Browston et al, Programming Expert systems in OPS5, Addison Wesley, 1986

6 BIOGRAPHY

Sehyeong Cho Received B.E. and M.Sc. from Seoul National University, Seoul, Korea in 1981 and 1983, respectively. He received Ph.D. in Computer Science from Pennsylvania State University, USA in 1992. He has been a Software Engineer at Electronics and Telecommunications Research Institute (ETRI) since 1984, and currently is a senior member of technical staff and project leader. His research interest includes Telecommunication Software Engineering and AI applications to Telecommunication.

Choong-Jae Im was born in Chungnam, Korea in 1968. He received the B.S. degree in computer science from Chungnam National University, Taejon, Korea in 1991 and 1993, respectively. Since 1993, he has been involved in Intelligent Network projects.

JeongHun Choi received the B.S. degree from KyeongBook University, and the M.Sc. degree from KAIST (Korea Advance Institute of Science and Technology), Korea, in 1985, and 1987, respectively. He is a Senior Member of Technical Staff of ETRI Intelligent Network Service Section. His current research interests include Service Creation Environment, Feature Interactions, and Formal Description Techniques.

Part Six
Multimedia and Mobility Services

18
IN AND MOBILITY

Terje Jensen
Telenor Research and Development
Instituttveien 23, Postboks 83,N 2007 Kjeller, Norway
Tel: +47 63 80 91 00, Fax: +47 63 81 00 76
E-mail: terje.jensen@tf.telenor.no

Abstract

Intelligent Network (IN) is a platform for supporting telecommunication services. Mobile services are expected to be a significant fraction of the service demand, counted in number of users and the number of calls. Therefore, the support of such services by IN are considered with interest by several organisations. This is also valid for the use of IN to allow for service differentiation by the mobile network operators.

In this paper, classes of mobile services are defined and the corresponding requirements from each of these stated on the network capabilities are given. It is then indicated how IN has the potential to handle these services.

1 INTRODUCTION

Mobile services are among the most intensive increasing markets in telecommunications. Several forecasts have predicted that these services will still be handling a larger part of the users as well as the number of calls in the future. Still, however, much have been invested in the telecommunications infrastructure. Therefore, as faced by the operators, the question of how to utilise the equipment in a better way becomes more pronounced. This is understood in the sense that several services and load can be handled by the installed network. In some ways the mobile services could be described as providing added value to the basic services. On the other hand, several basic aspects could be requested for such services leading to other sets of requirements for the telecommunication network capabilities.

Intelligent Network (IN) is a concept which seems to be introduced in most networks, mainly because of the possibilities for rapid service deployment and flexible allocation of functionality. Two other factors associated with IN are the separation of the control/service domain from the switching/transmission domain and the use of building blocks to compose services. Both of these aspects will support the construction of more services and service features. Basically, we may say that IN could be a platform for offering services when the focus has been placed on the operators/providers role.

A number of other roles can be identified as well. The main roles are the service subscriber, service user, access provider, network operator and service provider. The service subscriber makes a contract with a service provider and is economical responsible for one or more users covered by the subscription. The user is authorised by the subscriber to utilise certain services within specific limits. An access provider handles the equipment which enables the users to access their services. A network operator is, as a minimum, responsible for the exchanges and transmission equipment. More equipment could also be provided, like control nodes and data bases. A service provider is responsible for the subscriber handling and data base containing subscriptions and users. Several of these roles could be carried out by the same person/company. For instance, the service provider and the network operator could be the same company (vertical integration). A number of other roles can be identified as well, like the regulator, service designer and equipment manufacturer.

As seen from IN, mobility could be supported by defining appropriate sets of services. As will be described later, a number of other elements could also be required for some implementations of mobile services. Often, the phrase "any one , any where, any time and any form" has been seen when describing the future of such services. This do allow for increased flexibility for the users. On the other hand, as seen from the operators/providers, the combination of IN and mobility is also very interesting. A number of further applications could be introduced and exposed to the market.

In this paper the relationships between IN and mobile services will be examined. That is, the focus is placed on the fixed network side. On the other hand, mobile services are often associated with wireless access. Aspects related to that portion of the service provision are not treated here.

The objectives of this paper are to clarify a number of terms related to mobility and describe some possible implementations for supporting these services. As the title reveals, a specific weight is placed on the IN based implementation. Naturally, other ways of providing the services could also be thought of.

In Section 2 the mobility classes are described. In the following two sections the two main classes of mobility, terminal mobility and personal mobility, are treated in more detail. A few aspects of the existence of the various mobile services and the use of IN are looked at in Section 5, before some conclusions are given in the last section.

2 CLASSES OF MOBILITY

Mobile services can be classified into several main groups. Depending on the criteria used, these could be further divided into subgroups. This is only illustrated for terminal mobility.

2.1 Terminal mobility

Although terminal mobility often is interpreted as involving a radio connection, a broader categorisation is done in the next points, e.g., ref. [E.50301]:

- Fixed: That is, no mobility at all. In the widest sense, it is not possible to disconnect the terminal from the access. No requirements are imposed on the network capabilities after the initialisation of the configuration.
- Portable: The terminal can change access points and will assume the identity and the profile of the access point. No additional requirements are imposed on the network capabilities as the access points carry the identifications as seen from the network side.
- Movable: The terminal keeps its identify after it has been moved between access points. Access to the services is not possible while moving between access points. This implies that the network must have capabilities for mapping between the terminal identity and the access identity. In addition, more information associated with the terminal identity could be stored in the network.
- Mobile: The terminal keeps its identity and the possibility to access the services also while moving. In general, this requires that the network has capabilities for knowing the terminal whereabouts.

We may define the level of terminal mobility to increase from the fixed (no mobility) to mobile. Therefore, the mobile terminal includes the others as subsets considering the requirements stated on the network capabilities. Mobile terminals also imply the use of wireless access. That is, when the services are available while moving, a wireline can not connect the terminal to the network.

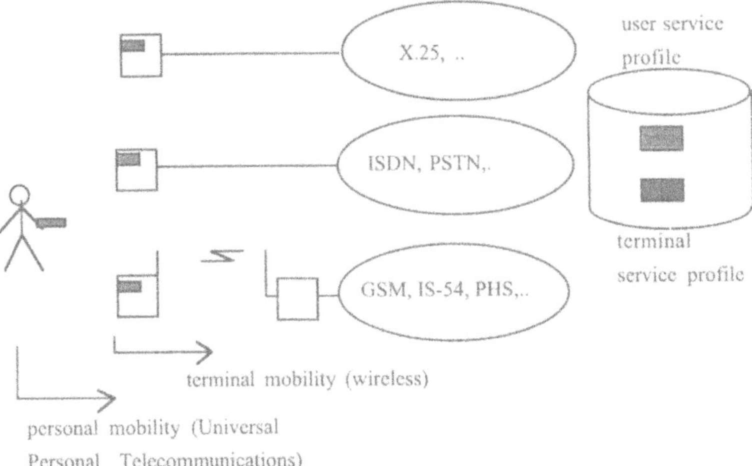

Figure 1 Illustrating different levels of mobility.

2.2 Personal mobility

In case of personal mobility, the user (person) has an identity associated. This allows, for instance, a call to be directed to a user and not to the terminal where the user is supposed to stay. Moreover, the user can utilise a number of terminals in order to access the telecommunication services. Terminals of all the mobility subclasses could support personal mobility.

In order for the network to provide this mobility, some capabilities for mapping between the personal identity to the terminal/access identity must be present. From this it also follows that information of the users' whereabouts must be available in the network. Personal mobility may also be defined with several levels, each adding requirements for functions implemented in the network.

2.3 Service mobility

The feature named service mobility is often seen to include the personal mobility for more than one network. In addition, the user has access to his/her own defined set of services from different access points (and networks). Depending on the information given in the service subscription, limited versions of the services could be available in some networks. Naturally, this could further be restricted by the capabilities of the network and the current terminal.

The requirements on the network capabilities are similar to those resulting from personal mobility when considering the mapping possibilities. In addition, the user defined services must be available through the networks which are covered by this mobility feature.

2.4 Relationships between mobility classes

The different mobility classes could be thought of as sets. Then, often, service mobility is included in the set called personal mobility. That is, service mobility implies that personal mobility is present. On the other hand, terminal mobility could be supported without implementing personal mobility and vice versa. This means that terminal mobility and personal mobility is partly overlapping, however, having non-overlapping parts as well.

A schematic illustration of the mobility classes is given in Figure 1. A few networks are depicted with belonging terminals and access points. As stated earlier, terminal mobility often is associated with wireless access. As personal mobility is associated with the user, several terminals could be utilised by the user in order to access the services. These terminals have wireless or wireline connections to the corresponding networks. In this sense, personal mobility could be thought of as a higher level of mobility compared to terminal mobility. A database is included to show that capabilities for storing and retrieving service profiles can be present. This may allow, for instance, the user to access similar set of supplementary services from the different networks. For example, when abbreviated dialling is subscribed to, the same abbreviated number could be used both from an GSM (Global System for Mobile communication) terminal and from a terminal connected to PSTN (Public Switched Telephone

Network). For terminal mobility, some information about the terminal itself could also be requested.

We may also say that terminal mobility is provided in close relationship with the systems, that is, it is a major characteristics of the system. On the other hand, personal mobility can be looked upon as a service provided by one or more systems.

One of the challenges by personal mobility is to support a consistent user interface over the different networks. Considering the example of abbreviated dialling, the sequence of dialled digits should be the same in the different networks. In most cases, this means that the different networks must have access to the data about the user. Personal mobility also introduces a number of other questions, like how to treat the fact that more than one user could be registered to the terminal at the same time. In particular, the problems raised by feature interaction must be looked into. Flexible and user-friendly solutions are requested for several issues.

3 TERMINAL MOBILITY

Although several terminal mobility subclasses were described in Section 2.1, only the one implying wireless access will be treated here. This is also regarded as the most interesting one in this context. Terminal mobility in this sense can also be called continuos mobility as the services may be accessed while moving.

3.1 Applications of terminal mobility

Terminal mobility could be applied in a number of environments each described by a set of characteristics. Typical applications can be domestic (private houses, flats, etc.), business (offices, storage rooms, production halls, etc.), vehicular (busses, air planes, trains, etc.) and public coverage. The listed environments are not exhaustive and not clearly separated. A set of other criteria could also be used to describe the areas where wireless systems can be used. Immediately, we recognize that these areas may have different users densities and different radio signal propagation characteristics. This invites for the introduction of different coverage areas for base stations. For example, a indoor base station may cover an area of 10 m radius while a base station in a satellite may cover an area of some hundreds of km in radius. In between these ranges there may be several other sizes depending on the traffic demand and the geography.

Recognizing the various environments also invites for the presence of a number of types of operators. Typically, a domestic area would be covered by a (set of) base station(s) provided by the persons living in that area, like a family could own the base station covering the house they are living in. In a similar way, a company could provide the set of base stations needed to allow for wireless access for the employee of that company. A third type of operators could be those providing public services. This service provision, however, could be for a restricted geographical area as well as limited in other ways. With all these potential operators, there must be corresponding lists of authorisation, that is, which users that can be allowed to use the operators' base stations. For example, the neighbour may not be allowed to utilise the base station owned by a family.

A number of classes of mobile terminals may also be present. These could be classified according to the usage of services and access to base stations (authorisation list). Service usage is, naturally, strongly correlated to the capabilities of the terminal, like what kind of audio/video functions that are implemented, the available battery capacity, and so forth.

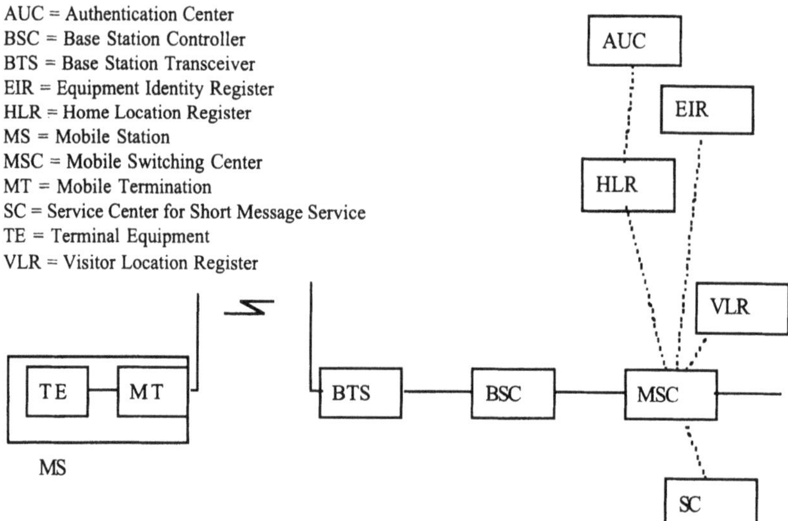

Figure 2 Example of entities present in a mobile system.

As the mobile terminal moves around between the different areas, often, the continuous use of telecommunications services is requested. This means that the functions supporting mobility in the different environments must interwork. As there may be several types of operators present, the definition of standard interfaces and procedures seems to be necessary. These are topics for the standardisation activities on future mobile systems.

The different wireless access systems present today support various sets of services. Therefore, it is simpler to compare the information bit rates the systems can provide. The more common digital mobile communications systems (e.g., GSM) can provide information bit rates of 9.6 kbits/s (data) or 13 kbits/s (voice). In some digital cordless systems (e.g., DECT, Digital European Cordless Telecommunication), several time slots could be utilised in order to arrive at information rates around a few hundreds of kbits/s. A single time slot in DECT, however, can support 32 kbits/s. In future systems, like FPLMTS (Future Public Land Mobile Telecommunications Systems), around 2 Mbits/s for information rates are examined, e.g., see [R.8.1.95]. Even higher bit rates are also studied for some applications, like MBS (Mobile Broadband Service), e.g., see [Fern95]. This means that most of the services present

BCF = Bearer Control Function
CCF = Call Control Function
MBCF = Mobile Bearer Control Function
MCCF = Mobile Call Control Function
MCF = Mobile Control Function
MSF = Mobile Storage Function
RACF = Radio Associated Control Function
RBCF = Radio Bearer Control Function
SCAF = Service Control Agent Function
SCF(M) = Service Control Function (Mobile)
SDF(M) = Service Data Function (Mobile)
SRF = Special Resource Function
SSF = Service Switching Function

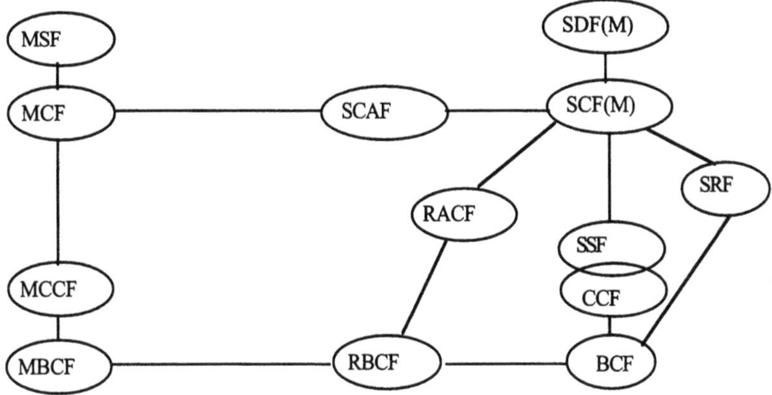

Figure 3 Functional model under examination for FPLMTS.

by wireline access also can be available by wireless access. In addition, mobile specific services, like location and navigation, could be supported. However, several topics must be further examined and a number of questions answered before such situations are reached. Naturally, all the services may not be available in all the environments outlined above. The highest bit rates may be provided in more local areas, while public access with traditional coverage areas may offer services supporting around 100 to 200 kbits/s.

3.2 Entities in mobile systems

Several of the mobile systems have defined similar entities in order to support terminal mobility, as illustrated in Figure 2. Beginning from the user side, a terminal and mobile termination may compose a mobile station. This can be connected to a base station by a radio interface. A base station subsystem can be composed of a base station transceiver and a base station controller. Several base station transceivers may be connected to one base station controller. A number of base stations are connected to a switching node (exchange). In addition to these basic entities, others including the necessary control and data base functions are usually present. Still more entities have been described as well, like the short message service center in GSM.

The functional model presented for systems supporting terminal mobility depends on the capabilities of the system. In Figure 3 one model proposed for FPLMTS is depicted.

Basically, we may recognize the IN functional model with SDF, SCF, SSF, CCF and SRF. In addition, the call and bearer (connection) may be separated. The bearer control functions are found in the lowest part in the illustration. Functional entities have been added because of the presence of a wireless access interface, like RBCF and RACF. There must also be functions capable of dealing with non-call related activities, like SCAF. These are typically used for mobility management. The different functional entities can be located in the terminal, access network and core network. However, a flexible allocation could be expected. Because of the capabilities of the mobile terminal, several of the functions in the network have a corresponding entity in the terminal. The control and data functions can be enhanced to support the mobile specific requirements. In the SSF we may find both a Basic Call State Machine (BCSM) and a Handover State Machine. The latter models the bearers in a way similar to the way the BCSM models the call and is, again, an effect of the mobility of the terminal.

3.2 Terminal mobility procedures

Because of the fact that mobile terminals can move around, a number of procedures must be enhanced or added compared to the non-mobile terminals. Some additional of such procedures are needed for:

- Call set-up: As the terminal may move, functions for locating the terminals before the call can be established are needed. This includes data bases and corresponding control functions.
- Handover: When the mobile, having an established call, finds that the quality of the connection towards the base station in use is below a specified threshold, a new base station may be taking over the call.
- Location update: A mobile without an established call may find that it has changed to a new geographical area (location area) and informs the network about this event.
- Paging: When a call is initiated towards a mobile terminal, the network must page for the terminal in order to locate it within the preregistered location area.
- Authentication: As the mobile terminal changes its connection point towards the network there is a need to make sure that the terminal and the network are the ones they pretend to be.
- Attach/Detach: In order to reduce the number of unsuccessful paging messages for terminals that are turned off, the terminal could send an indication of its change of status when turned on/off.

Each of these procedures implies additional requirements to the telecommunication network compared to when only wireline terminals are connected. Other mechanisms for interconnectivity could also be relevant as mobiles may access the network through different operators. In particular, arrangements for charging and accounting must be defined.

In future systems the service control and data are expected to become more distributed. Simultaneous procedures will also be required, like several services and mobility procedures could be active at the same time. In order to allow for improvement of the resource usage after handovers, bearer and call control should be more clearly separated. In addition, non-call related procedures must be allowed for. All these aspects have impact on the evolution of IN.

4 PERSONAL MOBILITY

Universal Personal Telecommunications (UPT) will be used for the discussion of personal (user) mobility. We may say that UPT arose from merging two telecommunications networks capabilities, that is, wireless access and intelligent network services. The first should allow for some mobility while the latter refers to UPT as a service based on a telecommunications platform. UPT can also be said to offer discrete mobility in the sense that specific actions are needed when changing connection points towards the network in order to still be accessible. For instance, the user may need to explicitly register on a terminal.

4.1 Personal mobility features

Some reasons for introducing UPT are to allow for increased flexibility, personification and availability (according to the user's requests). These may be achieved by utilising features like, e.g., ref. [T.F.850]:

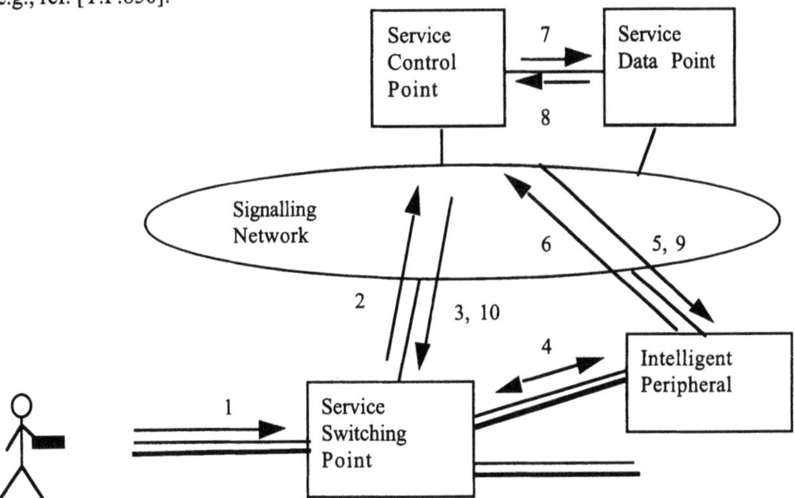

Figure 4. Sequence of interactions for registration to a terminal.

- Personal numbering: A unique number identifies the UPT user and is used by the caller to reach that UPT user. That is, the number is associated with the user and is transparent for the networks supporting UPT.
- Service profile: The profile can include a user defined set of services. That is, a list of services and facilities subscribed to by the subscriber on behalf of the user.
- Access from multiple networks: The service is available from several networks and there are common service control procedures for all these networks. However, support of the services in the service profile may be restricted by the terminal and the network operator/service provider as well as the subscription.
- Terminal independence: The user may dynamically register to terminals. A user may also be registered to several terminals for the different services.
- Security and privacy: The first term may include the use of passwords and other mechanisms for preventing that non-authorised persons are accepted to be the user. Also specific charging and barring facilities may be relevant. To take care of the privacy includes minimising the risk for revealing personal data and incorrect charging. In addition, third parties (e.g., terminal owners) should not be too disturbed by the UPT users' activities.
- Access device: Such a device may be used for atomising the interaction between the user and the network. The device may take different forms and be incorporated into equipment for other purposes. Programmable DTMF (Dual Tone Multi Frequency) transmitters are one way of implementing such devices.

4.2 Personal mobility procedures

In order to allow the UPT user utilising the services, a number of procedures have been defined. These are grouped into three categories named handling of personal mobility, UPT call and UPT profile management. Several procedures can be identified for each of these groups, like registration for incoming calls and registration for outgoing calls in the group named handling of personal mobility. The third group is special in the way that the service profile can be interrogated and modified. That is, how the service is provided by the network can be changed by the user. All these procedures involve the user. For several of them, activities not related to a user-to-user connection are defined, like between the user and the SCP, alternatively with an IP included in some phases of the activity.

The procedures are composed of a number of steps, starting from the access and finishing when the session ends. The access can be done by the user dialling the UPT access number. Then, the identity of the user should be given, e.g., by keying it on the pad of the phone set. In addition, the user authentication code must be provided by the user. After the combination of identity and authentication code has been verified, the user can continue by selecting the action to perform, which may be from one of the three groups described above. However, others or packages of actions could also be selectable. Ending that action, the session could be finished or a request for performing another action could be chosen (named Follow-on option).

A number of telecommunication network elements could be involved during these procedures, see Figure 4. The SSP would recognize the UPT access number and send a message to the SCP (2 in Figure 4). A reply could tell the SSP to establish a connection to an IP for translating the dialled UPT identity and authentication code from Dual Tone Multi-Frequency (DTMF) signals to digits. These digits can be sent from the IP to the SCP (6 in Figure 4) which composes a query to a SDP in order to verify the user provided information. The procedure can continue with the exchange of several messages and the involvement of more network elements.

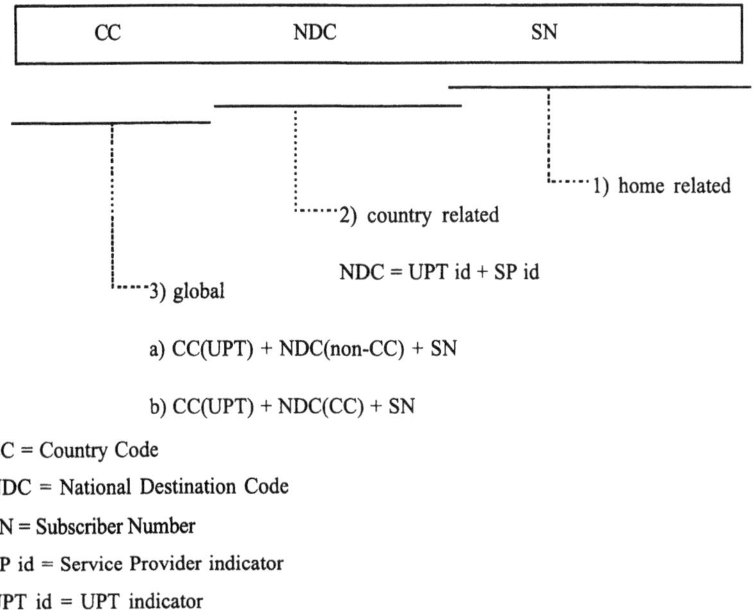

Figure 5 Possible numbering schemes for UPT.

Summing the number of digits provided by the user during such a session, we may find that this could be in the range of 25 - 30 digits just for making an outgoing call (without previous registration). In order to alleviate this procedure for the users, access devices seem to be requested. In addition, we see that a lot more signalling and processing are needed to serve the user requests. Data base look-ups could also be carried out during some phases of the call handling.

4.3 Network functions

The data related to a user could be stored in a user profile. In addition to keeping the identity and possibly the authentication code, other information for personalising the services could be given. This could, for instance, include routing tables, charging arrangements, customized voice announcements, and so forth. In particular, routing tables could allow for defining default accesses where the user usually stays at certain periods of the days. References to voice mail or other answering services could also be included.

Different numbering schemes for UPT are found in [T.E.168], see Figure 5. A number can be composed of a country code (CC), a national destination code (NDC) and a subscriber number (SN). Three schemes are defined based on these fields:

1) Home related: That is, the difference between the number of an UPT user and other users can only be found by looking at the subscriber number field. This implies that only the home network of the UPT user can recognize that it is a call to an UPT user.
2) Country related: That is, the NDC field can be composed of an UPT indicator and an service provider indicator. The first will only tell that this is an UPT user, while the latter identifies the provider of the UPT service.
3) Global: That is, an indication of the UPT number is found in the country code field. This number can then be recognized in all the networks supporting the UPT service.

As an alternative to these schemes, the use of a special prefix has been described. By dialling this prefix, it is directly indicated that the call is for an UPT user. Which scheme that is chosen, will influence the possible control architectures as returned to later.

The requirements stated by the UPT service on the network capabilities can be listed as:

- Recognizing the selected numbering plan: That is, the UPT numbers and the UPT procedures must be recognized.
- Dynamic registration of users to access points/terminals: This means that dynamic linking between the user and access points/terminals must be possible, and information about how to reach the user must be available. It also includes the use of authentication in order to ensure that the user is still present in case a call is presented at the given terminal.
- Access from multiple networks: This may imply the exchange of information between the visited network and the home network of an UPT user. Proper procedures for this have to be established. In addition, consistent user interface should be achieved.
- Storage of user specific service data: In particular, service profile should be allowed for in order to personalise the service.
- Flexible charging/billing arrangements: Special arrangements could be requested, like credit limits for a period, split charging, etc. Procedures for accounting between operators could also be needed.

These are basic requirements on the networks supporting the UPT service. Others could also be given that may increase the usability of the service, like the support of access devices. We may find that the use of IN could fulfill these requirements. However, a number of selections and arrangements must be established, in particular, when several networks and operators are involved.

5 SUPPORT OF MOBILITY SERVICES

Several services could be integrated in a telecommunications network in a number of ways. The most basic integration level may be common use of equipment, like the services are handled by the same exchange. Then, additional integration schemes like common signalling and functions can be defined. Although achieving a high level of integration may seem to be requested from an economical point of view, it could imply that several trade-offs must be done. This may again mean that the resulting solution might not be the better one for any of the services/applications.

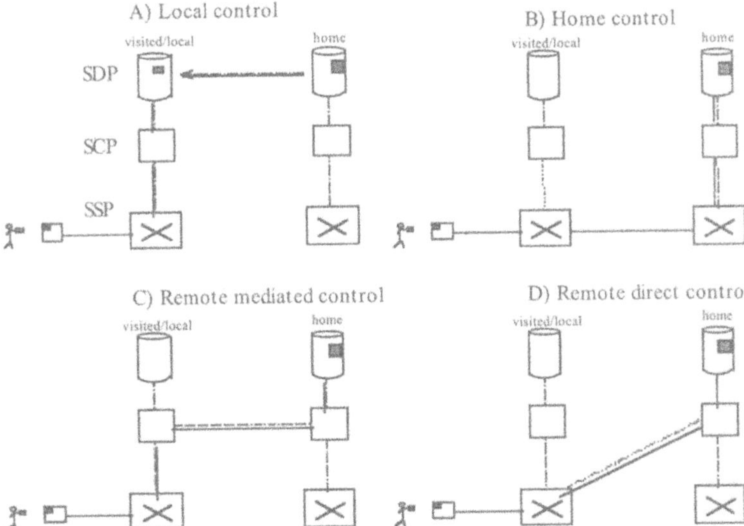

Figure 6 Some possible architectures for providing the service through a visited operator's network.

Another question is how the services are supported by a sequence of network operators, that is, how to handle the fact that users may want to access the services through the network handled by another operator. Some cases for enabling the support of services in several network have been defined. These could then be compared according to a set of evaluation criteria.

The use of standardised service profiles and service logic could be regarded as an advantage, but may also restrict the possibilities for operators to introduce better service features in the competition with others. This may also complicate the deployment of services when different versions of the network capabilities are present. Standardised profiles may allow them to be transferred between the different operators/providers. On the other hand, such transfers do also mean that customer information is revealed to the other operator/provider. The use of standardised interfaces is usually preferred. In case a solution requires the presence of an interface not treated in any standards, special arrangements must be agreed upon between the two parties. In particular, this must be treated in more detail when one operator is able to control the equipment provided by another operator.

The user may notice the interface, the set of services available and the quality of the services. The last one includes the delays for the procedures involved. Normally, the involvement of several elements leads to an increased delay. Longer signalling relations and connections may also imply a higher cost as seen from the operator side.

Different cases can be illustrated by looking at the configuration arising when a user, having a subscription in the home network, accesses the services through a visited network, see Figure 6. Some architectures for supporting the services over more network operators can then be classified like:

A) Local control: Where the service profile is transferred to the visited network when the user is registered in the visited network. That is, a relation between the two data bases is assumed.
B) Home control: Where the service is controlled in the home network. This implies that a connection to the home network must be established.
C) Remote mediated control: Where the service execution is controlled from the home network. The service control messages are mediated by the local service control.
D) Remote direct control: Where the service execution is controlled from the home network by direct accessing the switching functions in the visited network.

The different architectures can be compared according to the criteria outlined above. We find that local control could require that service profiles are standardised and that customer information will be revealed. Thus, allowing for special services and new versions could be more difficult. Home control may result in long connections although it may be better according to other criteria concerning independence and security. Remote direct control allows another operator to set parameters in the SSP, which implies that the corresponding call handling at that level must be defined.

6 CONCLUSIONS

Mobile services are among the telecommunication services with highest annual growth. As described above, these may be classified as terminal and personal/service mobility. These classes are not disjoint. However, as they provide mobility, we may face situations where they are competing in order to serve the demand. Some characteristics are relevant for each of these groups: As wireless access utilises radio propagation, the capacity is normally restricted (counted in kbits/s and number of users active simultaneously). Personal mobility is supposed to cover several networks including wireless. Personal mobility is associated with discrete mobility, therefore certain activities are requested from the user in order to achieve this mobility. For this, the use of access devices might be crucial, otherwise the procedures could be rather heavy. To a certain extent, this could be alleviated by the use of wireless access services for informing the network about the user's location. After knowing this, the network could have corresponding attributes given in service profiles and algorithms to select the most appropriate way of utilising the telecommunication network when a service is activated.

IN is a concept under standardisation which enables interoperability and provides an interface towards the user. In addition, management systems have been defined in order to support the services provided by the IN platform. UPT, which has been an implementation of personal/service mobility, can be regarded as a service executed on the IN platform. Several mobile networks have introduced a concept very similar to IN. Access to IN-based services through wireless is also regarded as a necessity in order to allow for further diversification for the operators.

As, by nature, mobile services allow the users to change locations, appropriate capabilities in the network must be defined to support these activities. This include the dynamic updating of location and routing. Furthermore, authentication procedures are necessary because the user is not identified by the access point that is used. Individualised service profiles and charging/billing arrangements are also requested. This implies that much data could be needed for providing the services, also data associated with each user.

Because an advanced level of service control is anticipated, the control activity could be allocated to separate nodes. According to the conceptual model, Intelligent Networks meet these requirements. However, some enhancements are foreseen to support the mobile services. How to include these features in the IN concept, is examined in the standardisation organisations.

It is expected that further services will be based on IN and that more systems will adapt to the IN concept. Naturally, other concepts may replace IN after a while, but the mobile services have to be taken into account for all the service control platforms.

7 REFERENCES

[E.50301] ETSI Technical Report Draft-ETR/SMG 50301: Framework of network architecture, interworking and integration for the Universal Mobile Telecommunications System (UMTS). Version 0.2.3. June, 1992.

[Fern95] Leandro Fernandes: Developing a System Concept and Technologies for Mobile Broadband Communications. IEEE Personal Communications. Feb. 1995, p. 54-59.

[R.8.1.95] ITU-R: FPLMTS/IMT-2000. Report of the Ninth Meeting of ITU-R Task Group 8/1. Tokyo, Japan. Sept. 1995.

[T.E.168] ITU-T rec. E.168: Application of E.164 numbering plan for UPT. Geneva, 1993.

[T.F.850] ITU-T rec. F.850: Principles of Universal Personal Telecommunication (UPT). Geneva, 1993.

8 BIOGRAPHY

Terje Jensen, Ph.D., is currently working as research scientist in the strategic network development group at Telenor Research and Development. The work interests include performance models and analyses of telecommunication networks and services. In particular network dimensioning issues, with main emphasis on Intelligent Networks, have been considered so far. Prior to receiving his Ph.D. degree he was also involved in administration and operational support for introduction of digital exchanges while working as senior engineer in Northern region of Norway. He has taken part in several international projects related to network dimension issues.

19
CALL PROCESSING MODEL FOR MULTIMEDIA SERVICES

Olli Martikainen
Telecom Finland Ltd.
P.O. Box 106, FIN-00511, Helsinki, Finland
Phone: +358 2040 3503, Fax: +358 2040 3251
E-mail: olli.martikainen@tele.telebox.fi

Valeri Naoumov, Konstantine Samouylov
Russian University of Peoples' Friendship
P.O. Box 9, 117419 Moscow, Russia
Phone: +7 095 955 {0863, 0956}, Fax: +7 095 952 2823
E-mail: {vnaoumov, ksam}@udn.msk.su

Abstract

To support a wide range of multimedia services a B-ISDN network capabilities should allow the distribution of service functions between the customer terminals and the network nodes. Intelligent networks (IN) can be considered as the bridging technology between the customer based service management and the service execution in the networks. Considering the mobile and broadband technologies it is a natural question, how the concept of intelligence will develop. To satisfy broadband mobile and multimedia service needs IN should support much more distributed operations. It will undoubtedly lead to the increase of the volume of signaling messages. To carry a large number of messages IN must use enhanced call processing model instead of current Basic Call Model. In the paper a distributed call processing model for multimedia services for broadband intelligent networks is proposed and argued for.

1 INTRODUCTION

With the research on multimedia and broadband mobile services there arises a need to control complex combinations of service components and resources. Examples of these services are digital interactive TV, video on demand services for banking, shopping and leisure, electronic press and publishing. The technological requirements for these services are cost effective broadband transmission and access technologies, flexible computer based management and control of networks, switching and service applications and the support of mobility. The broadband transmission and switching technology is also maturing and will provide a cost effective platform for service provision. When the broadband customer access will be available, then interactive business and consumer services based on video and multimedia will become possible. Common to all these developments will be the computer controlled structure of modern services, where protocols, application technology and resource management are key factors.

In modern telephony the most promising computer control and management concept is the Intelligent Network (IN). It facilitates the development and management of new services by service providers. Intelligent networks can be considered as the bridging technology between the customer based service management and the service execution in the networks. It seems to be clear that the service applications will be distributed over terminals and different service control and provider nodes. An example of distributed intelligence can be found in the specifications of Universal Mobile Telecommunications Services [UMTS] and the Advanced Intelligent Mobile Communications Network [AIMN]. Different possibilities to introduce and distribute intelligence in future networks have been considered in [FUTU] and [LOCA].

A number of modifications and improvements to IN and call processing are necessary before future mobile telecommunication systems can be implemented. In this paper we propose a distributed call processing model for multimedia services in broadband intelligent networks.

2 MULTIMEDIA SERVICES

Multimedia service is a service in which the interchanged information consists of more than one type (e.g. video, data, voice, graphics). Monomedia communications related to single information types are described in a parts of a multimedia service called service components. Multimedia services may offer several virtual connections, one virtual connection for each service component, and each of the virtual connections may have particular QOS attribute values [AUNI], [I.374].

The service provision mechanisms would become in the future different from those of today.

The subscriber subscribes to and pays for services offered by one or more service providers on behalf of the real user of the services (customer). It is possible that the customer and the subscriber are the same person. On the other hand various subscribers could be linked to the same customer.

The terminal can be available for the use by many users and provides a level of service that is customized to the user. This implies that each customer must be identified when accessing to the network. The content control allows the customer to influence the content of the service. The customer can manipulate on-line the quality of the receiving service component by the modification of QOS parameters during the connection life time. Different multimedia services may require a service negotiation by the involved terminal to check if they can serve the same type of multimedia service or not.

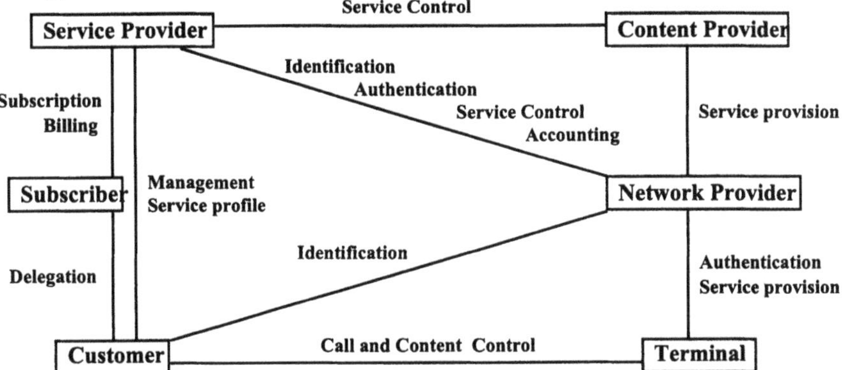

Figure 1 Multimedia service provision model.

The network provider provides standardized network services including security access to the network, individual connection links for a session and management of the service provider directory. The main function of a service provider is to design the interface for the customer. This includes the customer registration and identification, easy to use graphical interface, management of customer service profiles and billing/charging for the service. It performs also the session management function and end-to end service control. Service provider is responsible for agreements with network provider on network usage.

The primary function of a content provider is the storage and playback of multimedia content. For interactive services, it controls the playback content based on direct real-time interactions with the customer. This control may require access to and interaction with the network provided by the service provider.

3 MOBILITY AND IN

It is expected that the majority of new annual telephone subscriptions will be mobile-based after 1997. International Mobile Telecommunications 2000, IMT2000, formerly known as Future Public Land Mobile Telecommunication System, FPLMTS, should provide mobile telecommunications that support universal roaming and offer broadband multimedia services. To achieve in IMT2000 the Universal Mobile Telecommunication System, UMTS, is specified by ETSI [UMTS], and the Advanced Intelligent Mobile Communications Network, A-IMN, is developed by NTT [AIMN]. In mobile services the service mobility is supported by the user location updating, which is a non-call associated function, and by handovers during the call. In handover the bearer connections are transfered from a base station to another depending on the best quality available in the adjacent radio transmission cells.

The primary objective of IMT2000 is the service portability. It is obtained by terminal mobility and personal mobility. In addition to the basic IN Call Control Access Function (CCAF) for user call access, Call Control Function (CCF) for call processing control and Service Switching Function (SSF) for interaction between CCF and Service Control Function (SCF) [Q.1204], new functional entities have become necessary.

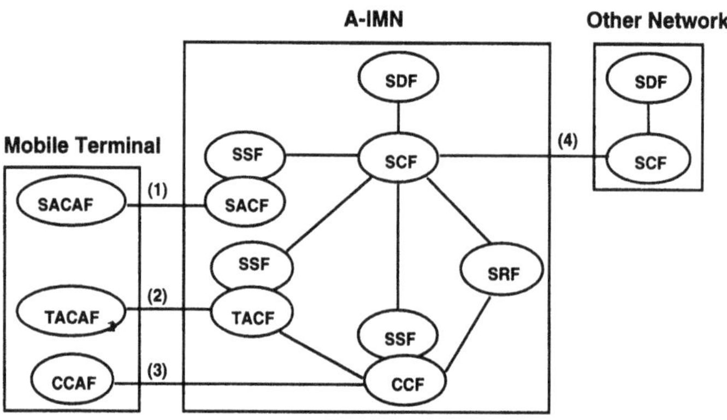

(1) - Non-Call/Non-Bearer Access Control Relationship
(2) - Terminal Access Control Relationship
(3) - Call Control Access Relationship
(4) - Inter-Network Control Relationship

Figure 2 IN Distributed Functional Plane Architecture introduced by A-IMN [AIMN].

Terminal mobility means the possibility to access services at anytime and anywhere. It implies terminal access control including terminal paging and terminal authentication for the authorization of the right to establish the bearer connection. For this control two new IN functional entities have been introduced: Terminal Access Control Function (TACF) and Terminal Access Control Agent Function (TACAF).

Personal mobility allows the identified customer to access services from any terminal. Several identification mechanisms can be used, e.g. manual PIN entry, biometric identification and smart card based identification [DATA].

Service profiles for service portability must be handled according to customer requests. Such non-call and non-bearer control includes terminal contation updating, customer registration, and service profile interrogation/modification. For this control Service Access Control Function (SACF) and Service Access Control Agent Function (SACAF) functional entities have been introduced.

4 MULTIMEDIA AND IN

Current IN call processing functions representation called the Basic Call Model must be enhanced to support much more distributed operations and events taking place outside a call.
A multimedia call model should support the ability to have several connections of the same or different types between any parties involved in a call. It has been widely agreed that future signaling protocols should support Call Control (CC) and Bearer Control (BC) separation [ISCP].

The Call Control is responsible for the establishment of an end-to-end communication configuration for users without allocating expensive network resources. No transit switches are involved in the Call Control. The Bearer (Connection) Control is responsible for establishing connections with the requested bandwidths. All transit switches have to be involved in the Bearer Control. The separation of Call and Bearer Control leads to the separation of Call Control Access Function (CCAF) and Bearer Control Access Function (BCAF).

Future multimedia services will require the allocation and on-line modification of special resources such as bridges and converters. Special resources may be both hardware devices and software entities providing functions such as compression, decompression, synchronization, security, multiparty branching and others. To control these specialized resources an additional Resource Control Protocol was proposed in [SIGN] and [COMP].

Quality of Service aims at producing a quantified assessment of user satisfaction. User satisfaction can be quantified according to the network's ability to fulfill service requirements, and the amount and flexibility of operational maintenance offered [E.800]. To maintain the QOS and to provide guaranteed QOS bearer and resource control are involved. For example degrading QOS caused by congestion can trigger rerouting of the call.

An often used example to create multimedia services is to use multi-channel or multi-party calls where different service components are obtained from different service nodes. Different service components can be displayed in different user interface windows which are created by the service application when needed. The question how to control these types of services has been studied in [ISCP] and [SUMM]. In [ISCP] the multimedia services are controlled by a Bridging/Branching node which resembles a present day switch with ISDN User Part. In [SUMM] an intelligent service control point called concept provider service node was introduced to control the multimedia service components.

5 CALL PROCESSING MODEL

In the mobile and multimedia services new control functions have been proposed. They are not necessarily user controlled or call related, but they trigger network intelligence depending on their triggering condition. We shall consider here different triggering conditions and the corresponding control functions and how they can be distributed in the network.

In the IN distributed functional plane architecture [Q.1204] the trigger event detection is centralized in the Basic Call State Model in CCF and the corresponding signalling interactions with SCF are managed by SSF. To include other types of triggering conditions and to make them communicate with corresponding control functions we propose the following new model.

Let us group trigger types and control function types and let us define signalling message interfaces provided by each control function. From the examples in previous chapters we can introduce the following trigger types:

- *Terminal access activated triggering* (TAT) - access rights to the network which can be in the terminal or in the network elements.
- *QOS activated triggering* (QAT) based on monitoring the service or bearer connection quality. Examples of this are
  - mobile location updating and handovers,
  - multimedia connection rerouting in case of degrading QOS.
- *Service component activated triggering* (SCAT) based on activating/deactivating or synchronising new service components with needed resource, e.g.
  - involvement of decompression resources in case of new service component stream,
  - synchronisation of video and audio components.
- *Service identifier activated triggering* (SIAT) based on the given service identifier (number), such as IN services with given service identifier (freephone with 0800).
- *User interaction activated triggering* (UIAT) providing control activated by user, such as
  - Calling card and voice services,
  - Universal Personal Telephony (UPT).
- *Service activated triggering* (SAT) controlled by the service logic in terminals or network nodes. For example the multimedia service uses another service as its part e.g. to open new window and to display there video segment from some other server.

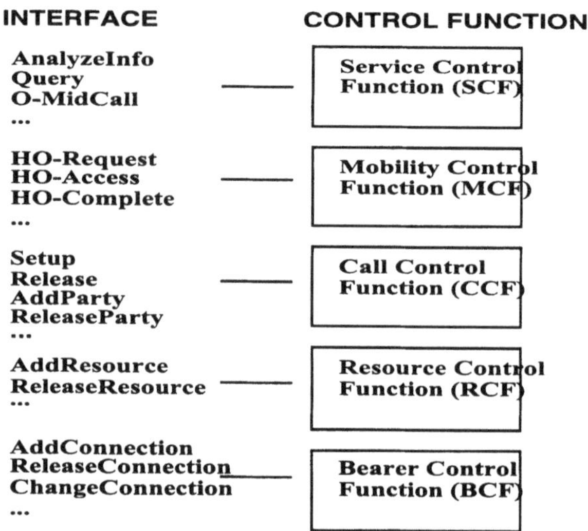

Figure 3 Example of Control Functions and their Interfaces

Each trigger type may communicate with the corresponding control functions:

- Service Control Function (SCF): SAT, TAT,
- Mobility Control Function (MCF): QAT, SAT, TAT
- Call Control Function (CCF): SIAT, UIAT, SCAT
- Resource Control Function (RCF): SCAT, QAT
- Bearer Control Function (BCF): SCAT, QAT

Here we have separated Mobility Control Function from SCF though in Chapter 3 they were both in SCF (Figure 2). The above control functions provide signalling message interfaces for the triggers (Figure 3). The SCF interface can be based on INAP operations [Q.1218] and the MCF interface can be based on GSM Handover procedures [ETSI].

Triggers may be considered as agents providing access to control functions. Hence we shall call entities providing triggering functions as trigger agents.

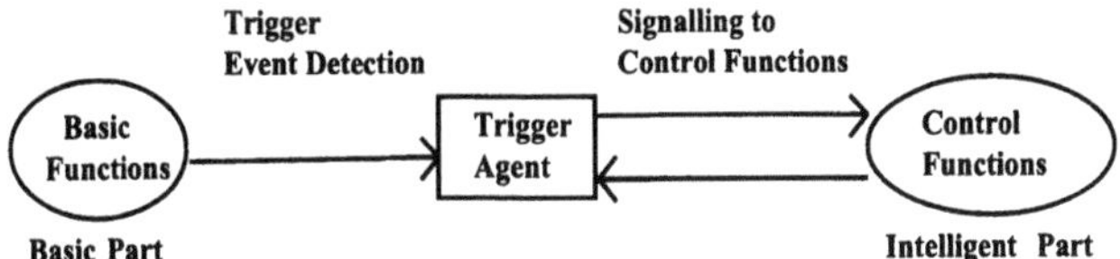

Figure 4 Trigger Agent

The trigger agents monitor the triggering conditions with triggering event detection. When some condition is met, the triggering agent sends signalling messages to the corresponding control functions provided by the network.

Trigger agents are mediators between the basic and additional (intelligent) functions in the network.

Intelligence is specified by giving the available control functions and the triggers which signal to them. The distribution of intelligence is described by placing the trigger and control functions in network entities. Network architecture specification describes how trigger agents and control functions are distributed in network entities.

A service can now be built on top of the basic (generic) control functions by introducing the trigger agents and their interactions with basic and additional control functions available (Figure 5). The trigger event detection can be distributed outside the CCF and the SSF can be considered as the interface specification for CCF based triggers and SCF.

There may be Mediation Points in the network which provide the control function signalling interfaces to Trigger Agents outside the teleoperator management domain and guarantee the separation of operator controlled network from user or service provider controlled network parts [MEDI].

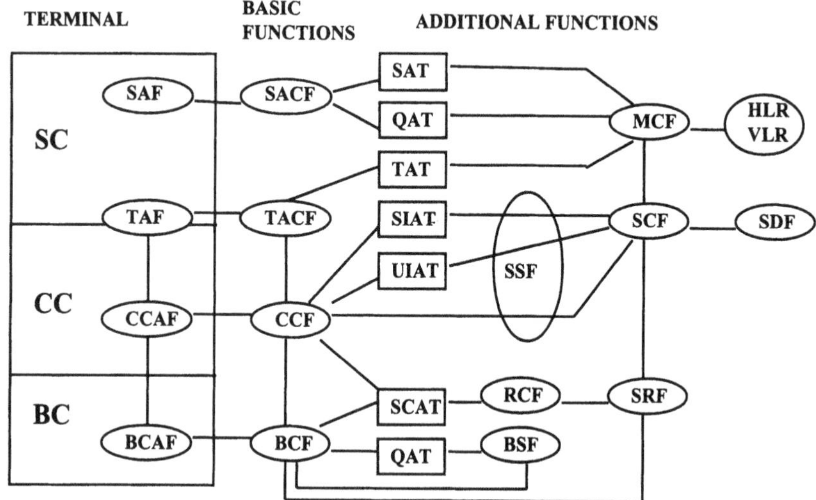

Figure 5 Examples of Control Functions and Triggers attached to them

6 EXAMPLES

With multimedia services the user application can have access to different service components with corresponding component SAT agent functions. In the user application the Trigger Agents can be represented by icons and they can be available from service providers (Figure 6).

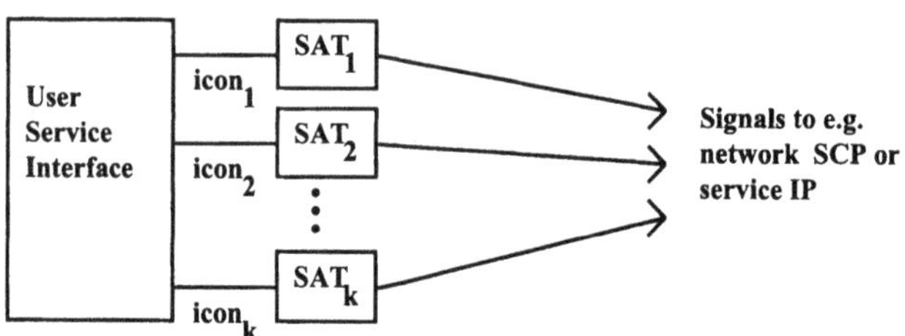

Figure 6 Trigger Agents as icons in user application

With connection less services the CCF is placed either in the user equipment or in the service nodes. In the next figure we present a connection less mobile service. The terminal in the figure contains some present day base station controller functions.

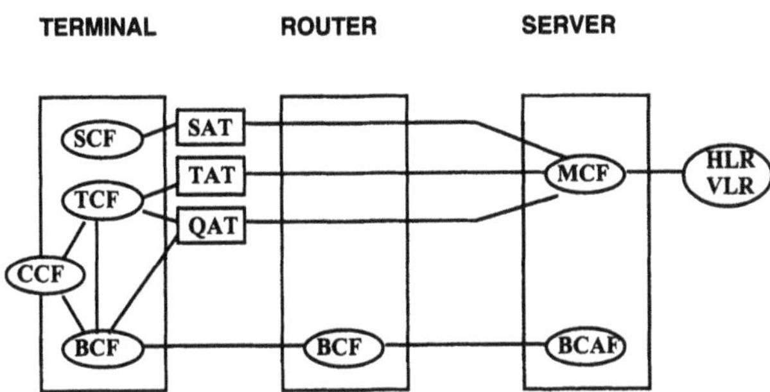

Figure 7 Terminal controlled connection less mobility

Let us finally consider a service provider operating a coordinating Service Control Point (SCP). The SCP is connected to the public network and it controls combinations of other services available in different service providers Service Nodes (SN).

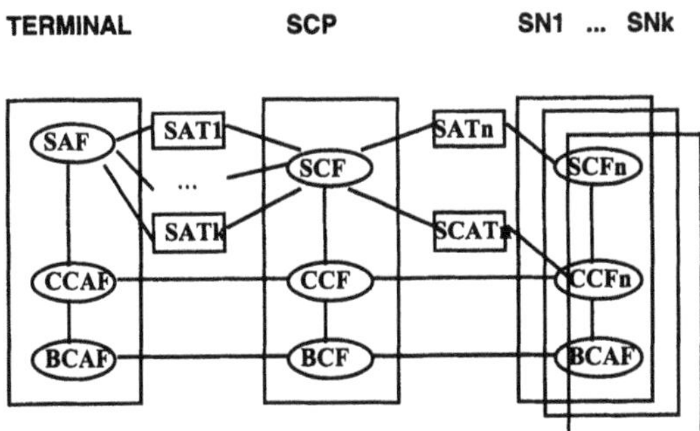

Figure 8 Service combined from several Service Nodes

The access to the separate services (j = 1, .., k) is provided by Trigger Agents SATj, which signal to the coordinating Service Control Function SCF. Each separate service SCFn can in turn invoke other services by Trigger Agent SATn signalling to the coordinating SCF. The coordinating SCP can provide the necessary agents SATj to the users and keep the user database for administration and billing. The use of services is measured by Agents SCATn in each

Service Node SNn, and the measurement results are signalled to the user database in the co-ordinating SCP.

7 REFERENCES

[AIMN] A. Nakajima, et. al. "Advanced Intelligent Mobile Communications Network", Proc. ISS '95, v. 2, April 1995, pp. 320-324.

[AUNI] ATM User-Network Interface Specification, Version 3.0, September 10, 1993.

[COMP] P. Hellemans, G. Reyniers "An Object-Oriented Approach to Controlling Complex Communication Configurations", Proc. ISS '95, v. 2, April 1995, pp. 72- 76.

[DATA] L. A. Craven, et. al. "Evolving Subscriber Data Management in Digital Switches to Provide Personal Communications Services", Proc. ISS '95, v. 2, April 1995, pp. 31-35.

[E.800] Terms and Definitions Related to the Quality Telecommunications Services, ITU-T Recommendation E.800, 1993.

[ETSI] European digital cellular telecommunications system (Phase 2); Handover Procedures, ETSI GSM 03.09, 1994.

[FUTU] D. Gaiti, G. Pujolle "Which Intelligence in Future Networks", Intelligent Networks, Chapman & Hall, 1995, pp. 280-290.

[I.374] Framework recommendation on "Network capabilities to support multimedia services", ITU-T Recommendation I.374, 1993.

[ISCP] ISCP Baseline Document, version 3.1, ITU-T Study Group 11 Report R10, June 1993.

[LOCA] H. Hämmäinen, P. Lahtinen "On Location of Service Control", Intelligent Networks, Chapman & Hall, 1995, pp. 246-253.

[MEDI] W. Heinmiller, et. al. "Solutions for Mediated Access to the Intelligent Network, Proc. ISS '95, v. 2, April 1995, pp. 217-221.

[Q.1204] Intelligent Network Distributed Functional Plane Architecture, ITU-T Recommendation Q.1204, 1993.

[Q.1218] Interface Recommendation for Intelligent Networks CS1, ITU-T Recommendation Q.1218, 1993.

[SIGN] S. Tohme', et. al. "Signaling Evolution for B-ISDN", Proc. Telcom 95, South-Africa, March 27—29, 1995.

[SUMM] O. Martikainen, et. al. "Comparison of Broadband Intelligent Network Architectures", Intelligent Networks, Chapman & Hall, 1995, pp. 253-269.

[UMTS] E. Buitenwerf, et. al. "UMTS: Fixed Network Issues and Design Options", IEEE Personal Communications, February 1995.

8 BIOGRAPHY

Konstantin E. Samouylov received the M.Sc. degree in mathematics from the Peoples' Friendship University of Russia (RPFU) in 1978 and the Ph.D. degree in probability and queuing theory from the Moscow State University in 1985. During 1985-1990 he held a position of associate professor at RPFU Faculty of Sciences (Information Technology laboratory). In 1991 he joined RPFU Computer Center where he is presently a vice-director. His current research interests are performance analysis of telecommunication networks, SS#7 network dimensioning and planning, IN signaling systems portable software development, broadband signaling and multimedia services.

Valeri Naoumov is Computer Center Director Research with People's Friendship University in Moscow, Russia. Areas of Expertise: queuing theory, computational algorithms, traffic models, intelligent network, signalling protocols, telecommunication protocols implementation. He has written many technical articles and two books. V. Naoumov holds Ph.D. in Applied Mathematics from Computer Centre, Russian Academy of Sciences (1979).

Olli Martikainen, Ph.D., from Helsinki University and M.Sc. from Helsinki University of Technology. He has been doing research in several positions in Helsinki University of Technology (1976 - 1982), Oxford University (1980 - 1981) and Technical Research Centre of Finland (1982 - 1985). Later he has been R&D manager at Nokia Electronics (1985 - 1986), department manager at Nokia Research Centre (1986 - 1988), professor and head of Datacommunications Institute at Lappeenranta University of Technology (1988 - 1989) and professor at Technical Research Centre (1989 - 1991). Since 1991 he has been research director at Telecom Finland. Currently he is vice president, R&D, at Telecom Finland Ltd., and professor at Helsinki University of Technology. His main areas of interest are telecommunication software methods and tools, network architectures, performance analysis and new industrial and economic structures in telecommunications.

20

A Generic Service Management Architecture for Multimedia Multipoint Communications

Constant GBAGUIDI, Simon ZNATY, Jean-Pierre HUBAUX
Swiss Federal Institute of Technology
Telecommunications Laboratory/Telecommunications Services Group
CH- 1015 Lausanne- Switzerland
Phone: +41 21 693 56 13, Fax: +41 21 693 26 83
E-mail: {gbaguidi, znaty, hubaux}@tcom.epfl.ch

Abstract

With rapid advances in multimedia processing technologies, and in high bandwidth network technologies, new kinds of multimedia applications are now emerging. These applications require real-time, multimedia, and multipoint interactions involving multiple communicating entities and leading to a need of more sophisticated communication control functionalities. In this paper, a generic integrated and flexible service management architecture for multimedia multipoint communications is proposed, which offers three main generic functions, namely Quality of Service (QoS) negotiation, QoS monitoring and control, and session management.

Keywords

Service Management.

1. INTRODUCTION

The availability of new technologies both in the network, where broadband transport facilities are being introduced, and at the user site, where terminals are providing for increasing intelligence and presentation capabilities, offers an unlimited number of opportunities for the introduction of new telecommunication and information networking services. Among these services are multimedia entertainment services, spanning from video on demand to interactive games, and communication services such as video conference and broadband virtual private networks.

A key issue to be tackled for providing such services is multimedia multipoint communications. For the time being these communications are not treated in an *integrated* way, that is, existing architectures do not cover the management of application resources, multimedia resources, and network resources altogether. As these resources cooperate to fulfil the users expectations, QoS requirements should not be defined statically for each given resource. For instance, when errors occur in the network resources, multimedia resources such as codecs must set their compression ratio to a low value. As a matter of fact, a low compression ratio introduces much redundancy so that the transmitted information can recover much easily from errors. The management architecture is expected to deal with this QoS adaptation between network and multimedia resources. Currently, this issue is not yet addressed.

In addition to integration, *flexibility* is another property to be required from management architectures. *Flexibility* allows for managing dynamically multimedia multipoint communications, in terms of communication session configuration (users participating in the session, involved media) and quality of service (QoS). Moreover, a *flexible* service architecture is not service-dependent and must be generic in order to be instantiated for any particular service.

To reach those objectives, namely *integration* and *flexibility*, both multimedia applications and broadband networks capabilities are to be considered. However, most of research works deal with either the former or the latter. Among these works, we can mention [2-5].

In [2], a classification of multimedia resources (also called special resources or media handling resources) is defined and a communication management architecture which exhibits two kinds of fabrics is proposed: SWFs (Switching Fabrics), concerned with classical switching functionalities (e.g., ATM switches), and MMSFs (Multimedia Multipoint Service Fabrics) concerned with value-added multimedia functionalities. Emphasis is put on meeting the users requirements, in terms of the possibility for the users to choose multimedia resources that best suit their communications needs. Although of prime importance, QoS issues mentioned above are not considered to provide the architecture with the *integration* property.

In [3], a generic multimedia service is described, in terms of service element (SE). Each SE is a "self-contained part of the service" and is composed of operations whose temporal ordering is specific. Two kinds of SEs are identified: confirmed SEs (SEs whose execution requires resource allocation) and unconfirmed SEs. Within a communication session eight service elements are outlined: call establishment SE, add users SE, add media SE, attach media SE, detach media SE, remove media SE, remove users SE and call release SE. Although it does not put emphasis on the methods for performing those service elements (e.g., which mechanisms should be achieved for adding media), that paper presents a broad view of the users requirements on a multimedia service.

In [4], the design of a QoS broker between multimedia applications and the network is presented. However multimedia resources are not mentioned, although they have a major impact on QoS. For instance, a high compression ratio decreases the needed bandwidth, and introduces more stringent constraints upon bit error rate than a lower compression ratio does.

Moreover, the Telecommunications Information Networking Architecture Consortium (TINA-C) has defined an "open" architecture for telecommunication services in the emerging broadband, multimedia and "information super-highway" era, based on distributed computing, object orientation, and integration of Intelligent Network (IN) and Telecommunications Management Network (TMN) [5]. Nevertheless, integrated multimedia multipoint communications are still under study. Even though TINA-C mentions the need of using "information converters" in a multimedia context, a QoS management framework taking into account the impact of these resources is not yet addressed.

In this paper, emphasis is put on multimedia service aspects, spanning from multimedia resources to communication management units. Undoubtedly, it is important to study carefully the service level before addressing networking issues particularly eventual effects of multimedia communications on connection management. On the contrary of the research works mentioned above, we first identify the generic functions to be held by call control, and then propose an integrated and flexible architecture to fulfil these functions. The structure of this paper is as illustrated in Figure 1.

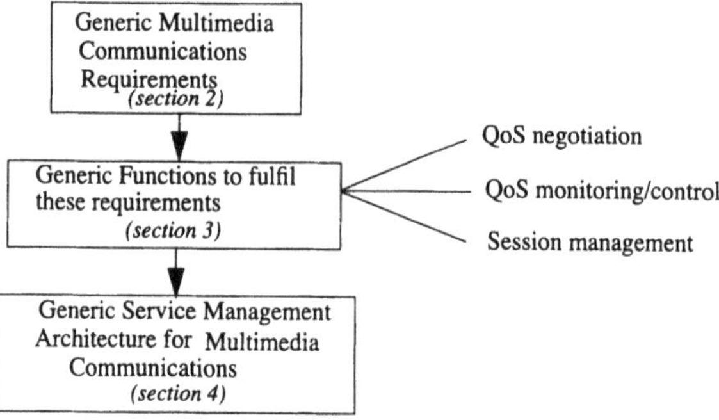

Figure 1 Structure of this document

In section 2, requirements over multimedia communications are outlined. Generic functions to be supported by the management architecture are derived from those requirements in section 3. This leads to the definition of a generic integrated and flexible service management architecture for multimedia multipoint communications in section 4.

2 THE GENERIC REQUIREMENTS OF MULTIMEDIA MULTIPOINT COMMUNICATIONS

Depending on the supporting service, constraints on multimedia multipoint communications are very different. For instance, telemedicine services have more severe requirements than classical videoconferencing services. However, both exhibit some common users expectations which are:

- The ability to manage dynamically the achieved QoS of each service component, may it be the audio, the video, the graphics, the picture information, and so forth. As QoS is highly subjective, it may be tuned by each party involved in the communication session. This requirement implies QoS adaptation, i.e., the possibility for end-systems to adapt the QoS to the behaviour of the network.
- The possibility for users to select some resources (network resources as well as media handling resources) that best meet their needs. As in a multimedia context, quality of service also depends upon factors such as coding techniques, quantization, and so on, the users must be proposed to choose resources according to their expectations and the capabilities of their terminal equipments.
- The possibility to interact with another user whose equipment has differing characteristics. The user expects the application he is running to deal with incompatibility problems any-

how. Hence, an MPEG video format emitter may transparently feed M-JPEG format receivers, despite the format incompatibility between the communicating parties.
- The communications network must allow any user to select the service components he wants: a user without any screen for visualizing a videoconference must be allowed to select the single audio component and interact with his partners anyway.
- The possibility for users to join in or withdraw from a communication session according to certain policies, such as agreement of all interacting partners before admitting a new user.
- The possibility to set such quality of service parameters as the synchronization tolerance between the service components, the echo sensitivity, the burst sensitivity, and so on.

The above outlined constraints require particular management schemes. The ensuing section analyzes these schemes and defines functions necessary for their achievement.

3 GENERIC FUNCTIONS NEEDED WITHIN THE SERVICE MANAGEMENT ARCHITECTURE

Service management may be defined as the intelligence, the set of mechanisms to be held in order to come up with the requirements over the provided services. Those mechanisms are interdependent, as they have to interact in order to fulfil the users exigencies. Therefore, mechanisms that are related can be grouped into *generic functions* that provide the service management with an organized view. As call control is a major building block that must be carefully studied when designing service architectures, we attempt herein to bring out the main functions that must be included in this important building block.

Three core *generic functions* are identified, that cover all requests to a multimedia network (this is the overall network that supports the multimedia communications). Those functions denote the meaning of the users requests to the multimedia network. They are *QoS negotiation, QoS monitoring and control,* and *session management*. Each of them consists of a set of mechanisms that should be achieved for performing a user request. For instance, a call establishment service element implies QoS negotiation, and then the related set of mechanisms. The following subsections focus on those mechanisms for each defined function.

3.1 QoS Negotiation

In this paper, we represent a communication session by a collection of LCGs (Logical Connection Graphs). LCGs are directed graphs that describe clearly and flexibly the connectivity among multiple logical resources[9]. These are logical and graphical views of physical resources. Examples are given in section 4.

The objective is to define completely all LCGs attached to a given session. Reaching this goal means negotiating the resources (network resources and special resources) needed for establishing the communication. Special resources are of three kinds[2]:
- *Mediation resources* are devices that carry out miscellaneous functions to deal with the interfaces' requirements between different resources. Encoding/decoding devices, synchronization devices, echo suppression or cancellation devices and multiplexing/demultiplexing resources belong to this group;
- *Heterogeneity handling resources* are devices providing useful services to overcome heterogeneity problems such as format type incompatibility and information type (audio, video, text, etc.) differences;

- *Multipoint communication resources* are used to provide multicasting, combining (or bridging) and selecting services. These elements are responsible for the duplication of data needed, e.g., in a videoconference, since such a service may require the user streams to be duplicated before any retransmission.

Thus, the first two kinds of resources deal with multimedia issues, whilst the last category mostly covers multipoint aspects.

However, prior to QoS negotiation, the application-defined QoS should be translated into resource-defined QoS, i.e., QoS meaningful to resources, by means of a QoS translation unit. Moreover, one LCG is defined per medium, such that, within a communication session, there are an audio LCG, a video LCG, a picture LCG and so forth. This separation of media enables a user without any video equipment to participate in a videoconference somewhat by receiving and paying for the single audio information. As a result, QoS negotiation partially solves equipment heterogeneity.

Nevertheless, selecting a resource is not so easy. Resources capabilities are to be considered. In this respect two appropriate tools, borrowed from [6] and [7], can assist in the estimation of resources capabilities versus the services these resources are able to carry out. Those tools are *schedulable region* (SR) for network resources and *multimedia capacity region* (MCR) for multimedia resources. SR and MCR respectively relate to traffic and service classes. A traffic class is characterized by some statistical properties and QoS requirements (e.g., processing duration and error rate). For instance, the most stringent traffic class is characterized by 0% as error rate, and no processing duration so that packets received at the resource interfaces are not subject to queuing. A service class is defined in the same terms for multimedia resources [6]. The schedulable region of a resource is "the set of points in the space of possible loads for which a given quality of service is guaranteed"[6]. The possible loads of interest for a network are traffic classes. The MCR is defined similarly to the SR with service classes as loads.

Schedulable region and multimedia capacity region are powerful tools, but the characterization of service classes, mainly based on encoding types, gives rise to many classes and does not facilitate the determination of the regions of interest. Even though the number of encoding types is currently huge, schedulable region and multimedia capacity region concepts could give a needful logical view of physical devices, which is helpful for performing call admission control for both networking and multimedia resources. The main open issue is the number of classes to be considered.

3.2 QoS Monitoring/Control

QoS monitoring/control is concerned with holding the achieved QoS up to the negotiated QoS as to guarantee the expected QoS. This can be done in two ways: *QoS compensation* and *distributed resource management*.

The purpose of *QoS compensation* is to find the way(s) QoS can be shared out among the network resources and those concerned with media handling. Indeed, the performance of the latter resources can help release the QoS requirements over the network. For example, using M-JPEG for video requires higher bandwidth than MPEG encoding technique does. The intelligence to be provided by QoS compensation then consists in having an integrated view of multimedia communications that encompasses network and multimedia technologies, so as to deal with the networking consequences after altering some multimedia resources parameters, and vice versa.

The *distributed resource management* function is essentially a fault management function. It may be performed using the L-E (Legislator-Executor) model [8] (Figure 2) for a collection of resources or each resource alone. The *legislator*, after consulting the *resource capacity* (which in fact is the schedulable region or the multimedia capacity region) and the *request intensities* data store, sets rules (*control policy*) on accessing the resource. The *executor* regulates this access conforming to the *control policy*. It also controls the *resource state* and, like a watch-dog, makes accurate decisions (e.g., failure notification) about using the resource. On top of the model *control parameters* are set to manage the control task.

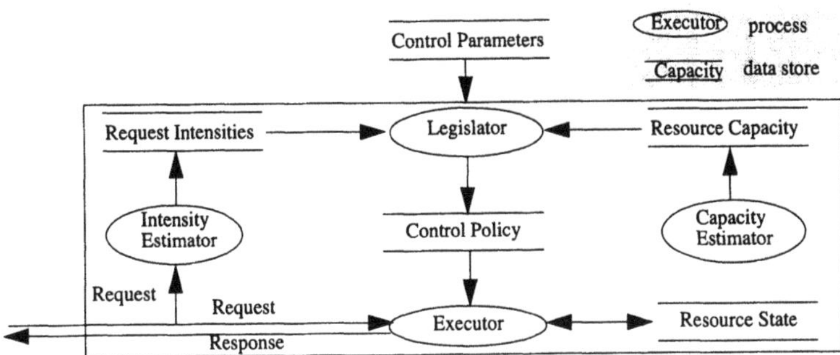

Figure 2 Resource control subsystem: the L-E model [8]

In short, QoS monitoring/control is performed in two manners: one, distributed, is the resource management function and the other, more centralized, is QoS compensation. This latter is more complex and may introduce undesirable transient effects when performed at runtime with a huge amount of resources. Transient effects are due to the fact that the set of resources involved in the communication has changed owing to a resource that failed in providing its QoS part. In the opposite of QoS compensation, the distributed feature of the resource management function makes it more robust against transients effects. Nevertheless, the scope of that distributed resource management is limited to "weak" failures. In cases where a resource cannot recover from its failure and cannot be replaced by any similar equipment, the only function available is QoS compensation which then initiates a new QoS translation.

3.3 Session Management

Session management provides mechanisms that enable the user to change the configuration and attributes of an on-going communication session. Among operations for changing those attributes and configuration are medium/user admission control, medium/user removal from a session and QoS modification. Most service elements described in [3] belong to this category of operations. Using medium-based LCGs, i.e, one LCG per medium, facilitates the achievement of management operations, since the session management function results in modifying existing LCGs (in case of admission or removal of a new user) or building new LCGs (in case of admission of a new medium). Admission control becomes a problem when the communication session could not admit a new user or medium for whom there is no more

resource available, unless the users participating in the session agree with decreasing their current QoS. Therefore, admission control mainly consists in modifying existing LCGs.
The generic functions described above are not just co-existing but they also need to co-operate. Thus, session management involves QoS re-negotiation and QoS monitoring/control. Yet, QoS negotiation presents a static aspect, as it consists in negotiating a required *constant* QoS, whilst QoS monitoring and control rather presents a dynamic feature as it keeps an existing configuration, especially the achieved QoS, by adjusting the resources efficiency. Session management presents a much more dynamic feature, since the QoS is dynamically changed as well as the communication session configuration. An open issue is to be studied is how to modify the achieved QoS without damaging the on-going communication.
As previously stated, the generic functions are necessary for meeting the users requirements. In the next section, an architecture is proposed with regard to these functions.

4 A GENERIC SERVICE MANAGEMENT ARCHITECTURE FOR MULTIMEDIA COMMUNICATIONS

The proposed architecture (Figure 3) enables the realization of the three generic functions described in section 3, namely QoS negotiation, QoS monitoring/control and session management. After specifying the roles taken by the architecture components, the way each function is carried out by the architecture is described, and the proof of the latter's flexibility brought up.
In order to allow for the sharing of multimedia resources among several users, a Customer Premises Equipment (CPE) is introduced between end-systems and public networks. Therefore users do not need to buy costly equipments before accessing multimedia services.

4.1 Architectural Units

Before addressing the overall functionality (provision of the generic functions) of the architecture components, we first describe these.
The `user agent` represents the profile of the service user, in terms of the latter's identity or some other features like the user's visual acuity, and so forth. Those features may be needful for making QoS user-specific, thus expressing the subjective nature of QoS.
The `application` provides the user with an interface for accessing the service.
The `user session management` unit is concerned with the management of the user session at the end-system; it is also in charge of selecting the terminal media handling resources needed to fulfil the QoS requirements. Moreover, the described unit interacts with the `service session management` instance for creating and managing the communication session.

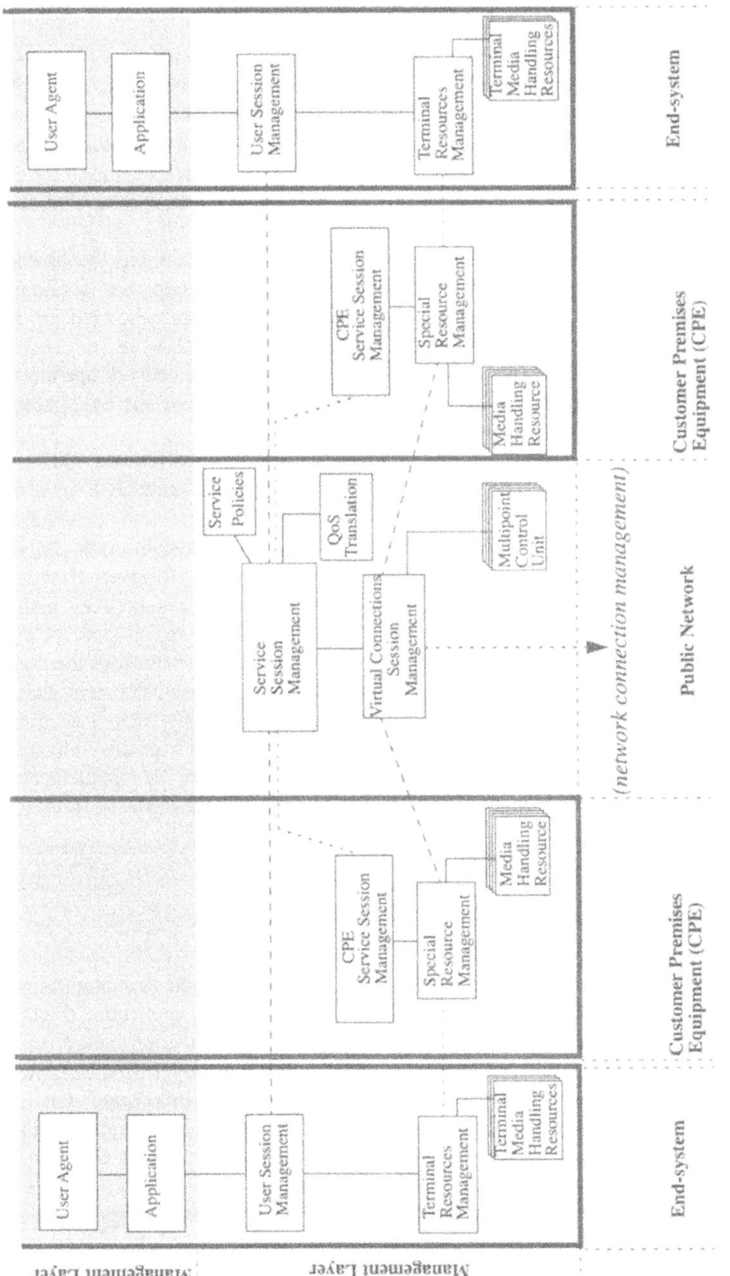

Figure 3 Generic Service Management Architecture for multimedia communications

The terminal media handling resources are the multimedia equipments embedded by the end-system.
The CPE service session management unit manages the session part within the CPE.
The CPE special resources management unit is responsible for selecting and managing the special resources within the CPE and for curing resource failures somehow.
The Terminal Resources management unit manages the special resources within the terminal.
The virtual connection session management unit manages the media LCGs used for the purpose of the communication. This management task necessitates interactions with the network and the CPE special resources management units to be involved.
The service session management unit manages the whole session, interacts with the network and the service policies to establish the bill, for instance. Another important decision it takes in concert with the service policies is how to select the networks to be used for the communication with the objective of reducing the communication cost. Service policies may serve to feature commitments between the service provider and some network operators.
The service session management unit is also responsible for creating the environment needed for managing the communication sessions.
The functionality of the QoS Translation unit is to translate application-required QoS into resource-defined QoS and eventually propose many issues for selecting the resources to involve in the communication sessions.
The multipoint control unit (MCU) is well described in [10] for bandwidths up to 2Mb/s, but this bandwidth restriction takes nothing off from the functionality. However, H series recommendations (e.g., [11]) do not consider sharing of special resources between many terminals and distinguish between primary terminals, whose capabilities match those of the used MCU, and secondary terminals whose capabilities are far different from those of the used MCU. For the sake of conciseness, the MCU is described as containing video combiners, audio mixers, multiplexing/demultiplexing resources, network interfaces, audio/video switching resources, echo handling unit, among the most important components. MCUs are discriminated by their processed audio/video features, their bandwidth, the number of simultaneous conferences they could handle, the number of participants in each conference, and so forth.

4.2 Realizing the Generic Functions

4.2.1 Realizing QoS negotiation

Considering the very first QoS negotiation that takes place during the communication establishment, a *user agent* creates a *user session* by running its *multimedia application* (Figure 3). The created session submits the desired LCG (Figure 4a) to the *service session management* instance. The latter is responsible for controlling the overall communication. It first translates the application-required QoS to resources-defined QoS using a *QoS translation* unit. A resource-defined QoS is a QoS expression made meaningful to the resources to be involved in the processing of the communication. The example illustrated below assumes that the session initiator negotiates for a video mixing. Thus, users are allowed to choose resources that best meet their needs. Note that initially the session initiator knows nothing about the capabilities of the terminals owned by its potential partners (this is depicted by the question marks in some end vertices in Figures 4a and 4b).

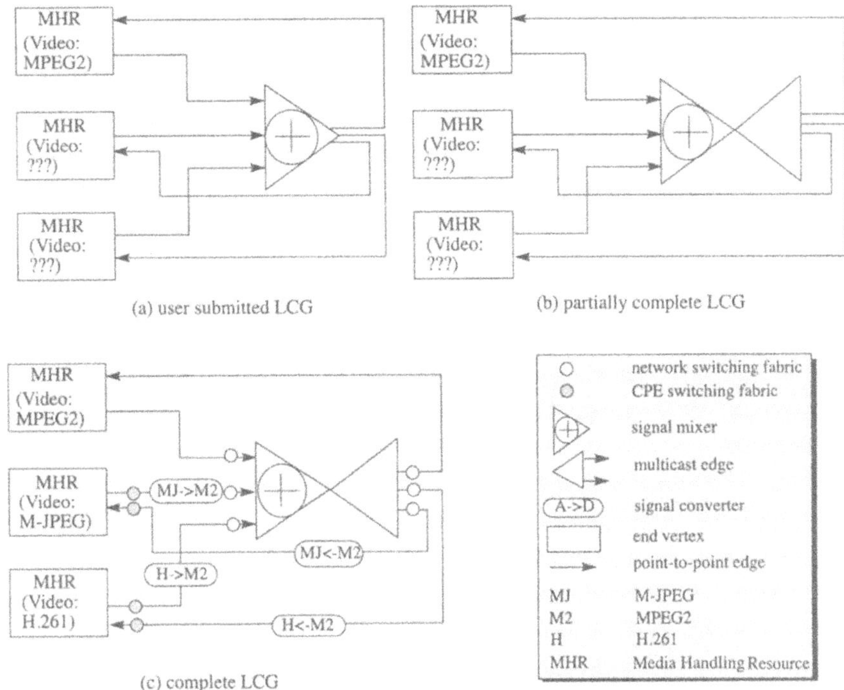

Figure 4 Evolution of a video LCG towards completeness

Once all resource-defined QoSs are calculated (thanks to the QoS translation unit) the *Virtual Connection Session Management* (VCSM) instance starts negotiating for the resources needed by interacting with the network and the *CPE special resources management* (SRM) instances. The VCSM is responsible of involving the networks selected according to the *service policies*. These also include billing and security policies to be applied to the communication messages. If any multipoint control unit (MCU) is needed, it is considered by the VCSM instance (Figure 4b, the signal mixer and the multicasting functionalities may be provided by a MCU). Networks stemmed from the *service policies* provide connectivity to support the multimedia communication. This connectivity adds information to the LCG (switching fabrics vertices).
However, the medium LCG is not yet complete at this stage. It still needs to check the capabilities of all partners in order to be managed by the VCSM. It is up to the latter to interact with the SRM instances within CPEs concerned with the communication session to select the special resources to use further on (Figure 4c). Those instances select the suitable resources to cope with the supplied resource-defined QoS using L-E models of resources (subsection 3.2). Before any selection the SRM instance first checks the capabilities of the terminal to connect. If these are not sufficient to support the communication, then the SRM instance chooses among

the resources it manages the adequate ones for processing the information.
Together, operations performed by the network, SRM and VCSM instances lead to a complete medium LCG. It is worth noting that the proposed framework makes no assumptions about the way networks to be involved are selected (that is up to the *service policies* unit). Moreover, it does not treat any particular service. That proves the *flexibility* of the architecture in Figure 3.
Throughout the above example, the strength of medium LCGs has been well stated. They permit to deal properly with media handling within a communication session and adds flexibility to the architecture.

4.2.2 Realizing QoS Monitoring and Control

QoS monitoring and control is concerned with keeping the achieved QoS close to the negotiated QoS. Monitoring is different from management in that its role is not to change the QoS, but to guarantee the quality of the communication. Hereafter is shown the way the two means (QoS compensation and distributed resource management) for performing the monitoring/control function are realized.
The QoS translation instance deals with the compensation function by suggesting different schemes to cater for the QoS requirements. For instance, when an encoding technique becomes unsatisfactory, due for example to network errors or low processing time at end-systems, another encoding type must be suggested. The compensation function is rather complex. The other method for QoS monitoring, namely the distributed resource management function, is simpler and should be first used before any compensation mechanism.
The resource management function is essentially a fault management function. In the proposed architecture, it is performed in end-systems (by *Terminal Resources Management* -TRM- units), in CPEs (by SRM) and in public networks (by network connection managers), i.e., every site whose resources are involved in the communication. The management units for network resources are not visible from application and service points of view. Therefore, they do not appear in the service management architecture. In the case of failure of the terminal resources, the TRM unit must inform the related SRM in order to find among special resources within the CPE those suitable for a replacement. In general, after any failure or recovery, the concerned medium LCG should be changed accordingly.

4.2.3 Realizing Session Management

Session management is mainly concerned with user/medium admission control, user/medium removal and QoS modification. Removal operations are generally not very difficult to perform in the opposite of admission or QoS modification operations. These have the particularity of modifying the media LCGs a great deal by implying QoS re-negotiation, and thereby re-negotiation of the resources to be involved in the communication session.
Medium admission requires the design of a brand-new medium LCG. On the other hand, user admission is usually expressed by the addition of new end-vertices to existing LCGs.
The main problem to be solved when managing a session is how to negotiate QoS in order to admit a new instance, medium or user as well. Two cases may occur: in the first, there are enough resources so that there is no problem, whilst in the second the lack of resources leads to implement a mechanism that may run successfully or not. In the latter situation, the participants in the communication session should be asked for agreement not only with decreasing their QoS

but also with the acceptable decrease amount. If this is acquired, a suggested mechanism is to find out the remaining capabilities of each involved resource (after decreasing) in order to "spread out" the new user's QoS over these resources. A solution from the proposed architecture, as described in a previous subsection, is for the QoS translation unit to suggest many different ways permitting to meet the QoS requirements. Thus, as a consequence of bandwidth decrease, a sufficiently high compression ratio can be recommended. The QoS translation unit should propose a suitable compression ratio. The way this is implemented is a matter of further research. However, the entities and concepts (mainly LCG, schedulable region, L-E model, multimedia capacity region) to use are visible in the generic functions.

5 CONCLUSION

The major achievements of this work are: identification of three generic functions that must be supported by any multimedia service management architecture, proposal of mechanisms for each generic function, and proposal of an *integrated* and *flexible* service architecture for multimedia communications featuring all proposed mechanisms. Although this architecture covers all users requirements, there still are many open issues not yet addressed. First, the QoS translation unit needs to be studied carefully, because it is the basic building block on which the architecture relies, especially when dealing with multimedia applications.

Another important issue is to define an information model of special resources in order for these to be manageable. The design of such a model requires the study of characteristics of those resources from architecture and management points of view. Moreover, synchronization issues need to be clarified. Synchronization is a relationship between one or more media, or between interacting parties. There are three kinds of synchronization: intra-medium synchronization (between information units pertaining to the same medium), inter-media synchronization (between two or more media, e.g., audio and video synchronization) and inter-party synchronization (between interacting parties). The latter kind of synchronization is intended to preserve the sequence of events occurring at the communicating parties.

Finally it is important to study how to hold an on-going multimedia communication while negotiating a new QoS. In other words, how to deal with transient effects? Once these issues are solved, the proposed architecture will be implemented in the context of the OAMS (Open management Architecture for Multimedia Services over ATM) project [14].

6 REFERENCES

[1] *Connection Management for an ATM Network* - L. CRUTCHER, A. GILL WATERS, IEEE Network, November 1992.

[2] *Towards Flexible Communications Management for Multimedia Multipoint Communications* - Y. KATSUBE, IECE Transactions on Communications, Vol. E77-B No. 11, Nov. 1994

[3] *Communications Systems Supporting Multimedia Multi-user Applications* - G.J. HEIJENK, X. HOU, I.G. NIEMEGEERS, IEEE Network, Jan./Feb. 1994.

[4] *The QoS Broker* - K. NAHRSTADT, J.M. SMITH, IEEE Multimedia, Spring 1995.

[5] *Building New Services on TINA-C Management Architecture* - J. PAVLON, L. RICHTER, M. WAKANO. 3rd International Conference on Intelligence in Networks, Bordeaux - France, 1994, pp259-263.

[6] *Real-Time Scheduling with Quality of Service Constraints* - J. M. HYMAN, A. A. LAZAR, G. PACIFICI - IEEE Journal on Selected Areas in Communications, vol. 9, Sept. 1991- Also available via: ftp.ctr.columbia.edu/CTR-Research/comet/public/papers.
[7] *Multimedia Networking Abstractions with Quality of Service Guarantees* - A. A. LAZAR, L.H. NGOH, A. SAHAI- Available via ftp.ctr.columbia.edu/CTR-Research/comet/public/papers.
[8] *An Architecture for Performance Management of Multimedia Networks* - G. PACIFICI, R. STADLER. Available via ftp.ctr.columbia.edu/CTR-Research/comet/public/papers.
[9] *A Connection Manager for Flexible Specification and Transparent Location of Special Resources* - S.L. MOYER & D.S. ROUSE, GLOBECOM'93, pp. 1526-1530.
[10] *Multipoint Control Units for Audiovisual Systems using Digital Channels up to 2 Mbit/s* - ITU-T Recommendation H.231.
[11] *System for establishing Communication between Audiovisual Terminals using Digital Channels up to 2 Mbit/s*- ITU-T Recommendation H.242.
[12] *Multipoint Audio and Video Control For Packet-Based Multimedia Conferencing* - F. GONG, ACM '94, San Francisco, CA, USA.
[13] *EXPANSE Software for Distributed Call and Connection Control* - H. BUSSEY et al., International Journal of Communication Systems, vol. 7, 146-160, 1994 by John Wiley & Sons, Ltd.
[14] *OAMS: An Open Management Architecture for Multimedia Services over ATM* - S. ZNATY, C. GBAGUIDI, J-P. HUBAUX, Internal Report, 1995.

7 BIOGRAPHY

Constant Gbaguidi obtained his MS degree from the Swiss Federal Institute of Technology in Lausanne (EPFL) in 95. He is currently a research assistant in the Telecommunications Laboratory at EPFL. His areas of research include service and network management, distributed systems and multimedia communications.

Simon Znaty obtained his Ph.D. degree in computer networks from Ecole Nationale Super-ieure des Telecommunications de Paris in 93. From 93 to 94, he was with the network archi-tecture laboratory at NTT Telecommunication Networks Laboratories, Tokyo, Japan, work-ing on TINA. In Oct 94, he joined the Telecommunications laboratory of the Swiss Federal Institute of Technology in Lausanne, where he leads the service engineering group.

Jean-Pierre Hubaux is professor with the Telecommunications Laboratory (TCOM) of the EPFL. Within TCOM, he leads the Telecommunications Services Group, which comprises around 15 people, most of them being Ph.D. candidates; his areas of interest cover service engineering, multimedia services, as well as teleteaching and telemedicine. In this position, he led the first negotiations which resulted in the creation of the Eurecom Institute and of the "Communications Systems" educational curriculum and diploma at EPFL. Prior to this activity, he spent 10 years in France with Alcatel, where he was involved in R&D activities, mostly in the area of switching systems architecture and software.

21

The Multimedia Reference Model: A Framework Facilitating the Creation of Multi-User, Multimedia Applications

Stephan Abramowski, Karin Klabunde, Ursula Konrads, Karl Neunast, Hermann Tjabben
Philips Research Laboratories
P.O. Box 1980, D-52021 Aachen, Germany
Tel.: +49 241 6003 535, Fax: +49 241 6003 518
E-mail: neunast@pfa.philips.de

Abstract

This paper focuses on the application control functionality required for future interactive multi-user, multimedia applications. Accordingly a new service control reference model is described which is part of a more general Multimedia Reference Model. This service control reference model introduces application building blocks that facilitate application creation and management in all kinds of multi-user, multi-point and multi-channel environments in interactive video and other multimedia broadband networks. The Multimedia Reference Model is derived by applying the conceptual modelling technique standardized for Intelligent Networks by ITU-T. Based on the service control reference model, a distributed functional architecture is developed which contains - among others - the functionality for application control, session management, and security handling. The functional design of the architecture enables it to be mapped in various ways onto the physical network components, for example, in a set-top box/ video server configuration. The new Multimedia Reference Model is applied to an interactive multi-user video games environment.

1 INTRODUCTION AND OVERVIEW

The advent of digital compressed video and fibre optic communication has opened up new avenues for advanced multimedia applications and services[1]. Much progress has been made concerning the standardization of video compression techniques (MPEG1 and 2), the development of new equipment such as video servers, digital set-top boxes etc. and the deployment of the

1. Throughout this paper the terms 'application' and 'service' are deliberately used as synonyms.

enabling network infrastructure (fibre/coax, FTTC). However, little effort has been spent up till now on the important issue of application control.
Some first aspects of application control, particularly as regards session control are dealt with in the ITU-T.120 series of recommendations on videoconferencing and multi-point communication [1]. Further elaboration of the session control concept can be found in the work carried out in the Bellcore Touring Machine system [2] and Expanse [3] projects.
Further standardization work for multimedia network and service control is being done in the ATM Forum and in the ITU-T study groups in the framework of the Broadband ISDN User Part (BISUP) [4] and ISCP ([5], [6]). However, Capability Set 1 (CS-1) of BISUP will be limited to setting up simple, single-connection calls; only CS-3 will add multimedia and broadcast connections (however, no date of completion of work has yet been specified [7]). Advanced control concepts are discussed in [8] and [9], also some preliminary discussion on service control is taking place in the DAVIC (Digital Audio-Video Council) project [10].
All the aforementioned activities do not yet meet the requirements of flexible multimedia application creation and management. Those requirements are:

- *The need for a reference model.* The complexity of networked multimedia systems makes a reference architecture indispensable, that treats the different views to the system (end user, application designer, provider, operator, etc.) and the relations between them. This also models the relationship between the application and the system infrastructure.
- *The need for abstraction.* Application designer need not to know all details of the underlying networked system. The provided system capabilities must be offered through adequate abstractions at application programming interfaces (APIs). The suitability of an API is given if no application designer sees the need to use lower level system interfaces, e.g. operating system calls for communication purposes.
- *The need for re-use.* The required APIs can be provided through re-usable building blocks offering well-tried solutions for the commonalities of networked multimedia applications, such as setting up and managing a multimedia conference.

The modelling techniques of the Intelligent Network Conceptual Model (INCM) by ITU-T (see [11]) proved to be a powerful means for the definition and communication of interfaces and protocols (as shown f.i. in the PHIDES [12] system for service creation and the Open Switching platform [13]), and turned out to be extensible for multimedia systems in general.
This paper further extends the scope of the Intelligent Network (IN) by describing a service control reference model which is part of a more general Multimedia Reference Model and can deal with all kinds of multi-user, multi-point and multi-channel environments in interactive video and other multimedia broadband networks. In section 2 the Multimedia Reference Model is derived by applying the conceptual modelling technique developed in IN. To maintain the INCM concept to construct the applications from application independent building blocks, we introduced Application Building Block (ABB) objects. The ABBs form an application construction set that facilitates application creation and management. Based on the service control reference model, a distributed functional architecture is developed which contains - among others - the functionality for application control, session management, and security handling. The functional design of the architecture enables it to be mapped in various ways onto the physical network components, for example, in a set-top box/video server configuration. This also includes a more flexible mapping of the cooperation protocols at the distributed functional

plane onto signalling protocols at the physical plane. In section 3 the new architecture is applied to an interactive multi-user video games environment.

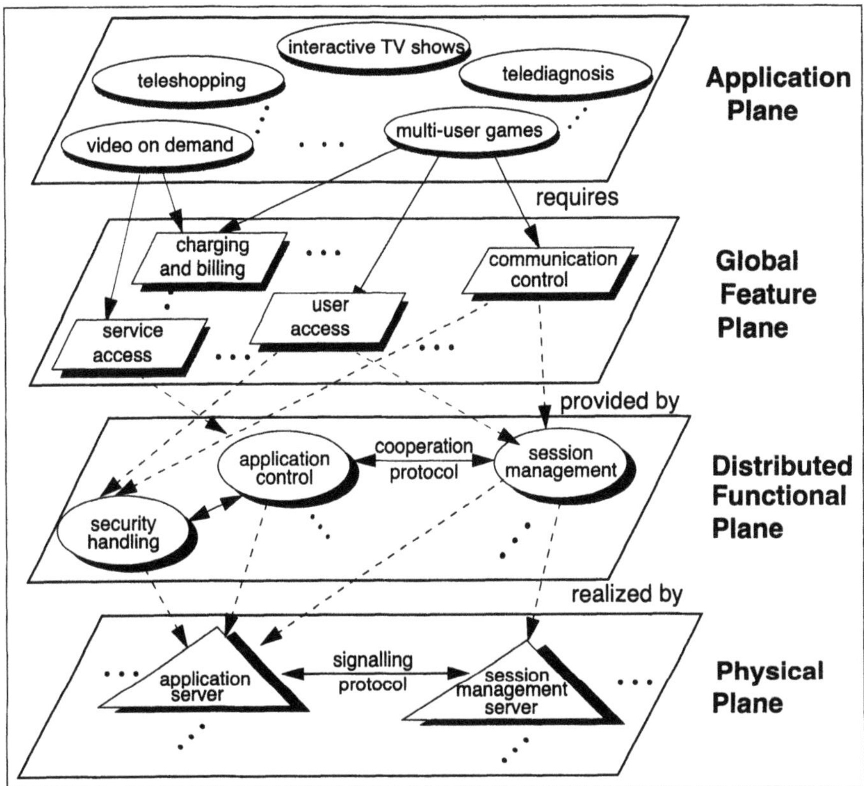

Figure 1 Multimedia Reference Model

2 MULTIMEDIA REFERENCE MODEL

For complex systems, such as networked multimedia systems, an overall conceptual framework is strongly needed. The knowledge and the interests of all concerned parties are too different to meet their needs with a single view to the system. Different views have to be offered to the end user, the application designer, provider, and the operator as well as to the network operator and the component manufacturer. Figure 1 shows the Multimedia Reference Model that we developed from the INCM, based on our experiences with the realization of IN concepts in form of the service creation environment PHIDES [12] and of the Open Switching platform [13].

The **Application Plane** describes the end user's view to the offered applications. Examples for such applications are telediagnosis, teleshopping, interactive TV shows, video on demand, or multi-user games. We replaced the INCM's term 'Service Plane' by 'Application Plane' because 'service' is a generic notion, (e.g. each of the seven layers of the OSI reference model offers its 'services' to the next higher layer).

The **Global Feature Plane** defines the application building blocks (ABB) of the networked, interactive, multi-user, multimedia applications described on the Application Plane. Those application building blocks are combined to construct the service logic of an application. The Service Control Reference Model is part of the Global Feature Plane. It is described in section 2.1 with special emphasis on the communication control ABB. In Figure 1 the features charging and billing, service access, user access, and communication control are shown exemplary. These application independent features take care of the commonalities among the different applications. Communication control refers to the logical ability to bring two or more parties together and to allow them to participate in a multimedia session.

We maintained the INCM idea to construct the applications from application independent building blocks, but dropped the CS-1 Service Independent Building Blocks (SIBs) and their chaining, which proved to be impractical.

The **Distributed Functional Plane** describes the functional entities and cooperation protocols between them, defining a functional architecture. Application control, security handling and session management are examples of important system functions to be provided by a distributed multimedia system. Other functions are e.g. network management, subscription and subscriber management and information access function. The functional architecture is defined such that it can provide the functionality needed for the application building blocks at the global feature plane.

The **Physical Plane** describes the possible distribution scenarios that can reach from centralized realization of single-vendor solutions to the maximum distribution where every functional entity of the Distributed Functional Plane is realized as a physical component by an independent organisation.

2.1 Global Feature Plane

The Service Control Reference Model of the Global Feature Plane offers a simplified view to the networked multimedia system, suitable for an application designer. Within the application logic many aspects of multimedia communication are hidden by abstractions which we call Application Building Block (ABB). These Application Building Blocks are software objects which offer the needed logical view via their operations. They are used by the application objects that are written by the application designer. We offer seven Application Building Blocks on the Global Feature Plane of our Multimedia Reference Model. These seven ABBs are shown on the right side of Figure 2, which outlines their use in a multi-user game application. This figure only indicates the 'use' relation between an application instance, i.e. multi-user game, and the ABBs. It does not give a complete specification of a service logic and of the timely order of the invocations.

In the following, we describe as an example the communication control object class which is a very important one. Before we present our solution - the *communication control ABB* - we give an overview of the state-of-the-art in communication control abstractions.

2.1.1 State-of-the-Art in Communication Control Abstractions

From the many activities in the field of multimedia communication control abstractions we selected those that influenced our approach. These are ITU-T's T.120 series for multi-point com-

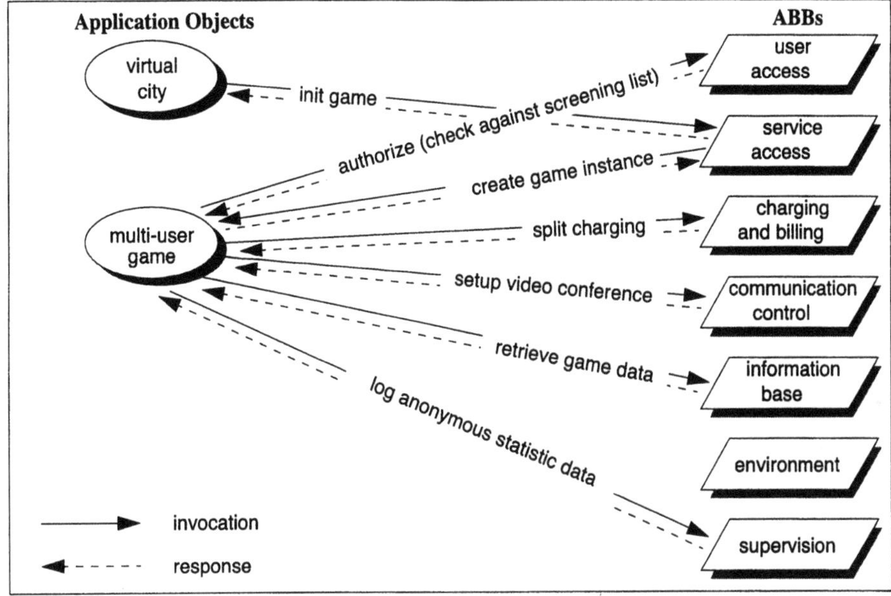

Figure 2 Sample interaction relation between the application objects and the ABBs

munication, Bellcore's Touring Machine and Twente's Service Elements. We will briefly describe them here.

Multi-point Communication Services And Generic Conference Control

During the last few years the ITU-T developed a set of data protocols for multimedia communications that permit group working between multiple – geographically dispersed – parties, see Figure 3. These protocols are published in the T.120 series [1]. Although these protocols were originally intended mainly for desktop videoconferencing it is now recognized that they are useful for a much wider range of application areas within the field of multi-point, multimedia communications.

There are two main characteristics of the T.120 standards that make them quite attractive. First, the T.120-protocols are applicable to a wide range of different networks profiles, e.g. PSTN, ISDN or PSDN. Second, within the T.120 system model certain standardized protocol entities are already offered that make use of the services defined in the lowest multi-point layer – the 'Multipoint Communication Services' and optionally of the services defined in the package 'Generic Conference Control'. Among these protocol entities are packages for 'Still Imaging' and 'Audio Visual Control'. However, it is also possible for users to develop their own non-standard application protocol entities that reside on top of MCS and GCC. End-user applications can then be developed that only use services that are defined in these dedicated protocol entities.

The MCS protocol defines primitives for connection, domain, channel and token management. GCC uses the MCS protocol to offer more enhanced services for conference establishment and

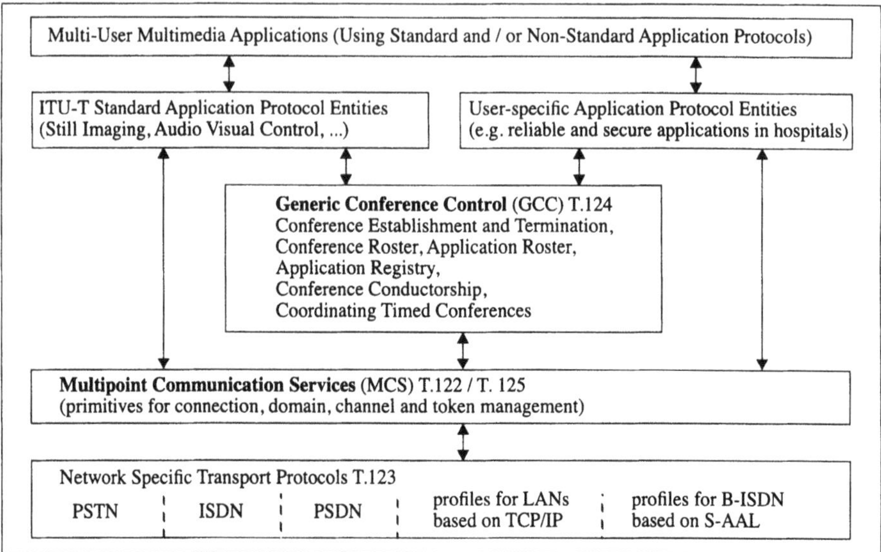

Figure 3 ITU-T's T.120 protocol stack

termination, conference and application roster, application registry, conference conductorship and coordinating timed conferences.
Because of the standardized interfaces and of the successive abstraction layers the T.120 protocols fit well into our concept of a multimedia reference architecture. The T.120 series is interesting with respect to the network transparency it offers. To reach a high degree of flexibility concerning the underlying networks we will consider to realize application independent building blocks on top of the MCS resp. GCC layer. However, the distribution of functionality and the identification of the various network components is not yet well addressed within the T.120 series. For the issue of an explicit representation of session topology we will now look at Bellcore's Touring Machine.

Bellcore's Touring Machine System
The idea behind Bellcore's Touring Machine[1] system is to provide a common platform which facilitates the development and operation of complex multimedia applications [2]. The applications can be developed using an Application Programming Interface (API) which hides the details of the underlying physical networks. The Touring Machine API offers capabilities to establish, and modify multiple concurrent multi-user multimedia communication sessions.
The transport topology of a communication session can be specified using a set of logical abstractions. Figure 4 shows the logical view on a three-party multimedia session. The abstractions used by the Touring Machine are: connectors, endpoints and ports.
A connector is an abstraction of a communication bridge and can have several types: data, audio and video. The termination points of a connector are logical ports which are called endpoints. Endpoints are characterized by the medium, the direction of flow and the connected client who provides respectively receives transport. Endpoints are mapped onto ports which

1. Touring Machine is a trademark of Bellcore.

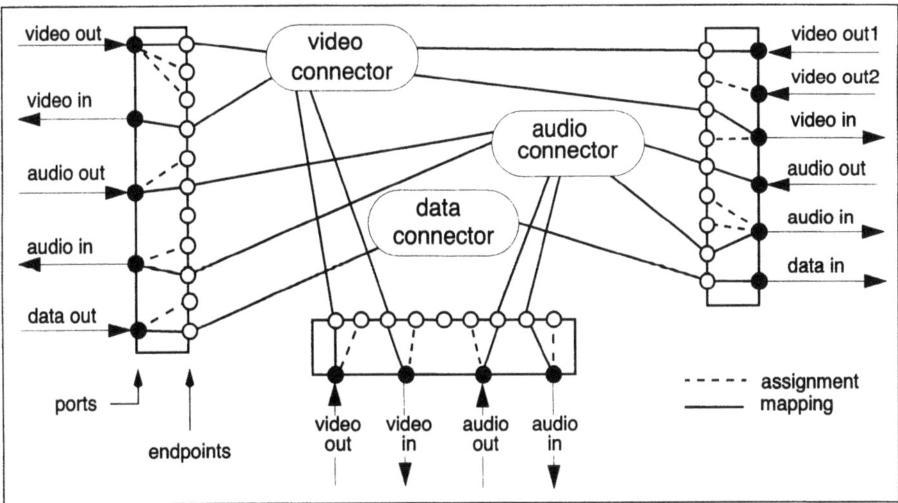

Figure 4 Touring Machine: three-party multimedia session

are network access channels to which a client can attach its data/audio/video equipment. Ports are characterized by medium and direction of flow. A client specifies how its endpoints shall be mapped to the ports. This assignment can be changed dynamically during a session thus allowing to share devices between concurrently running sessions.

The three-party multimedia session shown has three connectors, the video and audio connector have a multi-point connection whereas the data connector shows a point-to-point connection. All three parties have an in/out-connection to the video connector, i.e. they can all send and receive video. The same applies for the audio connector. The data connector is only attached to two parties, one party is sending data ('data out') and the other party is receiving data ('data in').

The logical view on the topology of a session as proposed by Touring is quite attractive although we think that possibly more abstractions are needed: the development of applications without knowledge about ports is desirable. Additionally, the collection of privileges, i.e. *in* and *out*, and the access mechanisms controlling them should be further developed. Furthermore, there is no representation of synchronization relations between different media.

Twente's Abstract Service Elements

Heijenk, Hou, and Niemegeers from the University of Twente described a set of abstract service elements [14] that enables to characterize networked multimedia applications. The Twente approach uses the abstractions of the Touring Machine API, but the connectors are called media. Additionally, an explicit synchronization relation between those media was introduced. Major aim of this work is to come to an agreement about a complete, explicit and precise description of the functional behaviour of a communication system supporting multi-user multimedia applications.

We consider this work as highly valuable – perhaps for a slightly different reason as it was originally intended by the authors – because the offered service description takes place on a very abstract level. The communication system is regarded as a whole, i.e. lower-layer details

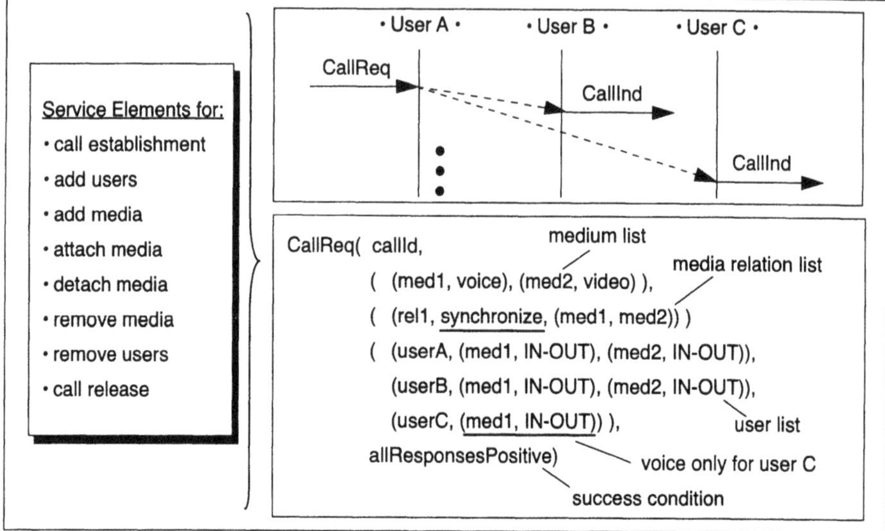

Figure 5 Call establishment with Twente's abstract service elements

remain hidden and no assumption is made about the distribution of the functionality in the network. It is explicitly stated in the article that only the required capabilities of a communication system are investigated but that nothing is said about the realization and how interactions between service primitives have to be resolved. Application developers benefit from such an approach because they can use the set of abstract service elements as building blocks for multi-user, multimedia applications without having in-depth knowledge of the underlying communication technologies.

We sketch the possible use of the proposed service elements which are listed in Figure 5. For each service element there exist several primitives (e.g. CallReq, CallInd). We do not go into details of the negotiation process, e.g. a request service is received as an indication message by the addressed parties and usually has to be responded. Parameters of the call establishment primitive CallReq are the list of media to be used in the call, the assumed synchronization relations between these media and the list of called users together with the media aimed at in that connection. The success condition prescribes when a call request is accepted. The call identifier is used to later add users or media to an existing call.

2.1.2 Communication Control ABB

The application designers need an abstract view on the capabilities of the multimedia communications system. Therefore details of the distribution of the functionality of the various components in a network and of their physical realization should be kept hidden as much as possible. Nevertheless it is important to visualize the constructs for the creation of multi-user multimedia applications that are offered by the network in a way that stimulates the process of creating new networked multimedia examples. In the following, we describe the session model and the API of the communication control ABB, our means to offer the required abstract view.

Session Model

The session model of the communication control ABB offers an abstract view to the session topology and to session management. This representation does not treat the state models for the different components of the session model. The state models introduce the detection points which offer the hooks for the event handling by the application.

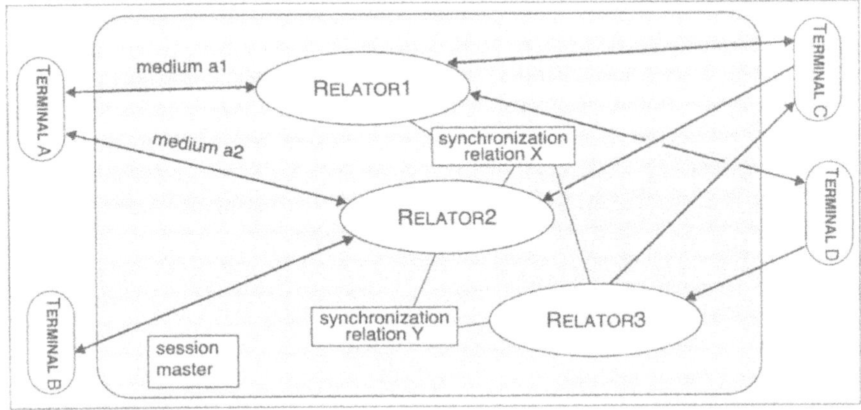

Figure 6 Topology of a four-party multimedia session

The following constructs are introduced by the session model: **relator, medium, synchronization relation, session master.**

The **relator** is a further development of the Touring Machine *connector* [2]. A relator supports sophisticated connect relations going far beyond the simple in or out of an audio/video bridge. Examples for those sophisticated connect relations are complex audio or video mixing or the support of the virtual city paradigm. The video mixing could combine input of a video camera (e.g. the head of a person) with application data (e.g. presentation of the body an artificial person) to produce a new actor of the application who is partially real and partially taken from a fantasy world. With the virtual city paradigm an application can diminish the volume of an input corresponding to a distance relevant in a virtual city application. This could be an audio conference of actors in a virtual pub where the audio input is diminished proportional to the actors distance to a virtual table. An example for the support of the virtual city paradigm is the product "The Parallel Universe" offered by the Virtual Universe Corporation.

The **medium**[1] represents the communication path between a terminal and the corresponding relator, e.g. of type audio. We introduced the **synchronization relation**, as proposed by Heijenk et al. [14] but not further described, to explicitly arrange synchronism between media, e.g. to keep video and audio of a movie lip-synchronous.

The **session master** is responsible for the composition and control of a session (e.g. granting privileges), and for the charging arrangements, e.g. splitting of charges between participants of a session. The application itself can also be the session master, e.g. for a meet-me-conference of the babble box type.

1. A *medium* is comparable to the Twente *attachment*, while a *relator* is comparable to the Twente *medium*.

Figure 6 shows the topology of a multimedia session with four terminals connected via three relators which are synchronized by two synchronization relations.

Communication Control API
The messages establishing the communication control API are:

- ***changeTopology*** which adds respectively removes a new construct to respectively from a session, i.e. a medium, a relator, a synchronization relation, or even creates a complete session.
- ***changePrivilege*** which supports a wide variety of communications privileges of the participants in an application. Besides the classic privileges 'read' (i.e. also 'listen' or 'watch') and write (i.e. also 'speak' or 'show'), sophisticated ones are provided like 'read but not copy'.
- ***changeQuality*** which is used for the modification of quality of service characteristics (such as bandwidth, delay, jitter, or synchronization). Additionally, the required confidence of communicated data is arranged with this operation.
- ***setDP*** which makes the explicit event handling for the multi-user, multimedia session possible. Detection points (DPs) can be activated for a collection of session events. Examples of possible session events are *(session) created* and *partyAdded* (see also Figure 10). If the corresponding event occurs, the DP fires and the communication control object informs the application objects. The basis of our DP mechanism are the ITU-T recommendations for IN [11].
- ***sessionEvent*** which informs the application objects about relevant session events. The relevance of a session event is determined by the application logic itself using the operation setDP.

2.2 Distributed Functional Plane

The service features of the Global Feature Plane have to be provided by functional entities (FEs) on the Distributed Functional Plane. These functional entities and their cooperation protocols construct the functional architecture of the multimedia system. Figure 7 shows the functional entities of the Distributed Functional Plane that provide the communication control ABB at the Global Feature Plane.

As it is indicated in Figure 7, the FEs security handling, subscription management, and several instances of session management at the Distributed Functional Plane provide communication control to the Global Feature Plane. In order to do so, these FEs cooperate with each other, and with other FEs (session management cooperates with transport network control, for instance). An example of typical cooperation protocols involved is given below (see Figure 10).

2.3 Physical Plane

For every physical component of a multimedia system the physical plane gives the functional entities that reside in the physical component. Figure 11 gives one possible mapping for a networked multi-user multimedia game system with set-top boxes, that are connected to a central

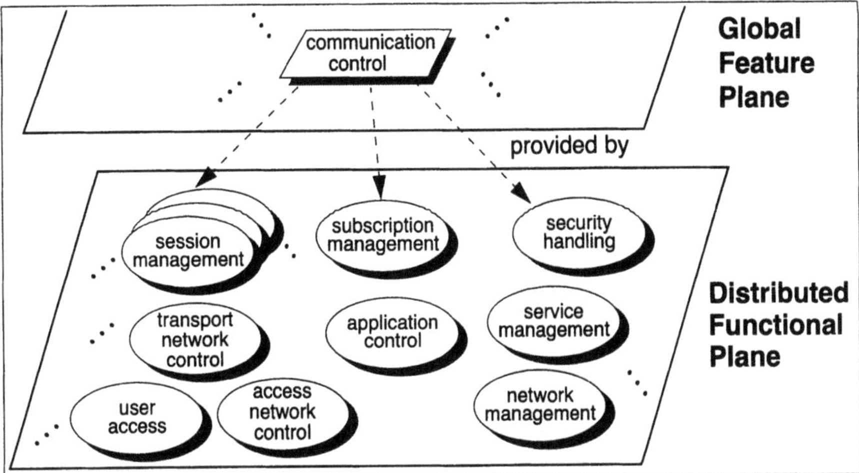

Figure 7 Mapping the system feature communication control onto functional entities

application server through a network. Service Creation and Management reside in their own physical components. The functional entities may be mapped and distributed in various ways into the physical components in the physical plane. It is the advantage of the IN conceptual model to provide a maximum of flexibility for such distribution scenarios.

3 VIEWS ON AN INTERACTIVE, MULTI-USER GAME

The four planes of the Multimedia Reference Model allow to present four different views to a multimedia system and its applications. In this section we illustrate this by a multi-user game example called Telepainter.

3.1 The Application Plane View

The Application Plane describes the end user's view to the offered application:
Telepainter is a networked multi-user game for two teams of players. In each round one player has to paint a term from one of six categories; the other players in his team have to guess the term from the painting as fast as possible. Meanwhile, the other team just watches how the active team is performing.
To offer a comfortable playing atmosphere, the players communicate via audio and video connections, in addition to the connections for the painting information and the control buttons.
Similar, non-electronic versions of this game idea are called *Images, Pictionary, Montagsmaler* etc.

3.2 The Global Feature Plane View

The ABB object classes of the Global Feature Plane form an application construction set that enables to easily create a new application by re-using these well-tried solutions for the commo-

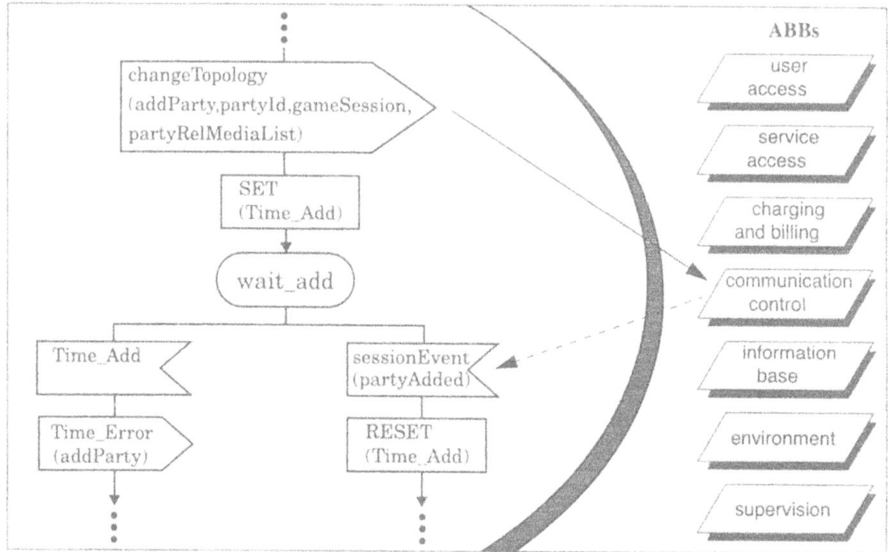

Figure 8 Sample of the Telepainter logic as SDL diagram using the ABB communication

nalities of networked multi-user, multimedia applications. The multi-user game object in Figure 2 represents the service control logic of the Telepainter game. There it is sketched how the application logic uses the Application Building Blocks (ABBs) to check the invited players, to setup the audio/video/data session, to arrange the charging among the players, to retrieve Telepainter terms, and to log some statistic data concerning the game.

In Figure 8 we outline how the application logic of the Telepainter game requests an operation of the communication control ABB and handles the resulting sessionEvent message. We use SDL[1] for the specification of the dynamic model of OMT [15]. The use of OMT results in an object-oriented design. The ABB object classes, such as the communication control ABB, can be used in every object-oriented development environment. Figure 9 outlines its possible use in a C++ environment. In this example the

1. ITU-T's Specification and Description Language (SDL)

service logic of a multi-user game application uses the communication control ABB to add a player to the multi-user, multimedia session.

```
// ===== Multiuser game

void main()
{
    communicationControl commCtrl();           // == define an object of class communicationControl
    sessionId gameSession;
    partyRelMediaListType partyRelMediaList; // list of media to be established between the party's
                                               // terminal and the corresponding relators
                                                 ⋮

    result = commCtrl.changeTopology (addParty, partyId, gameSession, partyRelMediaList);
    if (result != FAILED)    // == did the communicationControl object succeed in adding the party?
        ⋮
```

Figure 9 Sample use of the communication control object class in a C++ syntax

3.3 The Distributed Functional Plane View

The Distributed Functional Plane describes the functional entities and cooperation protocols between them.

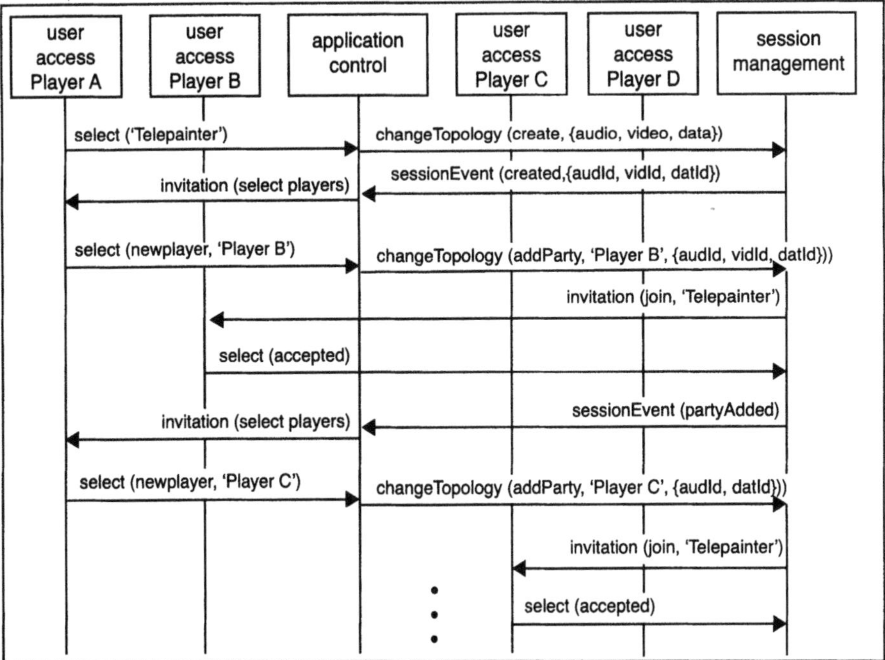

Figure 10 Sample message sequence for the start of a Telepainter playing round

Figure 10 outlines how some of those functional entities cooperate to start a Telepainter playing round. The involved functional entities are application control function, session management function, and four instances of user access function. The observed cooperation 'changeTopology' at the Distributed Functional Plane directly realizes the 'changeTopology' instruction the service logic gave in Figure 2 at the Global Feature Plane. We intentionally chose the same names for the operations at the DFP and at the GFP in this case. But in principal, operations at the DFP and at the GFP are different, and in general there is no one-to-one mapping between them.

3.4 The Physical Functional Plane View

A possible mapping to a physical architecture for multi-user games is shown in Figure 11. The session management functionality resides in this mapping example together with application control in the application server. Furthermore, the network must provide audio and video connections and highly reactive control channels for user control via control buttons. The mapping between the cooperation protocols at the distributed functional plane and the signalling protocols at the physical plane is not shown. The control information may be transported either via common channel signaling (i.e. SS7) including client-server control via user-user signaling, or partly via separate control channels between clients and servers. Another possibility currently being explored (see [4]) is to use a signaling protocol based on the OSI application layer service in an ATM environment.

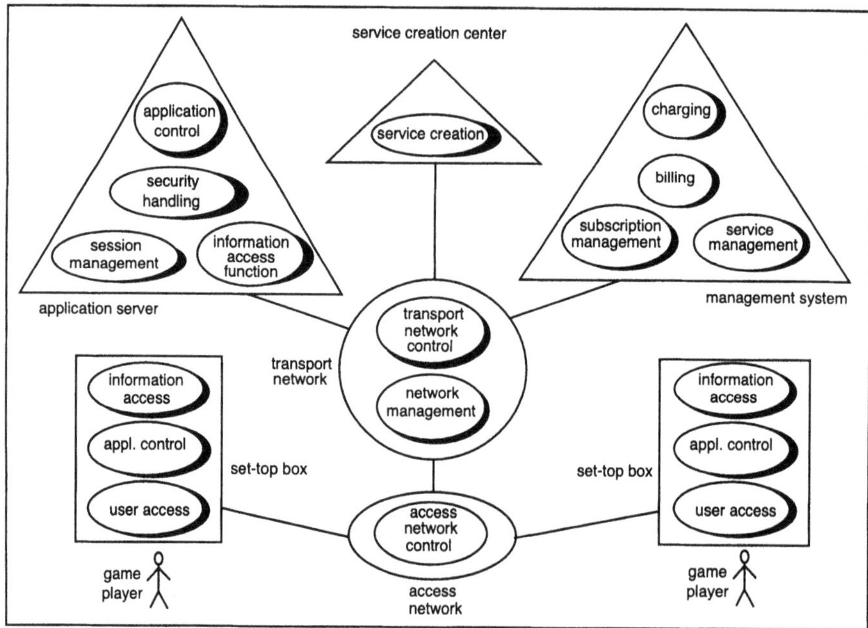

Figure 11 Possible mapping of functional entities to a physical architecture

4 CONCLUSIONS

The 'Intelligent Network Conceptual Model' [11] turned out to be a valuable tool in the development of 'intelligent networks', by establishing an accepted terminology and providing a framework for architectural discussions. A similar approach, that we called Multimedia Reference Model, can be followed in the field of multimedia, where layered architectural considerations are still largely missing (ATM, B-ISDN, DAVIC, ..), and may turn out to be equally valuable there. We showed the usefulness of the approach by modelling application control aspects, especially session management aspects. We introduced Application Building Blocks as the re-incarnation of the INCM idea to construct applications from application independent building blocks. We replaced the CS-1 SIBs and their chaining by a collection of object-oriented ABBs with their operational interface. These ABB object classes meet the requirements of flexible multimedia application creation and management: The need for a reference model, the need for abstraction, the need for re-use.

We applied our approach to an interactive multi-user, multimedia game, presenting the four different views to the Telepainter game.

5 REFERENCES

[1] ITU-T, Study Group 8; *T.120: Transmission Protocols for Multimedia Data*; 20 March 1995

[2] M. Arango et al.; *The Touring Machine system*; Communications of the ACM, January 1993, Vol. 36, No. 1

[3] S. Minzer; *A Signaling Protocol for Complex Multimedia Services*; IEEE Journal on Selected Areas in Communications, December 1991

[4] B. Law; *Signalling in the ATM network*; BT Technol. J., Vol.12, No.3, July 1994

[5] ITU-TSS WG 11/2 - COM 11-R 10-E; *ISCP Baseline Document*; Version 3.1, Geneva, June 1993

[6] A. Paglialunga, Biocca, A., Siviero, M.; *Signalling protocol for broadband ISDN*; Proc. ICC '92, 322.3

[7] P.J. Kühn, Pack, C.D., Skoog, R.A.; *Common Channel Signaling Networks: Past, Present, Future*; IEEE Journal on Selected Areas in Communications, Vol.12, No.3, April 1994

[8] T.F. La Porta, Veeraraghavan; *Design and Implementation of a Distributed Call Processing Architecture*; Proceedings of the IEEE International Conference on Communications, May 1-5, 1994

[9] T.F. La Porta, 5, M. et al.; *B-ISDN: A Technological Discontinuity*; IEEE Communications Magazine, October 1994

[10] N.N.; *responses to DAVIC's CFP*; DAVIC 5th Meeting, Tokyo, 4-7 December 1994, Parts 1-3

[11] CCITT; *Q.1211 - Q.1218 Draft Recommendations*; CCITT SG XI, March 1992

[12] S. Abramowski, M. Elixmann, H. Gappisch, U. Heister, U. Heuter, K. Klabunde; *A Service Creation Environment for Intelligent Networks*; Proceedings of the 1992 Int. Zurich Seminar on Digital Communication, March 16 - 19, 1992, Zurich, Switzerland

[13] M. Elixmann, B.L. de Greef, A.M.M. Lelkens, K.W. Neunast, H.Tjabben; *Open Switching – Extending Control Architectures to Facilitate Applications;* Proceedings of the ISS '95, Berlin, Germany, April 23 - 28, 1995

[14] G.J. Heijenk, X. Hou, I.G. Niemegeers; *Communication Systems Supporting Multimedia Multi-user Applications*; IEEE Network, January/February 1994

[15] J. Rumbaugh, M. Blaha, W. Premerlani, F. Eddy, W. Lorensen; *Object Modeling and Design*; Prentice-Hall, 1991

6 BIOGRAPHY

Karl W. Neunast studied Electrical Engineering at the Technical University of Aachen, Germany, where he graduated in Technical Informatics and received his doctor's degree for his thesis about security mechanisms of operating systems. Since 1990 Dr. Neunast has been with the Philips Research Laboratories in Aachen working on call control and service switch-ing in intelligent networks, especially for ITU-T IN and for ECMA CSTA. Actually Dr. Ne-unast is engaged in the architectural design of networked multimedia systems, in the protocol specification of their functional entities, and in the development of a creation methodology for networked, multi-user, multimedia applications.

Stephan Abramowski studied Computer Science at the Technical University of Karlsruhe, Germany, where he graduated in 1985 and received his doctor's degree in 1989 in the field of computer graphics and computational geometry. Since 1990 he has been with Philips Research Laboratories in Aachen, working in projects on service and network architectures, especially service creation in Intelligent Networks for value-added voice services and session management in multiuser, multimedia networks.

Karin Klabunde studied computer science and artificial intelligence at the University of Koblenz, Germany and University of Edinburgh, Scotland. She joined Philips Research Laboratories in Aachen in 1990. She is involved in research on Intelligent Networks since 1991, especially in the areas of service creation and database technology. Currently, she is working on networked multimedia systems with special emphasis on session management for net-worked multi-user multimedia applications.

Ursula Konrads studied Computer Science at the Technical University of Aachen, Germany, where she graduated in 1985 and received her doctor's degree in 1989 in the field of formal languages. Since 1990 she has been with Philips Research Laboratories Aachen working on service architectures, especially service creation in Intelligent Networks and service control for multimedia applications. She is currently the project leader of Architecture for Interactive Networked Multimedia Applications.

Hermann Tjabben received the degree Diplom-Mathematiker in 1989 from the University Hannover, Germany. Since 1989 he has been with Philips Research Laboratories. He is currently a member of the department Telecommunication Architectures and Multimedia Systems. His current research focuses on protocol architectures for networked multi-user multimedia applications on top of ATM-LAN networks.
He is especially interested in employing formal description techniques such as SDL to improve the quality of communications software.

22
Call Processing Architecture and Algorithms for Future Network

Geonung Kim, Sunshin An
Department of Electronic Engineering
Korea University
1, 5-ka, Anam-dong, Sungbuk-ku, SEOUL, 136-701, KOREA
Phone: +82-2-920-1556, Fax: +82-2-928-0179
E-mail: {kgu, sunshin}@dsys.korea.ac.kr

Abstract

This paper presents Service Call Processing Architecture 1 (SCPA1); a generic service call processing architecture that is well adaptable to emerging B-ISDN environment, especially to the Layer Network. We divide the call processing into two parts; one for negotiation among users and the other for management of lower layer resources. The Call Session Manager (CSM) is responsible for the negotiation among participants and the Resource Manager (RM) performs resource management functions. The recursive layering of CSM and RM makes it possible to model the Layer Network. We show SCPA1 is well suitable for the multiparty communication and the Layer Network. We describe the call processing algorithms of SCPA1 and depict the CSM functional model and RM functional model.

Keywords

Service Call Processing Architecture (SCPA1), Layer Network, Call Session Manager (CSM), Resource Manager (RM)

1 CALL AND LAYER NETWORK

The "call" is a unit of user's service request to service provider. The call in the B-ISDN environment has several requirements. Most of new services will be the multimedia multiparty services in future public networks, so the call should support these new services. To support multimedia services, it is necessary to define the basic components for each media that has its own property and they could be combined to cope with complex services. The multiparty services can not be viewed as just an extension of two-party services. The involvement of multiple parties in the same communication session may imply that the success of the establishment of a session is dependent on the existence of some specific party. Furthermore, each party can play different role. Therefore there should be capabilities to cope with these cases. There will be various quite different terminals in future networks. There will be telephones, cordless telephones, wireless telephones, workstations, personal computers, notebook computers, PDA (personal digital assistant)s, and so on. In this context, the problem of terminal and service capability will be significant and the call should support various terminals. There should be many resources to establish a call in B-ISDN environment, but there can be shortage of some resources, and it can cause another negotiation between user and service provider. So, it should be adapted to network circumstances and should be adapted for user's changing request during call setup phase.

Minzer introduced the "EXPANSE call model" that user can request multimedia multiparty services to network provider with extended transaction form. In the EXPANSE model, there are three kinds of elementary call objects. Those are *local objects* that are subject to control by a single user, *confirmed objects* that require concurrence with another user for construction and are subject to control by two users, and *virtual objects* that are created by network and may effect a user's behavior but are not directly subject to its control [1]. The EXPANSE call model is a basis of Bellcore's INA Communication Management Architecture [2]. There have been several researches of call model and resource model to provide multimedia multiparty communication [3][4][5].

The various functions comprising a telecommunications network can be divided into two broad classifications; delivery segment functions and service segment functions. Delivery segment functions are involved in the transfer of user's information. The delivery segment can be layered into a number of Layer Network (LN)s with a client-server relationship between adjacent layers. Each Layer Network represents the set of delivery segment that supports the transfer of a certain type of characteristic information. Examples of characteristic information include SONET/SDH VC-3 framed digital information, ATM cells, or Frame Relay blocks or fragments. Layering of the delivery segment offers several benefits to network providers in their tasks of operating and managing the network. A Layer Network can be partitioned according to several reasons; the network topology within a Layer Network, administrative boundaries between administrative authorities, domain boundaries, independent routing domain, etc. A Subnetwork can be recursively partitioned into a number of smaller Subnetworks interconnected by Links. There are many autonomous domains and the need for co-operation with other domains is increasing. The call that supports the layer networks should use resources of lower network efficiently and should hide the complexities of lower layer. [6]

TINA consortium is defining a software architecture for telecommunications systems. The mission of Connection Management functions in a TINA-C consistent network is to support telecommunication services in the need for connections. The TINA-C Connection Management Architecture (CMA) defines a target architecture for connection setup, control and release. The functionality of Connection Management is defined as a set of computational objects. The CSM (Communication Session Manager) provides services for the binding of computational stream interface. The clients of CSM will specify the references of stream

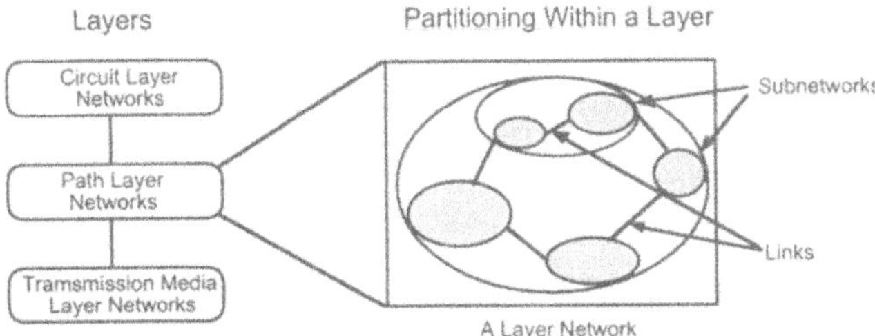

Figure 1 Layering and Partitioning [6].

interfaces they want to be interconnected, plus QoS parameters. The service of CC (Connection Coordinator) is the interconnection of physical nodes. The CC is responsible for creating connections between nodes. Connectivity through layer networks is provided by the collaboration of two types of computational objects: LNC (Layer Network Coordinator) and CP (Connection Performer). The LNC provides trails in a layer network. It also takes care of federation with other domains in a layer network. It requests the CP to set up the connection in its own domain. The CP provides connections in a single subnetwork. Since a subnetwork may be partitioned into smaller networks, each CP may request subordinate CPs to provide smaller subnetwork connections [7][8].

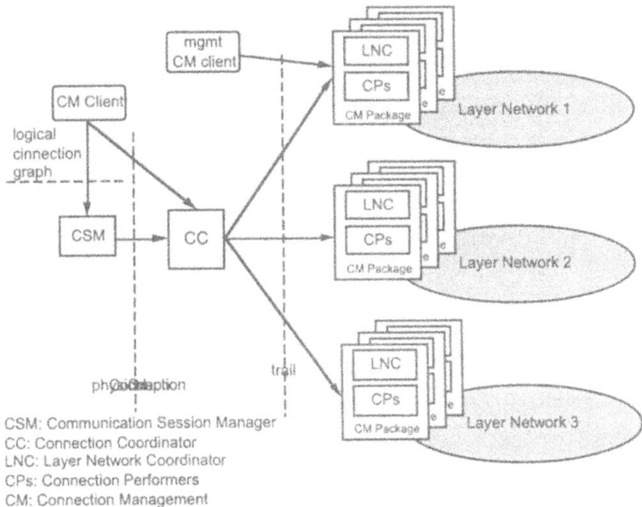

Figure 2 TINA-C's Connection Management Overview [8].

The TINA-C focuses on the communication management architecture and the communication management specification. However it does not consider the implementation of the CMA. In this paper, we propose a generic service call processing architecture: Service Call Processing Architecture 1 (SCPA1). The SCPA1 is well suitable for the Layer Network and it is a more practical architecture than TINA-C's CMA.

2 SCPA1 (SERVICE CALL PROCESSING ARCHITECTURE 1)

There are two functional elements to process a call in SCPA1. Those are elements to negotiate users' view and elements to manage resources of a lower layer. In the existing telephone service the user's requests are so simple that they can be mapped with the lower layer resources directly. When a user wants to use the telephone service he/she just picks up the phone and dials the destination telephone number. There is one connection per a call. However the user's requests in B-ISDN services will be very complex. He/she may want multimedia multiparty services. He/she can choose each medium's capabilities and each party's services.

There are the Call Session Manager (CSM) to receive and negotiate these user's complex requests and the Resource Manager (RM) manages lower layer resources. Figure 3 shows the SCPA1 that we propose.

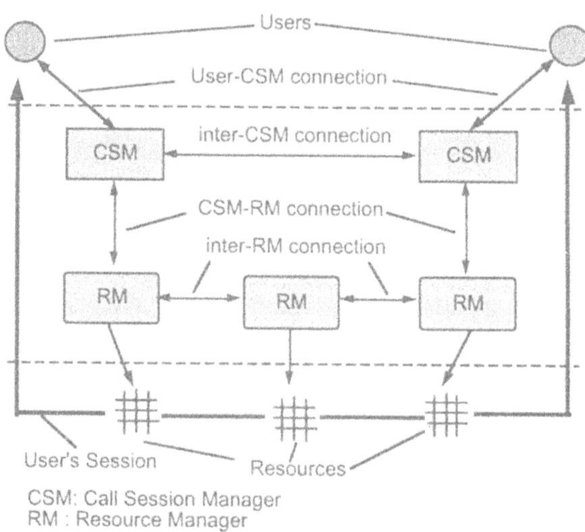

Figure 3 SCPA1: Service Call Processing Architecture 1

There are 5 kinds of connections in SCPA1. Those are (1) the user-CSM connection, (2) the inter-CSM connection, (3) the CSM-RM connection, (4) the inter-RM connection, and (5) the user's communication session.

When a user wants to use the service, he/she should register at the CSM first. The CSM maintains the information of each subscriber (user's address, user's profile, address of the RM that provides connection to user, billing information, etc.) It may give him/her the information such as the current status of communication, the cost of current call, etc. The user-CSM connection is a control channel between user and service provider. When a user wants to use B-ISDN services, he/she may request them via the user-CSM connection. Though there is only a bearer service in the existing public network, there can be other services like data conversion, video compression, and so on in future network. So, the user should negotiate not only on the quality of services of the bearer service but also on the quality of service of these non-bearer services.

When the source CSM receives a user's request, it negotiates with the destination CSMs via the inter-CSM connection. There are two kinds of requests; one for adjusting both sides' semantics into the same and the other for using the lower layer's resources. The CSM should support negotiation of these two kinds of requests. The negotiation of a user's request can be performed as an *atomic* action or an *interactive* action. In the atomic action, a user sends the information of his/her communication to a destination and then the destination user determines the request and makes decision whether the call is accepted or not. However in the interactive action, the receiving user may also send the information of its communication environment and call is accepted after several information exchanges.

When the CSM receives the request from its user, it creates *local* objects and *negotiation* objects, and transfers the replication of the negotiation objects to the destination CSMs. The *local* object is the object that can be managed by one local CSM and the *negotiation* object is the object that should be managed by co-operation of CSMs. Each destination CSM creates the local objects, replicates negotiation objects, and notifies user of them. The receiving user can modify the request by modification of the negotiation objects. The CSM modifies its negotiation objects and passes this modification request to source CSM. Then the source CSM modifies its negotiation objects and notifies a user of them.

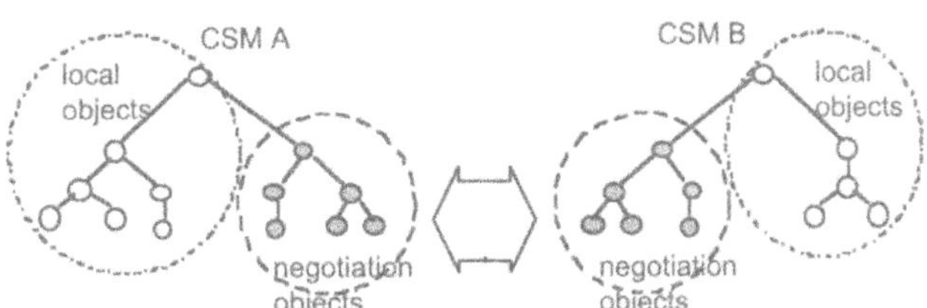

Figure 4 CSM objects

After the negotiation, the CSM translates the user's requests to resources and requests the RM to perform the resource management functions via the CSM-RM connection. There are *mediation* objects to maintain the relationship between these CSM objects and the resources that are managed by the RM. There can be various relationships among these CSM objects and the RM objects. One CSM object can be related to several RM objects. Several CSM objects can also be related on one RM object. The mediation objects perform a mapping between CSM objects and RM objects.

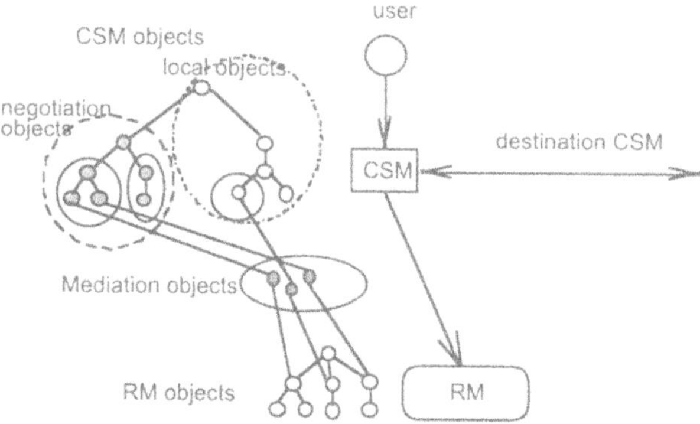

Figure 5 CSM objects, RM objects and Mediation objects

The RM can perform the resource management functions to its own resources. It can also request the tandem RM to perform the resource management functions. The RM is responsible for its domain. If a user wants to communicate with a user in an other domain, the RM should cooperate with other RMs. So, there can be one RM or several RMs in establishing one connection.

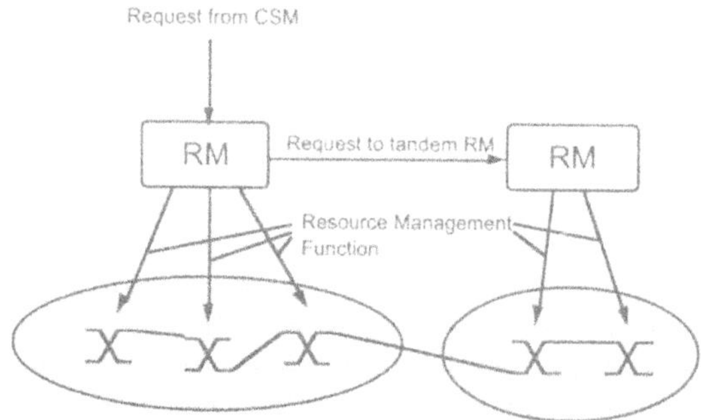

Figure 6 Role of RM

Before the resource allocation, the RM reserves the resources first. Because the multimedia multiparty communication needs many resources, there can be a shortage of some resources. At that time, the RM notifies the CSM that the requested connection cannot be established, and the CSM does the negotiation again.

When all the resources are reserved, the RM notifies the CSM that the connection can be established. When all required connections can be established the CSM requests the RM to perform resource management functions for resource allocation. After all resources are allocated, the RM notifies the CSM that the connections are established. Then, the CSM announces to the user and the destination CSM that the session is ready.

3 FUNCTIONAL MODELS OF THE CSM AND THE RM

The functions of CSM can be decomposed into a number of fundamental functional groups, such as call session management functions (CSMFs) and system management functions (SMFs). There are 5 kinds of CSMF objects in CSM. Those are the access CSMF object, the object manager CSMF object, the signaling CSMF object, the routing CSMF object, and the mediation object manager CSMF object. Figure 7 shows the functional model of the CSM.

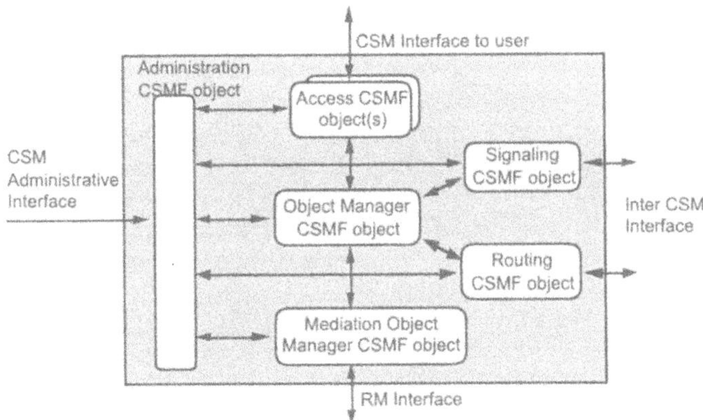

Figure 7 CSM Functional Model

The access CSMF object receives the user's requests and passes them to the object manager CSMF object. It also sends the notification from objects to upper users. There is one access CSMF object per one call. The object manager CSMF object processes the user's requests and it operates with the signaling CSMF object and the routing CSMF object to negotiate with other CSM's object manager CSMF objects. The signaling CSMF object terminates the network signaling between CSMs. The routing RMF object supports to create connections for signaling among CSMs. It determines a path between the source CSM and destination CSM. The mediation object manager CSMF object maps the CSM objects to RM objects.

The RM receives the request of the CSM and performs the lower layer's resource management function. One RM is responsible for its domain. Figure 8 shows the functional model of RM.

The access RMF object receives the CSM's requests and passes them to the object manager RMF object. It also sends the notification from resource object to the CSM. The object manager RMF object processes the CSM's request. It creates, deletes, modifies, and maintains logical resource objects. If the requested resources are in the RM's own domain, it performs lower layer resource management function via the lower layer resource RMF object. When the resource is not in its domain it co-operates with other RM. When the resource is not in the layer, RM plays the user's role of the lower layer's CSM via the lower layer network RMF object.

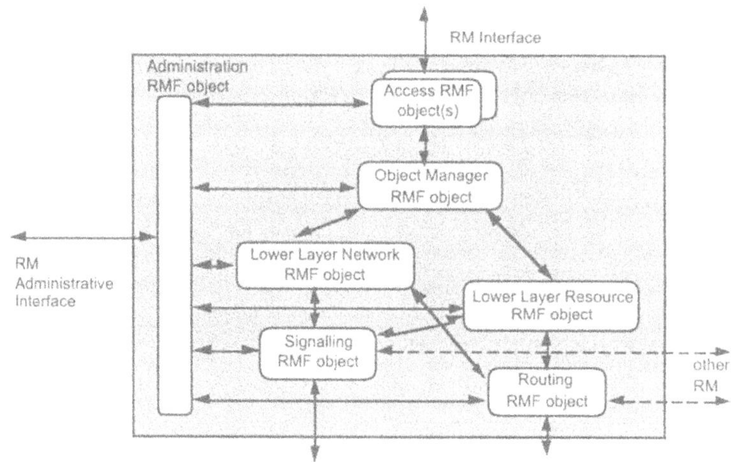

Figure 8 RM Functional Model

The signaling RMF object and the routing RMF object support the lower layer resource RMF object and the lower layer network RMF object. There are 3 kinds of signaling in the RM. One is for the lower layer resource management, another is for the cooperation between other RMs and the other is for the request to the lower layer network. The signaling RMF object terminates signaling between the RM and the CSM of the lower layer network and it also terminates signaling between the RM and the lower layer resources. Furthermore it terminates signaling between RMs. The routing RMF object determines 3 kinds of paths. It determines (1) paths between a source and destination address, (2) paths between RM and lower layer resources, and (3) paths between RMs. It also sends routing information to other RMs and receives it from others.

4 CALL PROCESSING ALGORITHMS OF THE SCPA1

We show the call processing procedure in this section. There are 4 phases in call processing in SCPA1. Those are the negotiation phase, the reservation phase, the allocation phase and the confirmation phase. The CSMs receive user's request and adjust both sides' views in the negotiation phase. After the negotiation the CSMs request the RMs to reserve resources. When all resources are reserved, the resource allocation phase is followed. Then the CSMs confirm to users that session is ready.

4.1 Normal call processing

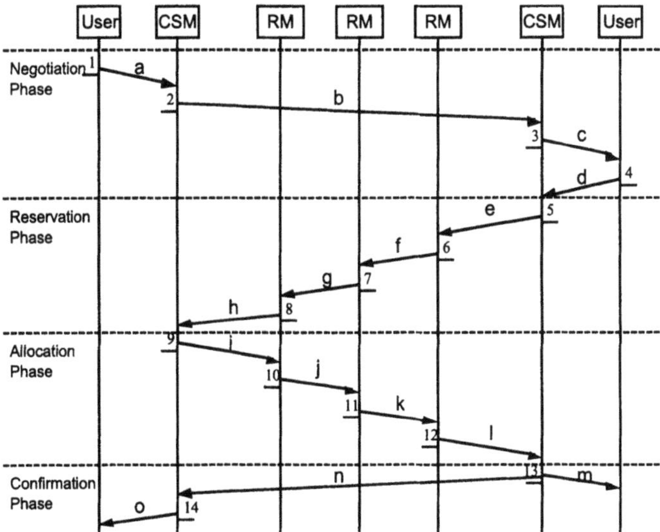

Figure 9 Normal Call Processing

The negotiation phase is from step 1 to step 4.

Step 1: User A who wants to communicate with user B transfers the request to the CSM A.

message a: { request to communicate with user B }

Step 2: The CSM A creates local objects and negotiation objects, and then transfers the request to the CSM B.

message b: { create negotiation objects }

Step 3: The CSM B creates local objects and negotiation objects, and notifies its user of them.

message c: { notify the creation of local objects and negotiation objects }

Step 4: When user B wants to accept the call, he/she sends acceptance message to the CSM B.

message d: { accept }

The reservation phase is from step 5 to step 8. In this phase, the CSM requests the RM to reserve the resources, and RMs do reservation.

Step 5: The CSM B requests the RM for resource reservation function.

message e: { do resource reservation }

Step 6,7: The RM reserves its own resources and requests the tandem RM to reserve the resources.

message f, g: { do resource reservation }

Step 8: The RM reserves its own resources and notifies CSM A that all resources for the call are reserved.

message h: { all resources are reserved }

The resource allocation phase is followed. In this phase, the RMs perform the resource management functions and request the tandem RM to perform it.

Step 9: The CSM A requests the RM to perform the resource allocation.

message i: { do resource allocation }

Step 10,11: The RMs perform the resource allocation functions and request the tandem RM to perform the resource management functions.

message j, k: { do resource allocation }

Step 12: The RM B performs the resource management function and informs CSM B that it has performed the resource management functions successfully.

message l: { all resources are allocated }

When all the resources are allocated, the CSMs notify users that the session is ready. We call this phase the confirmation phase.

Step 13: The CSM B notifies the CSM A and the user B that the session is ready.

message m, n: { session is ready }

Step 14: The CSM A notifies the user A that the session is ready.

message o: { session is ready }

4.2 Refused call

When a user requests unauthorized service to the CSM, the CSM can refuse the call. Because the destination CSM can do various functions such as authentication function, screen function, etc. it can also refuse the call. When the user B does not want to communicate with the user A, he/she can refuse it. Figure 10 shows the refused calls.

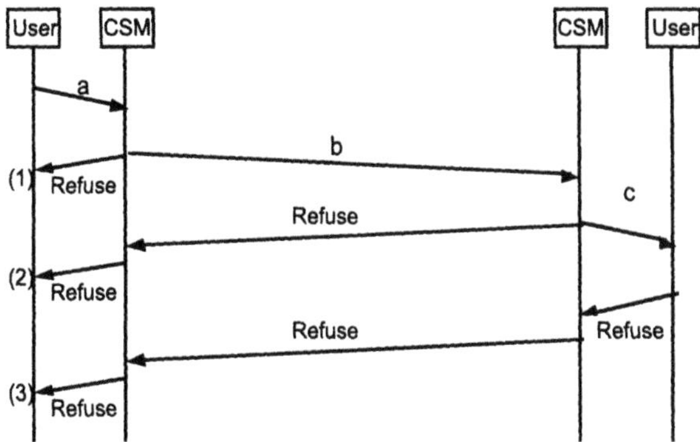

Figure 10 Refused Call

4.3 Renegotiation of call

If the user B wants to modify the attributes of negotiation objects, he/she requests the CSM to modify them. When the CSM receives this modification request, it modifies the negotiation objects and requests the peer CSM to modify them. After modification of its negotiation objects, the CSM notifies the user that the call is modified by the peer user. Figure 11 shows a 2-way negotiation.

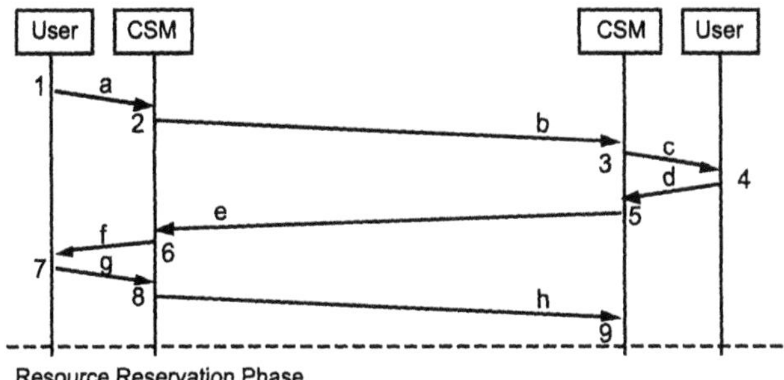

Figure 11 2-Way Negotiation

message d: { modify the negotiation objects }
message e: { modify the negotiation objects }
message f: { notify the user of the modifications }
message g: { accept }
message h: { accept }

This negotiation can be repeated. Figure 12 shows a 3-way negotiation. If the user A does not agree the modification, he/she can also request to modify the negotiation objects, he/she requests the CSM to modify them. When the CSM receives this modification request, it modifies the negotiation objects and requests the peer CSM to modify them.

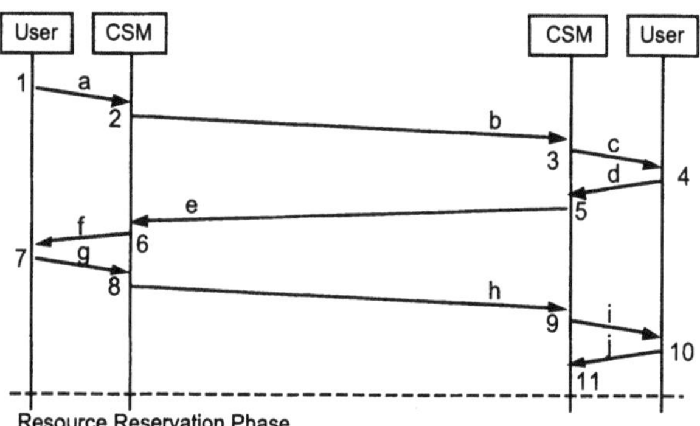

Figure 12 3-Way Negotiation

message d: { modify the negotiation objects }
message e: { modify the negotiation objects }
message f: { notify the user of the modifications }
message g: { modify the negotiation objects }
message h: { modify the negotiation objects }
message i: { notify the user of the modifications }
message j: { accept }

During the resource reservation phase, there can be a shortage of some resource. When it happens the RM notifies the requesting RM of it, and the notified RM announces to the requesting RM. When the CSM receives this notification message, it should negotiate again.

4.4 Call processing in a Layer Network

Figure 13 shows the call processing in Layer Networks. You should notice that the negotiation phase of the lower Layer Network is in the resource reservation phase of the upper Layer Network. When the negotiation of the lower layer Network is completed, the upper layer's RMs continue the resource reservation. The resource allocation of the lower Layer Network is performed after the resource reservation phase of the upper Layer Network.

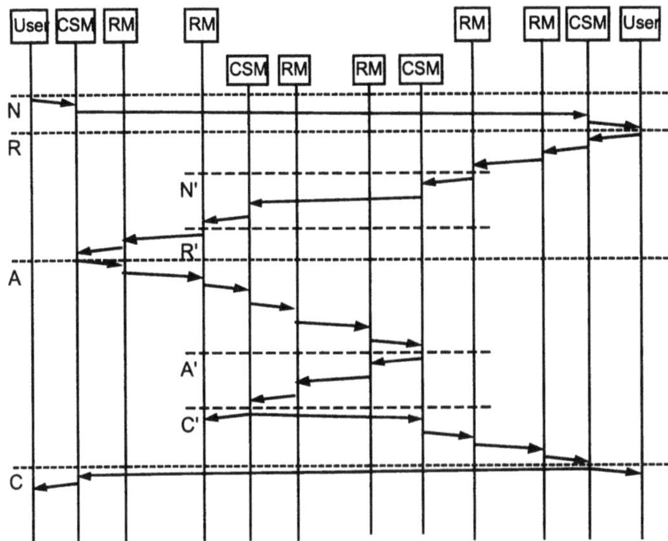

Figure 13 Call Processing in Layer Network

Figure 14 shows one example of the call processing. This shows the call processing in Layer Networks and the multiparty connections. There are 3 users and 3 CSMs in this figure, but one CSM is responsible for several users. There are 2 Layer Networks in this figure. You should notice that the negotiation process between the lower layer's CSMs. From the view-point of the upper layer's RM, the connection of the lower layer is a link. So, the RMs do not wait for the lower layer connection.

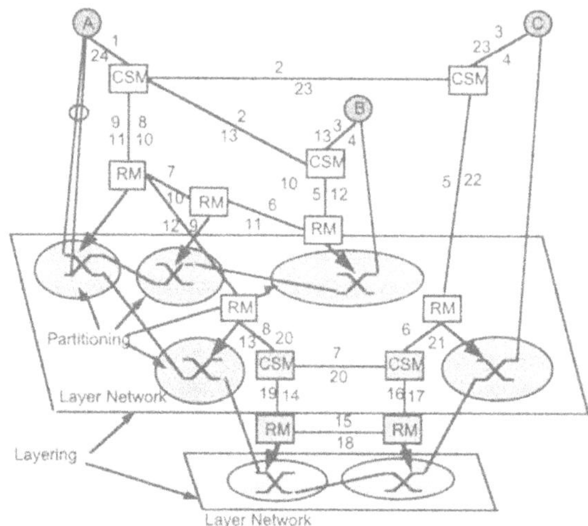

Figure 14 Example of Call Processing

5 SUMMARY

The Layer Network makes it possible to model the complex network composed of networks of several domains and several technologies. There will be new services in the future telecommunication network, and the most of them will be composed of existing services. The SCPA1 is a new call processing architecture for future networks, specially for the Layer Networks. We divide the call processing functions into two parts; one for negotiation of the upper layer's request and another for requesting the services of the lower layer. The CSM and the RM are responsible for performing those functions. We have presented the call processing algorithms and functional models of SCPA1. Now we have considered the generic resources of the service call and objects of each functional element.

6 REFERENCES

[1] S. Minzer, "A Signaling Protocol for Complex Multimedia Services," IEEE Journal on Selected Areas in Communications, Vol. 9, No. 9, Dec. 1991, pp. 1383-1394

[2] INA Document cycle 1 "Communication Management Architecture," Bellcore, 1992

[3] M. Jeffrey, A. Sarma, "A Signaling Architecture for Multimedia Services," Proc. of ICC'92 pp. 604-608

[4] M.E. Gaddis, R. Bubenik, J.D.Dehart, "A Call Model for Multipoint Communication in Switched Networks," Proc. of ICC'92, pp. 609-615

[5] S.W.Tu, G.A.Brenner, "An Object-oriented Resource Model for Supporting Signaling and Control of Broadband Services," Proc. of ICC'92, pp. 616-621
[6] CCITT Study Group XVIII/7, Recommendation G.803, "Architecture of Transport Networks Based on the Synchronous Digital Hierarchy (SDH)"
[7] Martin Chapman, Stefano Montesi, "Overall Concepts and Principles of TINA," TB_MDC.018_1.0_94, 1995
[8] Motoharu Kawanishi, Hiroshi Oshigiri, Juan Pavon, Mike Schenk, "Connection Management Architecture", TB_JJB.005_1.5_94, 1995

7 BIOGRAPHY

Sunshin An was born in Seoul, Korea, in 1950. He received the B.Eng degree from the Seoul National University, Korea, in 1973, and the M.S degree in electrical engineering from the KAIST(Korea Advanced Institute of Science and Technology), Korea, in 1975, and the Doctor degree in electric and information from the ENSEEIHT, France, in 1979. He joined the faculty of Korea University in 1982, where he is currently Professor of Electronic Engineering. Prior to joining Korea University, he was Assistant Professor of Electronic Engineering in Aju University, Suwon, Korea. He was with NIST(National Institute of Standards and Technology) in U.S.A, as a Visiting Scientist, in 1991. His research interests include the distributed system, communication networks and protocols, information network, intelligent network and multimedia communication system. Dr. An is the Chair of OSIA(Open System Interconnection Association) in Korea, and the Secretary of the Korea Information Science Society. He is a member of the ACM and IEEE.

Geonung Kim was born in Korea in 1967. He received the B.Eng degree and the M.S degree in electronic engineering from the Korea University, Seoul, in 1990 and 1994, respectively. He is currently a graduate research assistant in Super Highway Information Network Laboratory at the Korea University, where he is pursuing his doctorate in electronic engineering. His interests include network management, TMN, intelligent network and information network.

23
An Overview of the Telecommunications Information Networking Architecture

TINA Consortium
c/o Bellcore
331 Newman Springs Rd.
Red Bank, NJ 07701
USA
Tel: +1 908 758 2467
Fax: + 1 908 758 2865

1. INTRODUCTION

The Telecommunications Information Networking Architecture Consortium (TINA-C) is a worldwide consortium formed by network operators, and telecommunication and computer equipment suppliers. The consortium is aiming at defining and validating an "open" architecture for telecommunications services in the emerging broadband, multi-media and "information super-highway"/"information society" era.

The architecture is based on distributed computing, object orientation, and other concepts and standards from the telecommunications and computing industries, e.g., Open Distributed Processing (ODP), Intelligent Networks (IN), Telecommunication Management Networks (TMN), Asynchronous Transfer Mode (ATM), and Common Object Request Broker Architecture (CORBA).

The intention of this paper is to give an introduction to, and an overview of the TINA architecture. First, it presents an overall view of the TINA architecture, followed by a brief introduction to the main parts of the architecture, namely the computing, network, service, and management architectures of TINA. This is followed by a brief overview of interworking and migration scenarios for legacy systems. Finally, an overview of a tool-set to aid service specification and design is presented.

2. THE TINA-C VISION

The vision of the TINA efforts is, in short, to define and validate a consistent architecture for "open" telecommunications. TINA will address the needs of traditional voice-based services, future interactive multi-media services, information services, and operations and management type services, and will provide the flexibility to operate services over a wide variety of technologies [6][7]. This vision implies a **software architecture** that offers reusable software components, supports network-wide software interoperability, eases service construction, testing, deployment and operation, and hides from the service designer the heterogeneity of the underlying technologies and the complexity introduced by distribution.

The intention is to make use of recent advances in distributed computing (e.g., Open Distributed Processing (ODP) and Distributed Communication Environment (DCE)), and in object-oriented analysis and design, to drastically improve interoperability, re-use of software and specifications, and flexible placement of software on computing platforms/nodes. In addition, the consistent application of software principles to both services and management software is a primary goal. The TINA Architecture is furthermore defining how IN and TMN are "integrated" (related) within a common distributed framework. This will allow for a multi-supplier environment for telecommunications services and management systems.

Today, after three years of work with defining the TINA Architecture, the TINA-C Core Team is approaching a software architecture that seems to serve its initial purpose; a more suitable technique to handle distribution is taken into account, based on the standard to be RM-ODP [11]; end-user services and management services interwork smoothly and quite seamlessly; most are independent of the underlying technologies (computers, networks); while the most valuable concepts from the existing software architectures have been kept.

From this architecture, the steps to follow are to validate this architecture, to work out interworking scenarios with networks to be deployed and related interfaces (e.g., Intelligent Networks, Telecommunication Management Networks and ATM networks), to further develop related detailed specifications, and to further relate and to present the architecture to bodies and organizations working on similar issues (e.g., Network Management Forum, ATM Forum, Eurescom, RACE/ACTS and Object Management Group).

3. THE TINA OVERALL ARCHITECTURE

The technical goal of the TINA Architecture is to provide a set of concepts and principles to be applied in the design, processing, and operation of telecommunications software. The prevalent principle of TINA is that telecommunications services and management systems are considered software-based applications that operate on a distributed computing platform. The platform hides from applications the details of underlying technologies and distribution concerns, thus supporting the construction of portable and interoperable code.

3.1 Architecture introduction

The following outline three of the key foundations to the TINA architecture.

A TINA Service Component Model

A TINA service, as with any other telecommunications or computer applications, is provided as a set of interacting software components. These ***service components*** are structured according to a model called **USCM** (Universal Service Component Model) (Figure 1). The USCM states that each component should be composed of a **core** describing the nature of the object, regardless of how it is used or managed, a **usage** part describing its appearance to users and user access considerations, a **management** part providing operations, maintenance, and administration for the component, and a **substance** part representing the component's dependence upon other components [14].

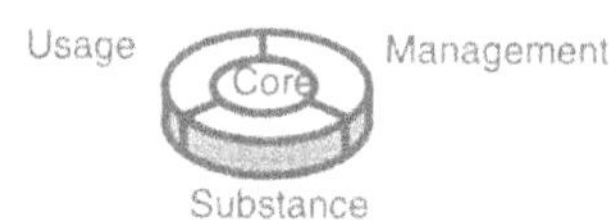

Figure 1. A service component model

Each service component can be divided into one or a number of software units, called ***computational objects***, as further described in Section 4 below.

A TINA Distributed Processing Environment

The communication (interaction) between the service components, or actually their computational objects, is supported in TINA by distributed computing mechanisms, e.g., Remote Procedure Calls (RPC). These mechanisms are provided by the core of a ***Distributed Processing Environment (DPE)***, as shown in Figure 2. The DPE provides a software sub-layer that operates above a native computing and communications environment (NCCE) and thus shields application programmers from technology dependant features.

In order to further enhance the components' capabilities to easily interact, generic support functions called *DPE Services* are defined. They are provided as additional services to a basic DPE. Examples of such generic servers are Traders and Name Servers, in order to help identify and locate relevant software units, Notification Servers, to help forward messages to the relevant software units, and Security Servers, to ensure only authorized interaction between service components takes place.

A TINA Component Categorization

In order to achieve good structure, modularity, and software re-usability, the service components are divided into three ***component categories*** - ***service components***, ***resource components*** and ***elements***, as shown in Figure 2. Components in the service category address the core functionality of services (service logic), and also covers access and service management considerations. Services can make use of common resources by interacting with components in the resource category. The resource components described (high-level) abstractions of the available resources, in a technology independent way, and provide facilities for the management of resources. Elements are software representations of individual resources such as transmission equipment, switch fabrics and computers.

3.2 Architecture subdivision

The TINA-C Overall Architecture is subdivided into four technical areas: the Computing Architecture, the Service Architecture, the Management Architecture, and the Network Architecture as shown in Figure 3.

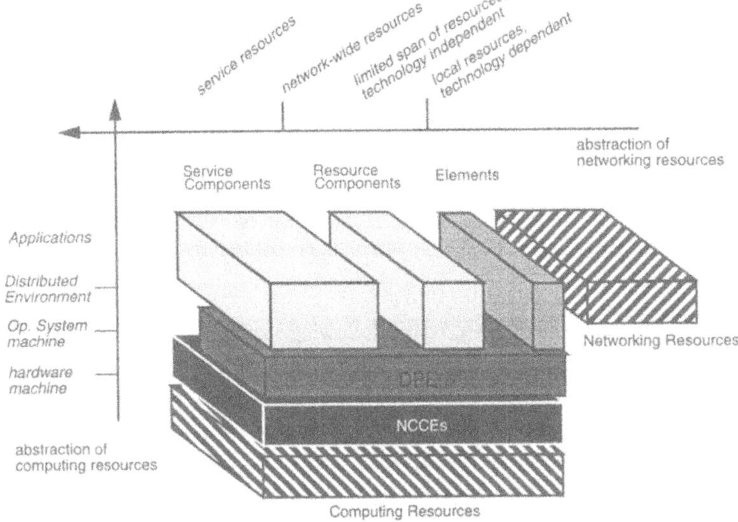

Figure 2. Overall Architecture

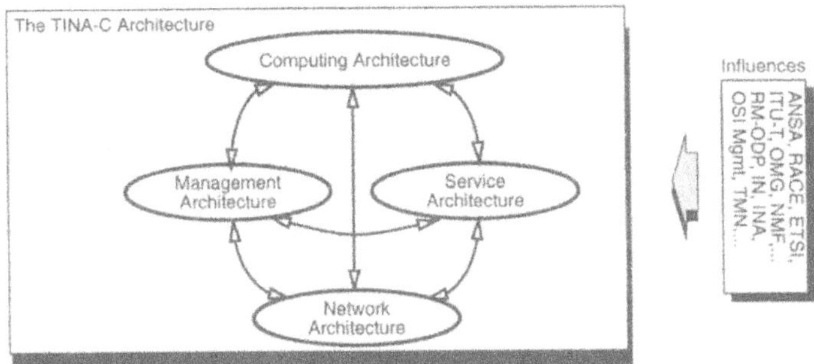

Figure 3. Partitioning of the TINA-C Software Architecture

- The *Computing Architecture* aims to provide a set of concepts and rules for how interacting software components (computational objects) are specified and how they communicate. The Distributed Processing Environment (DPE) defines the computer platform support for distributed execution of such components.
- The *Service Architecture* aims to provide a set of concepts and principles for specifying, analyzing, reusing, designing, and operating service-related telecommunications software components.

- The *Network Architecture* aims to provide a set of generic components for network resources. These components provide high-level abstractions of resources (e.g., connections) used by several service applications, and provides for the management of these resources.
- The *Management Architecture* aims to provide a set of generic management principles and concepts that are applied to the management of Service, Network, and Computing Architectures.

Components in the service, network and management architectures are structured according to the principles defined in the computing architecture, thus consistently applying the same software techniques.

Each architecture area is further discussed in the following sections.

4. THE TINA COMPUTING ARCHITECTURE

The computing architecture define the modelling concepts that should be used to specify object-oriented software in TINA systems. It also defines a distributed processing environment (DPE) that provides a support system that allows objects to locate and interact with each other. These concepts are based on the Reference Model for Open Distributed Processing (RM-ODP) [11][12].

The TINA computational modelling concepts defines the rules of how **computational objects** interact with one another. Computational objects are the units of programming and encapsulation. Objects interact with each other by sending and receiving information to and from **interfaces**. An object may provide many interfaces, either of the same or a different type. There are two forms of interface that an object may offer or use: **operational interface** and **stream interface**. An operational interface is one that has defined operations, that allow for functions of the offering (server) object to be invoked by other (client) objects. An operation may have arguments and may return results. A stream interface is one without operations (i.e., there is no notion of input/output parameters, requests, results, or notifications). The establishment of a **stream** between stream interfaces allows for the passing of other structured information, such as video or voice bit streams. Streams are established by interacting with service and network components, see below. Figure 4 depicts these concepts.

The interfaces are specified independently of any programming language by means of a specification notation. As none of the currently existing and relevant notations cover all the features of the TINA computational model, extensions to the Object Management Group's Interface Definition Language (OMG IDL) have been made. This notation is called TINA Object Definition Language (ODL)

Although the computational concepts are appropriate for describing how an application is structured into separate, logically distributed components, it is not appropriate for describing what this application is actually doing. This is the purpose of the **information modelling concepts**, from which an application designer can identify the information-bearing entities in the problem domain under study, show their relationships, and state the constraints directing their behavior. In TINA, information specifications are made using a quasi GDMO-GRM notation [13]. Stan-

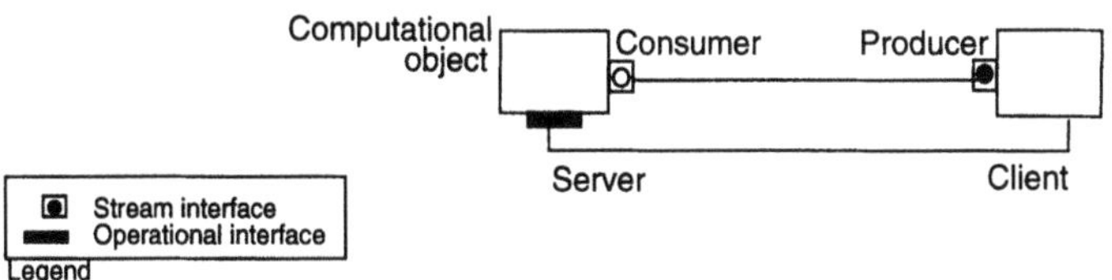

Figure 4. Computational modelling concepts

dard GDMO-GRM allows for many concerns to be expressed in a specification, including information and computational issues. TINA quasi GDMO-GRM uses only aspects of the language suitable for TINA information modelling.

A computational object can interact with another computational object without knowing on what computer the other object resides. There is an assumption that all computational objects reside on a platform, called a distributed processing environment (DPE). The DPE provides location and interaction mechanisms allowing computational objects to locate one another at execution time. **Engineering modelling concepts** define this DPE infrastructure.

From an application designer's standpoint, the DPE can be considered as one homogeneous infrastructure hiding the complexity and heterogeneity of the underlying network and computing resources. However, in reality a DPE consists of "kernels" implemented on top of different heterogeneous computing environments. The DPE kernels are interconnected by a specific logical network, called the **Kernel Transport Network** (KTN) [8]. The KTN facilitates the exchange of messages between computational objects. The DPE kernels are enhanced by additional generic software components, called DPE services, dealing with distribution, security, quality of service, operability, etc. An example is the trader which is a service that allows a computational object to locate interfaces of other computational objects.

A more detailed description of the DPE can be found in [2].

5. THE TINA NETWORK ARCHITECTURE

The purpose of the network architecture is to provide a set of generic concepts that describe transport networks in a technology independent way. The network architecture defines a set of abstractions that the resources layer can work with. At one end it provides a high level view of network connections to services. At the other end it provides a generic descriptions of elements, which can be specialised to particular technologies and products.

The network architecture has been defined by taking into account the layering principles of ITU-T Recommendation G.803 [9]. These principles separate the functions of a transport network into layers according to the characteristic information handled by the function. The **Network Resource Information Model** (NRIM) is an information specification of transmission and switch technologies, in which the technology dependent aspects have been extracted (e.g. hide differences between ATM and SDH switches). The model concerns how individual elements are related, topologically interconnected, and configured to provide and maintain end-to-end connectivity. The model therefore defines technology independent concepts that can be used to derive technology independent control and management functions. When designing and implementing a real network the technology dependant aspects must be taken into account as specializations of the generic model. The concepts found in this model include Layer Network, Sub-Network Connection (SNC), connection, topological link, and network termination points.). Figure 4 shows an example network.

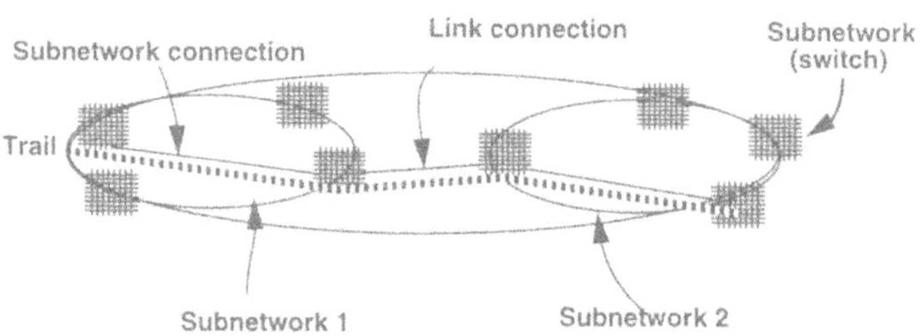

Figure 5. Example of abstract resource components

The network resource information model can be used to describe in detail a real network. However, it contains details which a user may not want to be, or should not be, aware of. Any use of a network, aside from management activity, must be as a result of using a service. It is important therefore to provide a service oriented view of connectivity. The concept of **connection graph** is used for this purpose. A connection graph depicts vertices with ports, where ports are connected by lines to represent connectivity.

There are two basic types of connection graph. In a **physical connection graph** the vertices represent physical nodes, the ports network access points and the lines connections. In a **logical connection graph** the vertices represent objects, the ports stream interfaces and the lines streams. The purpose of these two graphs is to provide two different views of a set of connections. Service software will request the establishment of streams between computational objects, and will build a logical connection graph to represent this. To build a stream will require the establishment of network connections and internal node bindings between the network access points and the stream interfaces. Network connections are expressed as physical connection graphs. Note that in each type of graph only the end-points of connections are represented. Intermediate nodes are abstracted away from since this information is not required by services.

A service can express connectivity requirements by building a graph of end-points (vertices and ports) and lines, and request a **communication session manager** to build an appropriate physical connection graph and intra-nodal bindings. The communication session manager accepts logical connection graphs as input from a service and produces a physical connection graph as output. It is responsible for negotiating with **connection management** to establish the physical connection graph (i.e. the connections), and to interact with nodes to perform internal network access point/stream interface bindings. The communication session manager is also responsible for monitoring and communicating any changes in each graph that could affect the other. The connection management components are derived from the NRIM definitions.

6. THE TINA SERVICE ARCHITECTURE

The service architecture provides means to build services and a service support environment [14]. The term "service" in TINA is used in a wide sense. It includes telecommunication services, management services and end-user services. Thus the service architecture is applicable to a wide range of service types, including management services, information services, transport services, and access services.

From the information viewpoint, the TINA service architecture provides a framework for describing what services do (their prime function), how they are managed, and how terminal and user mobility is achieved. Two main separation principles are defined in the service architecture; the separation of call and connection control, and the separation of user and terminal related issues. The separation of calls from connections, allows a more flexible approach to the handling of services, whereby the negotiation and involvement of users and their activities can be separated from the negotiation and involvement of communication resources. The separating of terminal and user issues allows for a flexible mobility model to be defined.

From the computational viewpoint, the TINA service architecture describes how a distributed service should be structured in order to be provide its function to a user. The service architecture identifies the run-time software components that should serve as a foundation for all services and therefore constitute a library of reusable components. The most important of these components are **user agents**, **terminal agents**, **service session**, **subscription manager**, and the **communication session manager**.

A user agent represents the user in the network. It handles, on a users behalf, incoming and outgoing service establishment, and maintains profiles of the user and any customized details associated with services. A terminal agent represents a terminal attached to a network. It provides a technology independent view of the terminal to the network, and manages terminal specific information. The power of the user and the terminal agent concept is that users and terminals can be located by finding the related agent. This avoids having to build in location knowledge and thus allows users and terminals to move around in the network.

A service session is an instance of an executing service. It maintains the state of a service, and provides interfaces for the joining and leaving of users to a service session. Embodied within a service session is the service core, or service logic. In response to commands from users, or service logic, connections may need to be established, modified, or 'dropped.' Service sessions do this by constructing logical connection graphs and instructing a communication session manager to build the connections (see Section 5).

The subscription manager maintains information related to customer and user subscriptions. It is use during the establishment of a users involvement in a service session to check access permission with respect to a subscription.

Figure 6 highlights these components operating above the DPE. More detail on these components can be found in [1][3].

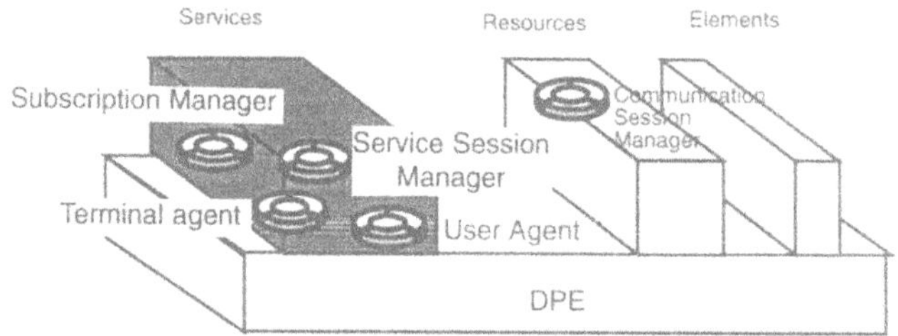

Figure 6. The Service Architecture reusable components

7. THE TINA MANAGEMENT ARCHITECTURE

The TINA management architecture [15] defines a set of principles for the management of the entities in a TINA system. There are two basic types of management in TINA systems: **computing management**, and **telecommunications management** (Figure 7).

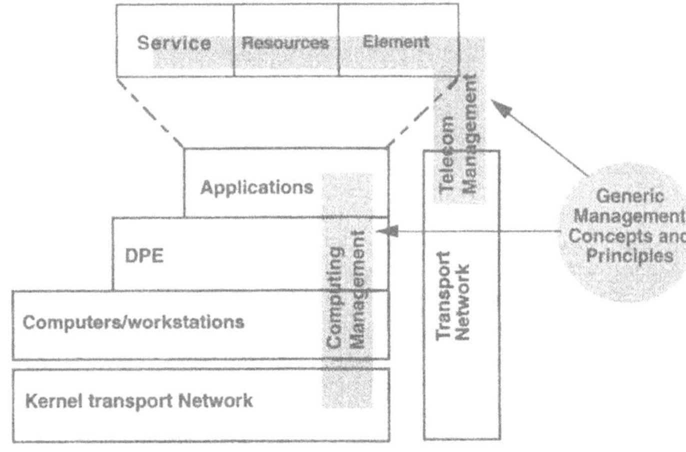

Figure 7. Two basic types of management.

Computing management involves the management of the computers, platform, and of the software (in general terms) that runs on that platform. Management here does not concern itself with what applications are doing nor on application specific management. The main concern is the deployment, installation, and load balancing of software.

In TINA the application software relates to telecommunications services, and software for the control and management of transmission and switching networks. These telecommunications specific software components require management. This is called telecommunications management.

The above two types of management are very broad, and therefore are divided into sub-types of management. Telecommunications management is broken down into service, network, and element management, in much the same way as TMN systems. Computing management can be broken down into generic software management, such as deployment, configuration, and instantiation of software, and management of the DPE and computer environments. The service, network and computing architectures therefore each contain management concepts and principles.

Even though different types of management have been identified, the TINA architecture defines a set of generic management concepts and principles.There are two basic sets of principles for generic management. The first deals with functional separations. This breaks down the problem of management into distinct areas of concern. The second deals with modelling of management systems. This provides principles for how to express management relationships and operations in terms of the TINA object models (information, computation and engineering). Five functional separations have been taken from OSI Systems Management, namely fault, configuration, accounting, performance, and security

Further details of management in TINA can be found in [1].

8. INTERWORKING SCENARIOS AND MIGRATION PATHS TOWARDS TINA

The TINA Architecture is being defined to suit the needs of a future marketplace of information services. However, the success of the architecture will largely depend on its ability to interwork with existing systems and on its applicability to migrate existing systems towards being consistent with the TINA Architecture. Thus work has begun on interworking and on migration issues.

Obviously, a number of different technologies are being deployed or are about to be deployed in today's networks - examples include IN, ISDN, ATM, TMN and OSI. It is not within the scope of the Core Team to examine all issues related to all these technologies. Instead a few characteristic examples will be chosen to be studied in **reference scenarios.**

A reference scenario shows, for a given technology and network environment, typical interworking and mapping possibilities. Interworking shows how a TINA system interacts with a legacy system in order to form a cooperation, or to reuse the existing technology. Mapping on the other hand shows how some of the TINA concepts can be used to evolve an existing system or architecture. Migration describes the development and realization of a series of reference scenarios.

It should be made clear that the Core Team is not working on migration scenarios per se, or on defining specific network solutions to be adopted and deployed by network operators. The work is instead intended as relevant "real-world" examples of how the TINA architecture might be used in different network environments. The results will focus TINA to better understand the **requirements** on the architecture which have to be satisfied before operational use of TINA can be achieved.

9. TINA METHODOLOGY AND TOOLS

Even though the intent and scope of the core team are not to build a service creation environment, some tools are currently used and customized by the core team. One is an Object Modelling Technique (OMT) tool that is used and customized in TINA-C to model information objects. From this modelling, quasi-GDMO-GRM specifications can be derived automatically as a text representation of the information model. We envisage expanding the OMT tool (into a skeleton of SCE) to help the designer to produce the TINA-C ODL specifications of computational objects. The other tool is the ODL specification pre-processor that produces IDL specifications suitable for compiling onto Common Object Request Broker Architecture (CORBA) platforms. Figure 5 depicts this tool chain.

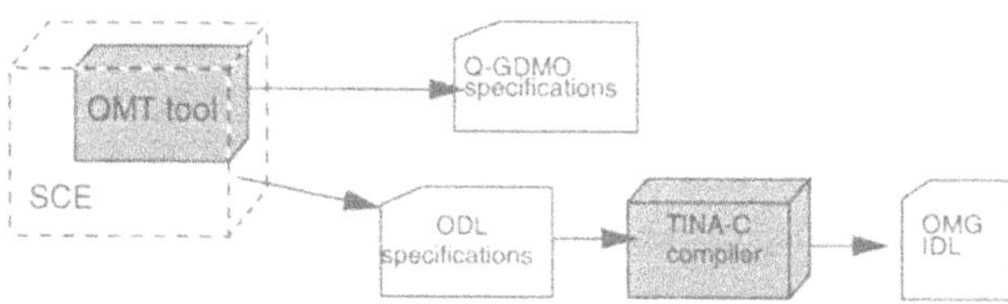

Figure 8. TINA-C Design Support Tools

10. CONCLUSION

Many telecommunications stakeholders have already shown their interest in the TINA architecture and expect it to be usable in 1996. Therefore, an important part of the TINA-C activities is devoted, prior to any standardization effort, to validating the current specifications through auxiliary projects, collaborative worldwide demonstrations, and implementations carried out by the TINA-C Core Team. It is also expected that typical migration scenarios from currently deployed intelligent networks and telecommunications management networks to TINA networks will be available soon. This will make it possible to proceed gradually and cost effectively towards a Telecommunications Information Networking Architecture.

11. REFERENCES

[1] H. Berndt, L. de la Fuente, P. Graubmann, *"Service and Management Architecture in TINA-C"*, TINA 95 Conference Proceedings.

[2] E. Kelly, N. Mercouroff, *"TINA-C DPE Architecture and Tools"*, TINA 95 Conference Proceedings

[3] H. Yagi, K. Ohtsu, M. Wakano, H. Kobayashi, R. Minerva, *"TINA-C Service Components"*, TINA 95 Conference Proceedings.

[4] J. Bloem, J. Pavon, M. Schenk, H. Oshigiri, *"TINA-C Connection Management Components"*, TINA 95 Conference Proceedings.

[5] V. Cohen, G. Chapman, N. Dathi, F. Ruano, *"TINA Validation through Prototyping"*, TINA 95 Conference Proceedings.

[6] William J. Barr, Trevor Boyd, Yuji Inoue, *"The TINA Initiative"*, IEEE Communications Magazine, March 1993.

[7] N. Natarajan, G.M. Slawsky, *"A Framework Architecture for Multimedia Information Networks"*, IEEE Communications Magazine, February 1992.

[8] G. Brégant, R. Kung, Y. Lepetit, J-B. Stefani, *"The evolution of networks: towards more integrated architectures"*, XIV ISS, October 1992.

[9] CCITT Recommendation G.803, *Architectures of Transport Networks Based on the Synchronous Digital Hierarchy (SDH),* June 1992.

[10] S. Rudkin, *"Templates, types and classes in open distributed processing"*, BT Technology Journal, July 1993.

[11] ISO 10746-2 RM-ODP Part 2, *"Descriptive model"*, Yokohama, June 1993.

[12] ISO 10746-3 RM-ODP Part 3, *"Prescriptive model"*, Yokohama, June 1993.

[13] ISO 10165-7, *"Information Technology - Open Systems Interconnection - Structure of Management Information - Part 7: General Relationship Model"*, January 1993.

[14] H. Berndt, P. Graubmann, *"Service Architecture, Service Session and Service Federation in TINA-C"*, submitted to ISS'95.

[15] L. A. de la Fuente, J. Pavon, J. C. Moreno, *"The TINA-C Approach to TMN"*, submitted to ISS'95.

[16] F. Dupuy, R. Minerva, J. Bloem, H. Hammainen, J. Moreno, *"The TINA-C Architecture as related to IN and TMN"*, 3rd International Conference on Intelligence in Networks, Bordeaux, October 1994.

INDEX OF CONTRIBUTORS

Abramowski, S. 265
An, S. 281
Belarbi, N. 51
Bredereke, J. 68
Capellmann, C. 37
Cho, S. 213
Choi, J. 213
Dyst, J. 95
Ericsson, L.M. 95
Fangchun Yang 201
Gaiti, D. 51
Gbaguidi, C. 252
Ginzboorg, P. 155
Haitao, T. 175
Hawkins, R. 191
Hermansen, K. 37
Hubaux, J.-P. 252
Im, C. 213
Jensen, T. 225
Junliang Chen 201
Kemppainen, P. 84
Kim, G. 281
Klabunde, K. 265
Kolehmainen, M. 107
Konrads, U. 265
Larsson, A. 143
Martikainen, O. 1, 84, 241
Melody, W.H. 7
Mudhar, P. 37
Naoumov, V. 241
Neunast, K. 265
Nørgaard, J. 37
Nyberg, C. 143
Raatikainen, K. 121
Samouylov, K. 241
Schmidt Jensen, T. 27
Schneps-Schneppe, M. 16
Simula, O. 175
Taina, J. 121
TINA Consortium, USA 296
Tjabben, H. 265
Wohlin, C. 143
Yuzhang Liu 201
Znaty, S. 252

GPSR Compliance
The European Union's (EU) General Product Safety Regulation (GPSR) is a set of rules that requires consumer products to be safe and our obligations to ensure this.

If you have any concerns about our products, you can contact us on

ProductSafety@springernature.com

In case Publisher is established outside the EU, the EU authorized representative is:

Springer Nature Customer Service Center GmbH
Europaplatz 3
69115 Heidelberg, Germany

www.ingramcontent.com/pod-product-compliance
Ingram Content Group UK Ltd.
Pitfield, Milton Keynes, MK11 3LW, UK
UKHW020105200726
13856UKWH00002B/381

* 9 7 8 1 4 7 5 7 5 5 4 4 2 *